Significant and Nonsignificant Risk Medical Devices

Prakash Srinivasan Timiri Shanmugam
Pugazhenthan Thangaraju
Thamizharasan Sampath
Indumathy Jagadeeswaran
Editors

Significant and Nonsignificant Risk Medical Devices

Springer

Editors
Prakash Srinivasan Timiri Shanmugam
Global Product Safety & Toxicology
Avanos Medical Inc.
Alpharetta, GA, USA

Pugazhenthan Thangaraju
Department of Pharmacology
All India Institute of Medical Sciences, Raipur
Raipur, Chhattisgarh, India

Thamizharasan Sampath
Department of Pharmacology & Toxicology, VMCGH
DR.YSR University of Health Sciences
Kurnool, AP, India

Indumathy Jagadeeswaran
Pediatrics
UT Southwestern Medical Center
Dallas, TX, USA

ISBN 978-3-031-52837-8 ISBN 978-3-031-52838-5 (eBook)
https://doi.org/10.1007/978-3-031-52838-5

This Springer imprint is published by the registered company Springer Nature Switzerland AG
The registered company address is: Gewerbestrasse 11, 6330 Cham, Switzerland

Paper in this product is recyclable.

Preface

The book *Significant and Non-significant Risks of Medical Devices* helps to identify the risks of medical devices with examples. The chapters discuss the devices, mechanism of devices, their intended use, single and/or multiple use, benefits and any side/adverse/toxicological effects on the patient, body contact, duration of contact, user error, etc. The sub-chapters discuss the description of the devices, reports of prior investigations conducted with the devices, basics of risk determination and the nature of harm that may result from the use of such devices. This book may be a helpful guide to understand significant and non-significant risks of medical devices.

Alpharetta, GA, USA — Prakash Srinivasan Timiri Shanmugam
Raipur, Chhattisgarh, India — Pugazhenthan Thangaraju
Kurnool, AP, India — Thamizharasan Sampath
Dallas, TX, USA — Indumathy Jagadeeswaran

Contents

Chapter 1
Introduction

Thamizharasan Sampath, Sandhiya Thamizharasan, Bharanidharan Indrakumar, Monisha Saravanan, and Prakash Srinivasan Timiri Shanmugam

Abbreviations

ALARA	As Low as Reasonably Achievable
CDRH	Center for Devices and Radiological Health
CFR	Code of Federal Regulations
CGM	Continuous Glucose Monitoring Sensors
CLABSI	Central LineAssociated Bloodstream Infections
CPR	Cardiopulmonary Resuscitation
CT	Computed Tomography
CVS	Catheters for Chorionic Villus Sampling
ECG	Electrocardiogram
ECMO	Extracorporeal Membrane Oxygenators
ECT	Electroconvulsive Therapy
EMA	European Medicines Agency
EU	European Union
FDA	Food and Drug Administration
FDC	Federal Food, Drug, and Cosmetic Act
HA	Hydroxyapatite
IEC	International Electrotechnical Commission

T. Sampath (✉)
Department of Pharmacology & Toxicology, VMCGH, DR.YSR University of Health Sciences, Kurnool, AP, India

S. Thamizharasan
The Tooth Doctor -Advanced Implant Centre, Vellapanchavadi, Chennai, India

B. Indrakumar
Veristatllc, South Borough, MA, USA

M. Saravanan
Bayshore Healthcare, Mississauga, Ontario, Canada

P. S. Timiri Shanmugam
Global Product Safety & Toxicology, Avanos Medical Inc., Alpharetta, GA, USA

P. S. Timiri Shanmugam et al. (eds.), *Significant and Nonsignificant Risk Medical Devices*, https://doi.org/10.1007/978-3-031-52838-5_1

IOL	Intraocular Lenses
ISO	International Organization for Standardization
IUD	Intrauterine Devices
IV	Intravenous
IVD	In Vitro Diagnostics
IVDR	In Vitro Diagnostic Medical Device Regulation
MDR	Medical Device Regulation
MRI	Magnetic Resonance Imaging
NSR	Non-Significant Risk
PORP	Partial Ossicular Replacement Prosthesis
SR	Significant Risk
SSI	Surgical Site Infections
TENS	Transcutaneous Electric Nerve Stimulation
TMJ	Temporomandibular Joint
TORP	Total Ossicular Replacement Prosthesis
UTI	Urinary Tract Infections
VAP	VentilatorAssociated Pneumonia

Medical devices have become indispensable tools in modern healthcare, ranging from simple instruments like syringes and thermometers to complex implantable devices and diagnostic equipment. "Medical device" means any instrument, apparatus, implement, machine, appliance, implant, in vitro reagent or calibrator, software, material, or other similar or related article, intended by the manufacturer to be used, alone or in combination, for human beings for one or more of the specific purposes of:

- Diagnosis, prevention, monitoring, prediction, prognosis, treatment, or alleviation of disease
- Diagnosis, monitoring, treatment, alleviation of, or compensation for an injury or disability
- Investigation, replacement, or modification of the anatomy or of a physiological or pathological process or state
- Providing information by means of in vitro examination of specimens derived from the human body, including organ, blood, and tissue donations, which does not achieve its principal intended action by pharmacological, immunological, or metabolic means, in or on the human body, but which may be assisted in its function by such means

1.1 Background

Medical devices play a crucial role in improving patient outcomes, enhancing diagnostics, and facilitating medical procedures. However, the use of medical devices also entails certain risks that need to be carefully evaluated and managed. Significant risks associated with medical devices refer to hazards that have the potential to cause severe harm or adverse events to patients or users. These risks can stem from various factors, including device malfunctions, design flaws, inadequate instructions for use, or adverse reactions to device materials. For example, a malfunctioning pacemaker can lead to life-threatening arrhythmias, while a faulty surgical instrument can cause serious injuries during a procedure. Identifying and mitigating these significant risks is paramount to ensuring patient safety and the overall effectiveness of medical devices [1].

Non-significant risks, on the other hand, are hazards that are relatively minor in nature and have minimal impact on patient health or device performance. These risks may result in temporary discomfort, local irritation, or non-serious adverse events. Examples of non-significant risks include mild skin irritation from a bandage or temporary drowsiness caused by a sedative medication. Although these risks may not pose immediate threats to patients, they still require consideration to ensure user satisfaction and overall device quality.

The classification and assessment of risks associated with medical devices are undertaken by regulatory bodies and standards organizations. These entities, such as the Food and Drug Administration (FDA) in the United States or the European Medicines Agency (EMA) in Europe, establish guidelines and regulations for the approval, monitoring, and post-market surveillance of medical devices. Through rigorous testing, clinical trials, and ongoing surveillance, these regulatory bodies aim to ensure the safety and effectiveness of medical devices, as well as provide guidance to manufacturers and healthcare professionals.

Understanding the distinction between significant and non-significant risks is vital for manufacturers, regulators, healthcare providers, and patients. It allows stakeholders to make informed decisions regarding device selection, usage, and patient counseling. By conducting thorough risk assessments, implementing appropriate risk management strategies, and adhering to regulatory guidelines, stakeholders can strike a balance between the potential benefits and risks of medical devices. This knowledge and proactive approach contribute to improved patient outcomes, enhanced healthcare quality, and increased confidence in the use of medical devices.

1.2 Why Risk Assessment Is Important?

The importance of risk assessment in the context of medical devices cannot be overstated. Risk assessment is a critical process that helps ensure the safety and effectiveness of medical devices throughout their lifecycle, from design and development to manufacturing, deployment, and use. Here are some key reasons why risk assessment is essential in the field of medical devices:

Patient safety: The primary objective of risk assessment in medical devices is to protect the safety and well-being of patients. By identifying and evaluating potential risks associated with the use of a medical device, manufacturers can implement necessary measures to mitigate those risks and prevent harm to patients. This includes identifying potential hazards, assessing their severity and likelihood, and implementing risk control measures to minimize or eliminate them.

Regulatory compliance: Regulatory bodies, such as the US Food and Drug Administration (FDA) and the European Medicines Agency (EMA), require medical device manufacturers to conduct risk assessments as part of the regulatory approval process. Compliance with these regulations is crucial for market access and ensuring the device meets the necessary safety standards. Failure to perform adequate risk assessment can lead to regulatory non-compliance and the removal of the device from the market.

Product development: Risk assessment plays a vital role in the product development phase of medical devices. It helps identify potential risks and hazards early in the design process, allowing manufacturers to incorporate appropriate risk control measures and design features to mitigate those risks. By proactively addressing safety concerns during the development stage, manufacturers can enhance the overall safety profile of the device and reduce the likelihood of adverse events in real-world use.

Post-market surveillance: Risk assessment is not limited to the pre-market phase but continues throughout the lifecycle of a medical device. It forms a crucial component of post-market surveillance activities, enabling manufacturers to monitor and evaluate the performance of their devices in real-world settings. By collecting and analyzing data on adverse events, complaints, and other safety-related information, manufacturers can identify any new risks or hazards that may arise and take appropriate actions to address them, such as issuing safety alerts, recalls, or product modifications.

Liability and legal considerations: Conducting thorough risk assessments and implementing appropriate risk management strategies can help protect manufacturers from potential liability claims. In the event of an adverse incident or harm caused by a medical device, having a well-documented risk assessment process demonstrates that the manufacturer took reasonable steps to identify and mitigate risks. This can be crucial in legal proceedings and risk mitigation.

Overall, risk assessment is essential for ensuring the safety and effectiveness of medical devices. It helps protect patients, ensures regulatory compliance, informs product development, enables post-market surveillance, and safeguards manufacturers from liability. By systematically identifying and managing risks, medical device manufacturers can improve the quality of their products and promote patient safety [2].

1.3 Classification of Medical Devices

The Food and Drug Administration (FDA) has established classifications for approximately 1700 different generic types of devices and grouped them into 16 medical specialties referred to as panels. Each of these generic types of devices is assigned to one of three regulatory classes based on the level of control necessary to assure the safety and effectiveness of the device.

Class 1 – Low Risk

There are currently approximately 780 Class I devices on the market. General Controls include Adulteration/Misbranding, Electronic Establishment, Registration, Electronic Device Listing, Premarket Notification, Quality Systems, Labeling, and Medical Device Reporting (MDR).

Examples: Corrective glasses and frames, manual wheelchairs

Type of Certification: Self Certification/Self Declaration

Class 1s – Low Risk (Sterile)

Examples: Personal protection kits, sterile urine bags, etc.

Type of Certification: Notified Body

Class 1m – Low Risk (Measuring Body attributes)

Examples: Stethoscopes, weighing Balance

Type of Certification: Notified Body

Class 1r – Low Risk (Reused Device)

Examples: Surgical forceps (all types of SS/Tit surgical equipment sterilized and reused by hospitals)

Type of Certification: Notified Body

Class IIa – Medium Risk

Examples: Orthodontic wires, surgical gloves, lancets

Type of Certification: Notified Body

Class IIb – Medium to High Risk

Most devices are classified as Class II, an intermediate-risk device that is subject to "special controls" to assure safety. The majority of Class II devices are subject to premarket review and clearance by the FDA through the 510 (k)-pre-market notification process and may have rigorous review requirements in line with a Class III device. There are currently over 800 Class II devices on the market.

Examples: Orthopedic nails and plates, intra-ocular lens, pregnancy test kit, incubators for babies

Type of Certification: Notified Body

Class III – High Risk

The devices are subject to the most rigorous review process that includes general controls, special controls, and premarket approval. There are fewer than 120 Class III devices currently on the market.

Examples: Pacemakers, prosthetic heart valves, cardiovascular sutures, brain spatulas, drug-device combination products

Type of Certification: Notified Body

1.4 Medical Device Classification: 21 CFR 862–892

"Most medical devices can be classified by finding the matching description of the device in Title 21 of the Code of Federal Regulations (CFR), Parts 862–892. FDA has classified and described over 1700 distinct types of devices and organized them in the CFR into 16 medical specialty panels such as cardiovascular devices or in vitro diagnostics."

862 = Chemistry/Toxicology
864 = Hematology/Pathology
866 = Immunology/Microbiology
868 = Anesthesiology
870 = Cardiovascular
872 = Dental
874 = Ear, Nose and Throat
876 = Gastro/Urology
878 = General Plastic Surgery
880 = General Hospital
882 = Neurological
884 = Obstetrical/Gynecological
886 = Ophthalmic
888 = Orthopedic
890 = Physical Medicine
892 = Radiology Regulations

For each of the devices classified by the FDA, the CFR gives a general description including the intended use, the class to which the device belongs (i.e., Class I, II, or III), and information about marketing requirements. Your device should meet the definition in a classification regulation contained in 21 CFR 862–892.

1.5 NSR and SR Devices

The following examples may help sponsors and IRBs in making SR and NSR determinations. The list includes many commonly studied medical devices. Inclusion of a device in the NSR list is not a final determination because the evaluation of risk must reflect the proposed use of a device in a study [3].

Non-significant Risk Devices

- Caries Removal Solution
- Contact Lens Solutions intended for use directly in the eye (e.g., lubricating/rewetting solutions) using active ingredients or preservation systems with a history of prior ophthalmic/contact lens use or generally recognized as safe for ophthalmic use
- Conventional Gastroenterology and Urology Endoscopes and/or Accessories
- Conventional General Hospital Catheters (long-term percutaneous, implanted, subcutaneous, and intravascular)
- Conventional Implantable Vascular Access Devices (Ports)
- Conventional Laparoscopes, Culdoscopes, and Hysteroscopes
- Daily Wear Contact Lenses and Associated Lens Care Products not intended for use directly in the eye (e.g., cleaners; disinfecting, rinsing, and storage solutions)
- Dental Filling Materials, Cushions, or Pads made from traditional materials and designs
- Denture Repair Kits and Realigners
- Digital Mammography
- Electroencephalography (e.g., new recording and analysis methods, enhanced diagnostic capabilities, measuring depth of anesthesia if anesthetic administration is not based on device output)
- Externally Worn Monitors for Insulin Reactions
- Functional Non-Invasive Electrical Neuromuscular Stimulators
- General Biliary Catheters
- General Urological Catheters (e.g., Foley and diagnostic catheters) for short-term use (<28 days)
- Jaundice Monitors for Infants
- Low Power Lasers for treatment of pain
- Magnetic Resonance Imaging (MRI) Devices within FDA-specified parameters
- Manual Image Guided Surgery
- Menstrual Pads (Cotton or Rayon, only)
- Menstrual Tampons (Cotton or Rayon, only)
- Nonimplantable Electrical Incontinence Devices
- Nonimplantable Male Reproductive Aids with no components that enter the vagina
- Ob/Gyn Diagnostic Ultrasound within FDA-approved parameters
- Partial Ossicular Replacement Prosthesis (PORP)
- Total Ossicular Replacement Prosthesis (TORP)
- Transcutaneous Electric Nerve Stimulation (TENS) Devices for treatment of pain (except for chest pain/angina)
- Ureteral Stents
- Urethral Occlusion Device for less than 14 days
- Wound Dressings, excluding absorbable hemostatic devices and dressings (also excluding Interactive Wound and Burn Dressings that aid or are intended to aid in the healing process)

Significant Risk Devices

General Medical Use

- Catheters for General Hospital Use, except for conventional long-term percutaneous, implanted, subcutaneous, and intravascular
- Collagen Implant Material for use in ear, nose, and throat, orthopedics, plastic surgery, urological and dental applications
- Surgical Lasers for use in various medical specialties
- Tissue Adhesives for use in neurosurgery, gastroenterology, ophthalmology, general and plastic surgery, and cardiology

Cardiovascular

- Annuloplasty Rings
- Aortic and Mitral Valvuloplasty Catheters
- Arterial Embolization Devices
- Atherectomy and Thrombectomy Catheters
- Cardiac Assist Devices: artificial hearts, ventricular assist devices, intra-aortic balloon pumps, cardiomyoplasty devices
- Cardiac Bypass Devices: oxygenators, cardiopulmonary blood pumps, axial flow pumps, closed chest devices (except Class I cardiovascular surgical instruments), heat exchangers, catheters/cannulae, tubing, arterial filters, reservoirs
- Cardiac Mapping and Ablation Catheters
- Cardiac Pacemaker/Pulse Generators: antitachycardia, esophageal, external transcutaneous, implantable
- Cardiopulmonary Resuscitation (CPR) Devices
- Cardiovascular Intravascular (vena cava) Filters
- Coronary Artery Retroperfusion Systems
- Distal Embolic Protection Devices
- Extracorporeal Counterpulsation Devices
- Extracorporeal Membrane Oxygenators (ECMO)
- Implantable Cardioverters/Defibrillators
- Intravascular Brachytherapy Devices
- Intravascular Stents
- Laser Angioplasty Catheters
- Organ Storage/Transport Units
- Pacing Leads
- Percutaneous Conduction Tissue Ablation Electrodes
- Percutaneous Transluminal Angioplasty Catheters
- Replacement Heart Valves
- Transcatheter Cardiac Occluders for atrial and ventricular septal defects, patent foramen ovale, and patent ductus arteriosus
- Transmyocardial Revascularization, Percutaneous Myocardial Revascularization Devices
- Ultrasonic Angioplasty Catheters
- Vascular and Arterial Graft Prostheses
- Vascular Hemostasis Devices

Anesthesiology

- Breathing Gas Mixers
- Bronchial Tubes
- Electroanesthesia Apparatus
- Epidural and Spinal Catheters
- Epidural and Spinal Needles
- Esophageal Obturators
- Gas Machines for anesthesia or analgesia
- High Frequency Ventilators greater than 150 BPM
- Rebreathing Devices
- Respiratory Ventilators and new modes of ventilation
- Tracheal Tubes

Dental

- Absorbable Materials to aid in the healing of periodontal defects and other maxillofacial applications
- Bone Morphogenic Proteins with and without bone, e.g., Hydroxyapatite (HA)
- Dental Lasers for hard tissue applications
- Endosseous Implants and associated bone filling and augmentation materials used in conjunction with the implants
- Subperiosteal Implants
- Temporomandibular Joint (TMJ) Prostheses

Ear, Nose, and Throat

- Absorbable Gelatin Sponge
- Auditory Brainstem Implants
- Cochlear Implants
- Endolymphatic Shunt Tubes with or without valve
- ENT Cements/Adhesives
- Implantable Bone Conduction Hearing Aids
- Implantable Middle Ear Hearing Device
- Injectable Teflon Paste
- Laryngeal Implants
- Synthetic Polymer Materials
- Tissue Autofluorescent Devices
- Vocal Cord Medialization (Augmentation) Devices

Gastroenterology and Urology

- Anastomosis Devices
- Balloon Dilation Catheters for benign prostatic hyperplasia (BPH)
- Biliary Stents
- Components of Water Treatment Systems for Hemodialysis
- Dialysis Delivery Systems
- Electrical Stimulation Devices for sperm collection
- Embolization Devices for general urological use

- Extracorporeal Circulation Systems
- Extracorporeal Hyperthermia Systems
- Extracorporeal Photopheresis Systems
- Femoral, Jugular and Subclavian Catheters
- Hemodialyzers
- Hemofilters
- Implantable Electrical Urinary Incontinence Systems
- Implantable Penile Prostheses
- Injectable Bulking Agents for incontinence
- Lithotripters (e.g., electrohydraulic extracorporeal shock-wave, laser, powered mechanical, and ultrasonic)
- Mechanical/Hydraulic Urinary Incontinence Devices
- Penetrating External Penile Rigidity Devices with components that enter the vagina
- Peritoneal Dialysis Devices
- Peritoneal Shunt
- Plasmapheresis Systems
- Prostatic Hyperthermia or Thermal Ablation Devices
- Retention Type (Foley) Balloon Catheters for long-term use (>28 days)
- Suprapubic Urological Catheters and accessories
- Urethral Occlusion Devices for greater than 14 days use
- Urethral Sphincter Prostheses
- Urological Catheters with anti-microbial coatings
- Urological Stents (e.g., urethral, prostate)

General and Plastic Surgery

- Absorbable Adhesion Barrier Devices
- Absorbable Hemostatic Agents
- Artificial Skin and Interactive Wound and Burn Dressings
- Breast Implants
- Injectable Collagen
- Implantable Craniofacial Prostheses
- Repeat Access Devices for surgical procedures
- Sutures

Neurological

- Electroconvulsive Therapy (ECT) Devices
- Hydrocephalus Shunts
- Implanted Intracerebral/Subcortical Stimulators
- Implanted Intracranial Pressure Monitors
- Implanted Spinal Cord and Nerve Stimulators and Electrodes
- Neurological Catheters (e.g., cerebrovascular, occlusion balloon)
- Transcutaneous Electric Nerve Stimulation (TENS) Devices for treatment of chest pain/angina

General Hospital

- Implantable Vascular Access Devices (Ports) – if new routes of administration or new design
- Infusion Pumps (implantable and closed-loop – depending on the infused drug)

Obstetrics and Gynecology

- Abdominal Decompression Chamber
- Antepartum Home Monitors for Non-Stress Tests
- Antepartum Home Uterine Activity Monitors
- Catheters for Chorionic Villus Sampling (CVS)
- Catheters Introduced into the Fallopian Tubes
- Cervical Dilation Devices
- Contraceptive Devices:
  - Cervical Caps

    Condoms (for men) made from new materials (e.g., polyurethane)
    Contraceptive In Vitro Diagnostics (IVDs)
    Diaphragms or Female Condoms
    Intrauterine Devices (IUDs)
    New Electrosurgical Instruments for Tubal Coagulation
    New Devices for Occlusion of the Vas Deferens
    Sponges
    Tubal Occlusion Devices (Bands or Clips)

- Cryomyolysis
- Devices to Prevent Post-op Pelvic Adhesions
- Embryoscopes and Devices intended for fetal surgery
- Endometrial Ablation Systems
- Falloposcopes and Falloposcopic Delivery Systems
- Fundal Pressure Belt (for vaginal assisted delivery)
- Gamete and Embryo Surgical Systems
- Intrapartum Fetal Monitors using new physiological markers
- New Devices to Facilitate Assisted Vaginal Delivery
- Operative Hysteroscopy and Laparoscopy
- Uterine Artery Embolization

Ophthalmics

- Aniridia Intraocular Lenses (IOLs) or Rings (for iris reconstruction)
- Capsular Tension Rings
- Class III Ophthalmic Lasers Contact Lens Solutions intended for direct instillation (e.g., lubrication/rewetting solutions) in the eye using new active agents or preservatives with no history of prior ophthalmic/contact lens use or not generally recognized as safe for ophthalmic use
- Corneal Storage Media
- Extended Wear Contact Lens (i.e., including a single overnight use)

- Glaucoma Treatment Devices (e.g., trabeculoplasty devices, devices that treat ciliary bodies, devices that raise or lower intraocular pressure, aqueous shunt/drainage devices)
- Implants for Refractive Purposes (e.g., intraocular lenses, corneal implants, scleral expansion bands)
- Intraocular Lenses (IOLs)
- Keratoprostheses
- Refractive Surgical Devices (e.g., lasers, electrical current devices, thermal and non-thermal keratoplasty devices, ablation devices, expansion rings, treatment of ciliary bodies)
- Retinal Disease Treatment Devices (e.g., electrical stimulation devices to treat macular degeneration, lasers to ablate epiretinal membranes and vitreous strands)
- Retinal Prosthesis (implant)
- Retinal Reattachment Devices (e.g., fluids, gases, perfluorocarbons, perfluorpropane, silicone oil, sulfur hexafluoride, balloon catheter for retinal reattachment)
- Viscosurgical Fluids (viscoelastics)

Radiology
- Boron Neutron Capture Therapy
- Hyperthermia Systems and Applicators

Orthopedics and Restorative
- Anti-Adhesion Gels
- Bone Growth Stimulators
- Bone Morphogenetic Proteins/Biodegradable Scaffolds combination products, with or without allograft/autograft combinations and with or without metallic implant
- Bone Void Fillers (hydroxyapatite and other materials)
- Bovine Collagen Meniscus Implants
- Computer Guided Robotic Surgery
- Implantable Peripheral Neuromuscular Stimulators
- Implantable Prostheses (ligament, tendon, hip, knee, finger)
- Implantable Spinal Devices
- Injectable Sodium Hyaluronate

Significant Risk
In the context of medical devices, there can be various significant risks that need to be assessed and managed. FDA regulations state that for studies involving use of an investigational device, the investigator (or sponsor) must obtain either a "significant risk" Investigational Device Exemption (IDE) from the FDA, or a determination of "non-significant risk" from the IRB. The specific risks associated with a particular medical device will depend on its intended use, complexity, and potential interactions with patients.

Biocompatibility: Medical devices that come into direct contact with the human body, such as implants or devices with prolonged skin contact, need to be biocompatible. The risk assessment should evaluate the potential for allergic reactions, inflammation, or other adverse reactions due to the materials used in the device.

Mechanical hazards: Some medical devices involve mechanical components or moving parts, which can pose risks such as mechanical failure, entrapment, or injury. Risk assessment should identify potential hazards, evaluate their severity, and implement appropriate safety mechanisms or design features to prevent harm to patients or users.

Electrical safety: Medical devices that utilize electrical components or energy sources can present electrical safety risks. These risks include electric shock, burns, or interference with other devices. Risk assessment should include evaluation of electrical safety standards and implementation of measures to mitigate the potential hazards.

Software and cybersecurity: With the increasing integration of software and connectivity in medical devices, the risks associated with software malfunctions, data breaches, or unauthorized access become significant. Risk assessment should consider vulnerabilities, privacy concerns, and potential impacts on patient safety to implement adequate cybersecurity measures.

Use-related hazards: Risks can arise from the use or misuse of medical devices by healthcare professionals or patients. Factors such as inadequate training, improper maintenance, or failure to follow instructions can lead to errors or adverse events. Risk assessment should identify potential use-related hazards and consider human factors to design devices and accompanying instructions that minimize the risks.

Sterilization and infection control: Medical devices intended for reuse or that come into contact with bodily fluids need to undergo effective sterilization procedures to prevent the spread of infections. Risk assessment should address potential risks associated with inadequate sterilization or ineffective infection control practices.

Performance and accuracy: Some medical devices, such as diagnostic instruments or monitoring devices, need to provide accurate and reliable results. Risk assessment should evaluate potential risks related to inaccurate measurements or incorrect interpretation of data and implement measures to ensure device performance and accuracy.

The actual risks will vary depending on the specific device and its application. Thorough risk assessment should be conducted for each medical device to identify and manage these risks effectively [4].

1.6 Significant Risks Associated with Medical Devices

1.6.1 Death and Serious Complications

Death is indeed one of the most significant risks associated with medical devices. While medical devices are designed to improve patient outcomes and save lives, there are instances where device-related factors can lead to severe complications or even death.

Device failure: Malfunction or failure of a medical device can have life-threatening consequences. For example, a malfunctioning defibrillator may fail to deliver a life-saving shock during cardiac arrest, or a ventilator malfunction may result in inadequate oxygenation, leading to respiratory failure and death. Device failures can occur due to design flaws, manufacturing defects, or improper use.

Surgical complications: Medical devices used during surgical procedures, such as implants or surgical instruments, carry inherent risks. Surgical errors, such as incorrect device placement or improper technique, can lead to severe complications, including internal bleeding, organ damage, or systemic infections. These complications can result in significant morbidity or mortality.

Infections: Infections related to medical devices can be life-threatening. Implantable devices, such as prosthetic joints or cardiac devices, carry a risk of device-related infections. These infections can lead to sepsis, endocarditis, or deep-seated abscesses, which can be difficult to treat and may result in death if not promptly addressed.

Medication delivery devices: Certain medical devices, such as infusion pumps or automated medication delivery systems, play a critical role in administering medications to patients. Malfunctions or programming errors in these devices can lead to medication errors, including incorrect dosages or rates of administration, which can have lethal consequences, such as drug overdoses or adverse drug reactions.

Radiation therapy devices: Medical devices used in radiation therapy, such as linear accelerators or brachytherapy devices, are essential for treating cancer. However, errors in radiation dosage calculations or delivery can cause significant harm, including severe tissue damage, radiation burns, or radiation-induced organ failure, potentially leading to death.

User errors: Human factors play a significant role in the safe and effective use of medical devices. User errors, such as incorrect device setup, programming errors, or misinterpretation of device readings, can result in adverse events, including life-threatening situations. Inadequate training, lack of familiarity with the device, or high-stress environments can contribute to these errors.

Implantable device complications: Implantable medical devices, such as pacemakers, defibrillators, or neurostimulators, carry risks associated with the implantation procedure and long-term use. Complications such as infection, device migration, lead fractures, or battery failures can occur, leading to serious health consequences, including death.

It is important to note that the occurrence of death as a result of medical device-related factors is relatively rare compared to the vast number of successful device uses. However, the potential for severe outcomes underscores the critical need for rigorous risk assessment, ongoing monitoring, and continuous improvement of medical devices to ensure patient safety and minimize these significant risks. Regulatory bodies, healthcare professionals, and manufacturers collaborate to establish safety standards, conduct post-market surveillance, and address potential risks to mitigate the occurrence of device-related deaths [5].

1.6.2 Infection or Contamination

Medical devices that come into contact with patients, such as catheters or surgical instruments, can pose a risk of infection if not properly sterilized or maintained. Contamination can lead to serious infections, including bloodstream infections, surgical site infections, or device-related infections. Infections related to medical devices can be life-threatening. Implantable devices, such as prosthetic joints or cardiac devices, carry a risk of device-related infections. These infections can occur when microorganisms, such as bacteria, viruses, or fungi, colonize or invade the site where a medical device is inserted or applied. These infections can lead to sepsis, endocarditis, or deep-seated abscesses, which can be difficult to treat and may result in death if not promptly addressed.

Types of device-related infections: Device-related infections can occur in various scenarios, including:

- Surgical site infections (SSIs): These infections occur at the surgical incision site or in the tissues surrounding the surgical site. Examples include infections following implant surgeries (e.g., joint replacements) or after the placement of catheters or prosthetic devices.
- Central line-associated bloodstream infections (CLABSIs): These infections arise when microorganisms enter the bloodstream through central venous catheters or other intravascular devices. CLABSIs can be particularly severe, leading to bloodstream infections and sepsis.
- Urinary tract infections (UTIs): Urinary catheters, commonly used in hospitals or long-term care settings, can introduce bacteria into the urinary tract, resulting in UTIs.
- Ventilator-associated pneumonia (VAP): Patients on mechanical ventilation are at risk of developing pneumonia due to the presence of the endotracheal tube or tracheostomy tube.

Risk factors: Several factors contribute to the risk of device-related infections:

Duration of device use: Prolonged use of medical devices increases the risk of infection. The longer a device remains in place, the more opportunity bacteria have to colonize and form biofilms on the device surface.

Poor aseptic technique: Inadequate adherence to proper infection prevention measures during device insertion, maintenance, or removal, such as hand hygiene, sterile technique, or disinfection protocols, can increase the risk of infections.

Immunocompromised patients: Individuals with weakened immune systems, such as those undergoing chemotherapy or organ transplantation, are more susceptible to device-related infections.

Contamination: Contaminated devices or improper handling, storage, or cleaning practices can introduce bacteria or other microorganisms, increasing the risk of infection.

Prevention and control: Preventing device-related infections is a multifaceted approach that involves various strategies, including:

- Adherence to infection prevention guidelines: Healthcare facilities follow infection prevention guidelines and best practices to minimize the risk of device-related infections. This includes appropriate hand hygiene, sterile technique during device insertion, and maintenance protocols.
- Catheter selection and maintenance: Proper selection of catheters and devices with infection prevention features, such as antiseptic-coated catheters, can help reduce the risk of infections. Regular assessment, monitoring, and timely removal of devices are also crucial.
- Education and training: Healthcare professionals receive training on infection prevention protocols, aseptic techniques, and device handling to ensure proper practices are followed.
- Surveillance and monitoring: Healthcare facilities implement surveillance systems to monitor and track device-related infections, enabling early identification, intervention, and quality improvement initiatives.

Antibiotic resistance: Device-related infections are increasingly complicated by the emergence of antibiotic-resistant microorganisms. Infections caused by multidrug-resistant bacteria can be challenging to treat, leading to prolonged hospital stays, increased healthcare costs, and poorer patient outcomes.

Addressing infections related to medical devices requires a comprehensive and multidisciplinary approach involving healthcare professionals, infection prevention specialists, manufacturers, and regulatory bodies. By implementing evidence-based practices, adhering to infection prevention guidelines, and continuously monitoring and improving processes, healthcare facilities can mitigate the risk of device-related infections and enhance patient safety.

Medication delivery devices: Certain medical devices, such as infusion pumps or automated medication delivery systems, play a critical role in administering medications to patients. Malfunctions or programming errors in these devices can lead to medication errors, including incorrect dosages or rates of administration, which can have lethal consequences, such as drug overdoses or adverse drug reactions.

1.6.3 Radiation Exposure

Medical devices that use radiation for imaging purposes, such as X-ray machines or CT scanners, carry the risk of overexposure to radiation. Inadequate shielding, incorrect settings, or prolonged exposure can lead to harmful radiation doses and potential long-term health effects.

Radiation therapy devices: Medical devices used in radiation therapy, such as linear accelerators or brachytherapy devices, are essential for treating cancer. However, errors in radiation dosage calculations or delivery can cause significant harm, including severe tissue damage, radiation burns, or radiation-induced organ failure, potentially leading to death. Radiation exposure from medical devices is an important consideration in healthcare settings. While many medical devices use various forms of radiation for diagnosis or treatment, it's crucial to manage and minimize the risks associated with radiation exposure. Here are some significant risks of radiation exposure from medical devices:

Direct radiation exposure: Healthcare professionals and patients who are in close proximity to radiation-emitting devices, such as X-ray machines, fluoroscopy systems, or nuclear medicine equipment, may be at risk of direct radiation exposure. Prolonged or repeated exposure to ionizing radiation can increase the risk of radiation-related health issues, including cancer.

Scatter radiation: When radiation is used in medical imaging procedures, such as X-rays or computed tomography (CT) scans, some of the radiation scatters and can reach areas outside the intended target area. This scatter radiation can expose healthcare workers and patients to additional radiation doses.

Cumulative radiation doses: Patients who undergo multiple diagnostic or therapeutic procedures involving radiation over their lifetime may accumulate higher radiation doses. This cumulative exposure can increase the long-term risk of radiation-related complications.

Inadequate shielding or protection: In some cases, inadequate shielding or lack of proper protective measures around radiation-emitting devices can lead to unnecessary radiation exposure. Shielding materials, such as lead aprons, leaded glass, or protective barriers, are crucial to reducing radiation exposure to individuals in the vicinity.

Operator error: Improper use or incorrect settings of radiation-emitting devices can result in higher radiation doses than necessary. Human error, such as prolonged exposure or incorrect positioning, can contribute to increased radiation exposure risks.

To mitigate these risks, healthcare facilities and professionals must prioritize radiation safety protocols and guidelines, including:

- Proper training and education of healthcare professionals operating radiation-emitting devices.
- Implementing appropriate shielding and protective measures, such as lead aprons, thyroid collars, and protective barriers.

- Adhering to ALARA (As Low as Reasonably Achievable) principles, which aim to minimize radiation doses during procedures.
- Regular calibration and maintenance of radiation-emitting devices to ensure accurate and controlled radiation output.
- Follow-up and monitoring of patients' cumulative radiation exposure to avoid excessive doses.
- By implementing strict radiation safety measures and maintaining a culture of awareness, healthcare facilities can significantly reduce the risks associated with radiation exposure from medical devices while ensuring effective diagnosis and treatment.

1.6.4 Implantable Device Complications

Implantable devices are medical devices that are designed to be placed inside the body for various therapeutic purposes. While these devices have significantly improved patient outcomes and quality of life, they are not without risks. Implantable medical devices, such as pacemakers, defibrillators, or neurostimulators, carry risks associated with the implantation procedure and long-term use. Complications such as infection, device migration, lead fractures, or battery failures can occur, leading to serious health consequences, including death. Complications related to implantable devices can occur, and it's essential to understand these potential issues. Here are some common implantable device complications:

Infection: Infection is a significant complication associated with implantable devices. The presence of a foreign object in the body can create an entry point for bacteria, leading to localized or systemic infections. Common signs of infection include pain, redness, swelling, discharge, or fever. Prompt diagnosis and appropriate treatment, which may involve antibiotics or device removal, are necessary to manage implant-related infections.

Device migration: Implantable devices may occasionally migrate or move from their intended position. This can occur due to factors such as improper surgical placement, insufficient fixation, or anatomical changes in the patient. Device migration can result in loss of function, tissue damage, or the need for surgical intervention to reposition or remove the device.

Device failure: Implantable devices may experience mechanical or functional failures over time. This can include issues such as fractures, disconnections, battery depletion, or loss of electrical stimulation. Device failure can lead to a loss of therapeutic efficacy, necessitating device revision or replacement.

Allergic reactions: Some patients may develop allergic reactions to the materials used in implantable devices, such as metals or polymers. Allergic reactions can manifest as localized inflammation, pain, or systemic allergic responses. In severe cases, allergic reactions may require device removal and alternative treatment options.

Tissue damage or irritation: The presence of an implantable device can sometimes cause tissue damage or irritation. This can occur due to factors such as device movement, friction, or foreign body reaction. Tissue damage can lead to pain, inflammation, erosion, or ulceration in the surrounding tissues.

Hematoma or bleeding: Surgical implantation procedures carry the risk of hematoma formation or excessive bleeding at the surgical site. Hematomas can lead to pain, swelling, or even compromise the function of the implanted device. Surgical techniques and postoperative management strategies aim to minimize the risk of hematoma formation.

Thrombosis or embolism: Some implantable devices, such as stents or artificial heart valves, carry the risk of thrombosis (blood clot formation) or embolism (the detachment of a clot that can travel and block blood vessels). Anticoagulant therapy or antiplatelet medications may be necessary to mitigate the risk of these complications.

Psychological and emotional impacts: Implantable devices, particularly those related to cosmetic or functional enhancements, can have psychological and emotional impacts on patients. These may include body image concerns, psychological adjustment issues, or changes in lifestyle and self-perception. Adequate patient counseling and support are important to address these aspects of implantable device use.

It's important to note that the occurrence of complications associated with implantable devices is relatively low, and the benefits of these devices generally outweigh the risks. Healthcare professionals work closely with patients to assess individual risks, select appropriate devices, and provide post-implantation care to minimize complications. Rigorous adherence to surgical techniques, patient monitoring, and timely management of complications are crucial for optimizing patient outcomes when using implantable devices.

1.6.5 Device Failure or Malfunction

There is a risk of device failure or malfunction, which can have serious consequences for patients. This can include issues with electronic components, mechanical failures, or software glitches that affect the device's functionality or accuracy. Malfunction of medical devices can indeed pose significant risks to patients and healthcare providers. These risks can vary depending on the type of device and its intended use, but some potential consequences include:

Patient harm: Malfunctioning medical devices can directly harm patients by delivering incorrect or inappropriate treatments, providing inaccurate readings, or causing physical injury. For example, a faulty infusion pump could administer an incorrect dosage of medication, leading to adverse effects or inadequate treatment.

Delayed or incorrect diagnosis: Diagnostic devices that malfunction can lead to delayed or incorrect diagnoses, potentially resulting in delayed treatment or unnecessary procedures. This can have serious implications for patient outcomes and overall healthcare costs.

Infection and contamination: Malfunctioning devices, such as improperly sterilized surgical instruments or contaminated implantable devices, can increase the risk of healthcare-associated infections. These infections can lead to prolonged hospital stays, additional treatments, and even life-threatening complications.

Operational disruptions: When medical devices fail, it can disrupt the workflow and operations of healthcare facilities. This may result in delays in patient care, increased workload for healthcare professionals, and potentially compromised patient safety.

Financial implications: Device failures can lead to additional costs for healthcare providers, including the need for device replacement, repairs, or legal repercussions. These financial burdens can have ripple effects on the healthcare system as a whole.

Regulatory compliance: Medical device failures can trigger investigations from regulatory agencies, such as the US Food and Drug Administration (FDA). If a device failure is determined to be due to design flaws, manufacturing defects, or inadequate quality control, it may result in regulatory actions, including warnings, recalls, or even bans on device usage.

To mitigate the risks associated with device failures or malfunctions, healthcare providers should implement rigorous quality assurance and maintenance programs, conduct regular device inspections and testing, stay updated on manufacturer recalls and safety alerts, and ensure proper staff training on device usage and troubleshooting.

1.6.6 Adverse Tissue Reaction

Implantable medical devices, such as artificial joints or pacemakers, carry the risk of adverse tissue reactions or implant rejection. These reactions can manifest as inflammation, pain, or other complications that may require revision surgeries or removal of the device.

Allergic reactions: Some individuals may have allergies or hypersensitivity to certain materials used in medical devices, such as metals (e.g., nickel, cobalt) or latex. When these materials come into contact with the body's tissues, they can trigger an immune response, leading to symptoms like inflammation, redness, itching, pain, or even systemic allergic reactions.

Inflammatory response: The body's immune system may react to the presence of a medical device as a foreign object, leading to an inflammatory response. This response can cause tissue damage, swelling, pain, and impaired device function.

In some cases, excessive or prolonged inflammation can result in chronic conditions or complications.

Infection risk: Medical devices can provide a surface for bacterial or fungal colonization, increasing the risk of infections. Infections can cause local tissue damage, delay healing, and lead to systemic infections that can be life-threatening. Inadequate sterilization during device implantation or improper maintenance can contribute to infection risks.

Wear and degradation: Over time, some medical devices, especially those with moving parts or that experience mechanical stress, can degrade or wear down. This degradation can result in the release of particles or debris, which may trigger an inflammatory response or damage surrounding tissues.

Corrosion and leaching: Certain medical devices made from metals may corrode or leach substances into the surrounding tissues. Corrosion can lead to the release of metal ions or particles, causing localized tissue damage and potential systemic effects. This is particularly relevant in implants such as joint replacements or cardiovascular stents.

Fibrous encapsulation: In response to the presence of a medical device, the body may form a protective fibrous capsule around it. While this encapsulation is a natural healing process, excessive or abnormal capsule formation can result in device migration, mechanical interference, or compromised device function.

To mitigate the risks of adverse tissue reactions, manufacturers must ensure the biocompatibility of materials used in medical devices and conduct thorough premarket testing. Healthcare providers should follow proper sterilization protocols, monitor patients for allergic reactions or signs of infection, and promptly address any complications or adverse events associated with medical devices.

In cases where patients have known allergies or sensitivities to certain materials, healthcare professionals should carefully consider alternative device options or take appropriate precautions to minimize the risk of adverse tissue reactions.

1.6.7 Electrical Shocks or Burns

Medical devices that utilize electricity, such as defibrillators or electrosurgical devices, can pose risks of electrical shocks or burns if not properly designed, maintained, or used. Improper grounding, faulty wiring, or user errors can lead to these hazards.

Improper grounding or wiring: Faulty electrical grounding or wiring of medical devices can result in electrical shocks or burns. If the electrical current is not properly grounded or insulated, it can pass through the patient's body, causing harm. This can occur during device use, such as during surgery or when using electrical stimulation devices.

Malfunctioning or defective devices: Malfunctioning or defective medical devices can generate excessive heat or deliver electric shocks to patients. This can occur due to manufacturing defects, design flaws, inadequate quality control, or wear and tear over time. Examples include faulty defibrillators, electrocautery devices, or electrically powered surgical instruments.

Inadequate maintenance or inspections: Failure to perform regular maintenance, inspections, and calibration of electrical medical devices can increase the risk of electrical shocks or burns. Worn-out cables, frayed wires, or damaged insulation can expose patients to electrical currents or create sparks that can cause burns.

Inappropriate use or user error: Incorrect use of medical devices, such as improper connection or incorrect settings, can lead to electrical shocks or burns. User error, lack of training, or inadequate understanding of device operation can contribute to these incidents.

Sensitivity of certain patient populations: Some patient populations may be more susceptible to electrical shocks or burns, such as those with compromised skin integrity, reduced sensation, or pre-existing medical conditions. These individuals may have a diminished ability to perceive pain or respond to electrical stimulation, making them more vulnerable.

To mitigate the risks of electrical shocks or burns from medical devices, healthcare providers should adhere to the following measures:

- Regularly inspect and maintain electrical devices according to manufacturer guidelines.
- Train healthcare personnel on proper use and handling of electrical devices.
- Ensure proper grounding and electrical safety measures are in place, including appropriate electrical outlets, grounding pads, and surge protectors.
- Follow established protocols for device setup, use, and maintenance.
- Monitor patients closely during device use, and promptly address any signs of electrical complications or adverse events.
- Encourage patients to report any discomfort or unusual sensations during device usage.

Manufacturers play a crucial role in ensuring the safety of medical devices by adhering to quality standards, conducting thorough testing, and addressing any identified electrical safety issues. Regulatory bodies, such as the FDA, also have regulations and guidelines in place to monitor and mitigate the risks associated with medical device electrical hazards.

1.6.8 User Errors

The risk of user errors should also be considered. Healthcare professionals or patients may misinterpret device instructions, use the device incorrectly, or make mistakes during device setup or calibration, leading to incorrect readings, dosages, or treatment decisions.

Incorrect device operation: Users may not fully understand the proper operation of a medical device, leading to errors in its use. This can include errors in setting up the device, adjusting settings, or interpreting and responding to device readings or alarms.

Improper device selection: Choosing the wrong device for a particular procedure or patient condition can result in errors. Different medical devices have specific indications, limitations, and patient requirements. Failure to select the appropriate device can lead to ineffective treatment, inadequate monitoring, or adverse events.

Inadequate training: Insufficient training on the proper use of medical devices can contribute to user errors. Healthcare professionals need comprehensive training on device operation, safety precautions, troubleshooting, and potential complications. Lack of training may result in improper device handling, misinterpretation of data, or failure to recognize and respond to device malfunctions.

Misinterpretation of device readings: Medical devices often provide data and readings that require interpretation. User errors can occur when healthcare professionals misinterpret or misread device data, leading to incorrect clinical decisions or inappropriate treatment interventions.

Failure to follow manufacturer guidelines: Manufacturers provide detailed instructions for the proper use and maintenance of medical devices. User errors can occur when healthcare professionals fail to adhere to these guidelines, including failure to perform regular maintenance, calibration, or use of incompatible accessories or consumables.

Inadequate communication: Poor communication among healthcare professionals can result in user errors with medical devices. This can include incomplete or inaccurate handoff of information, failure to communicate device-related concerns, or lack of clear protocols for device use in interdisciplinary settings.

To minimize user errors with medical devices, healthcare providers should prioritize the following:

- Comprehensive training programs that cover device operation, safety precautions, troubleshooting, and potential complications
- Implementation of standardized protocols and checklists for device use and maintenance
- Clear communication channels and effective interdisciplinary collaboration
- Regular competency assessments and ongoing education to ensure healthcare professionals' proficiency in device operation
- Encouraging an environment that promotes reporting and learning from near misses or incidents related to device use

Manufacturers can contribute to reducing user errors by designing devices with user-friendly interfaces, clear instructions, and intuitive features. Providing comprehensive training materials and offering customer support for device users can also help minimize errors. Collaboration between healthcare providers and manufacturers can lead to the development of safer and more user-friendly medical devices.

1.6.9 Cybersecurity Threats

With the increasing connectivity of medical devices, there is a growing concern regarding cybersecurity risks. Hacking or unauthorized access to medical devices can potentially compromise patient data, disrupt device functionality, or even cause harm to patients.

Unauthorized access: Cybercriminals may attempt to gain unauthorized access to medical devices, such as implantable devices, infusion pumps, or diagnostic equipment. Once inside the device's network, they can manipulate or disrupt its functionality, potentially causing harm to patients or compromising data integrity.

Data breaches and privacy risks: Medical devices collect and transmit sensitive patient data, including personal health information. If these devices are compromised, there is a risk of data breaches, leading to the exposure of patients' private medical information. This can result in identity theft, insurance fraud, or other malicious activities.

Malware and ransomware attacks: Medical devices connected to the internet or hospital networks can be targeted with malware or ransomware attacks. Malicious software can infiltrate the device, disrupt its operation, steal data, or even encrypt critical patient data and demand ransom for its release. Such attacks can disrupt healthcare operations and compromise patient safety.

Lack of patching and vulnerability management: Medical devices often run on specialized software or operating systems, which may not receive regular security updates or patches. This leaves them vulnerable to known exploits and makes it challenging to address newly discovered vulnerabilities promptly. Without proper vulnerability management, cyber attackers can exploit these weaknesses.

Insider threats: Insider threats, including disgruntled employees or individuals with authorized access to medical devices, pose a significant cybersecurity risk. These individuals may intentionally misuse or compromise the devices, either for personal gain, revenge, or other malicious purposes.

Supply chain risks: The complex supply chain involved in the development and distribution of medical devices introduces potential vulnerabilities. A compromised component or a compromised update from a third-party supplier can compromise the overall security of a medical device.

To address cybersecurity threats in medical devices, various stakeholders must take action.

Manufacturers: Device manufacturers should prioritize cybersecurity throughout the entire lifecycle of a medical device, including secure design practices, regular security updates, and proper vulnerability management. They should also implement robust authentication mechanisms, encryption, and intrusion detection systems.

Healthcare providers: Healthcare facilities should implement comprehensive cybersecurity protocols and practices. This includes network segmentation to isolate medical devices, regular security assessments, staff training on cybersecurity best practices, and incident response plans.

Regulatory bodies: Regulatory bodies play a crucial role in setting and enforcing cybersecurity standards for medical devices. They should establish robust regulations and guidelines to ensure the security and privacy of medical devices, as well as mandate regular security assessments and updates.

Collaboration: Collaboration among stakeholders, including manufacturers, healthcare providers, cybersecurity experts, and regulators, is crucial to effectively address cybersecurity threats. Sharing information, best practices, and lessons learned can help enhance the security posture of medical devices.

By acknowledging the cybersecurity risks associated with medical devices and implementing proactive security measures, the healthcare industry can work towards mitigating these threats and ensuring patient safety and privacy.

1.6.10 Device Materials

While medical devices are designed to improve patient outcomes, there can be instances of serious adverse reactions to the materials used in these devices. Adverse reactions can occur due to a variety of factors, including allergies, hypersensitivity, or incompatibility with the body's tissues. Here are some examples of serious adverse reactions to medical device materials:

Allergic reactions: Some individuals may develop allergies to specific materials used in medical devices, such as metals (e.g., nickel, cobalt), latex, or certain plastics. Allergic reactions can range from mild irritation to severe immune responses, including hives, rash, swelling, or even anaphylaxis.

Hypersensitivity reactions: Hypersensitivity reactions can occur when the body's immune system reacts adversely to the presence of a medical device or its components. This immune response can lead to inflammation, tissue damage, or systemic symptoms, depending on the individual's sensitivity.

Inflammatory response: In certain cases, the body may react to a medical device by initiating an inflammatory response. This can result in local inflammation, pain, redness, and swelling around the site of device implantation or contact.

Infections: Although rare, medical devices can potentially increase the risk of infection. Bacteria or other microorganisms may adhere to the surface of a device, leading to localized or systemic infections. The risk of infection can be influenced by various factors, such as the material's surface properties, the presence of biofilms, or the duration of device use.

Implant failure or complications: In some cases, the body may exhibit a negative reaction to a medical device, leading to implant failure or complications. This can occur due to factors like material degradation, wear and tear, mechanical stress, or the body's response to the device, such as fibrosis or rejection.

Chemical toxicity: Certain materials used in medical devices can release toxic substances over time, leading to adverse effects. For example, leaching of harmful chemicals from certain plastics or metals can cause systemic toxicity or organ damage.

To address and minimize the risk of serious adverse reactions to medical device materials, it's important for manufacturers to conduct thorough biocompatibility testing and follow regulatory guidelines. Healthcare professionals should also be vigilant in assessing patients for potential allergies, sensitivities, or prior adverse reactions to specific materials. Individual patient factors and medical history should be carefully considered when selecting and implanting medical devices to minimize the risk of adverse reactions.

These examples highlight the diverse range of risks associated with medical devices. It is crucial for manufacturers, healthcare professionals, and regulatory bodies to identify and mitigate these risks through comprehensive risk assessments, rigorous testing, adherence to safety standards, and ongoing monitoring of device performance in real-world settings.

1.7 Non-significant Risks of Medical Devices

Non-significant risks of medical devices refer to risks that are considered to have minimal or negligible impact on patient safety or health. These risks are typically associated with medical devices that have undergone rigorous testing, evaluation, and regulatory approval processes. While no medical device is entirely risk-free, non-significant risks are generally outweighed by the benefits provided by the device [6]. Some examples of non-significant risks may include:

Temporary discomfort: Some medical devices, such as adhesive bandages or compression stockings, may cause temporary discomfort or irritation at the application site. However, these effects are usually mild and transient.

Allergic reactions: Certain individuals may experience minor allergic reactions, such as skin redness or itching, in response to certain materials used in medical devices, such as latex. However, the prevalence of these reactions is relatively low, and alternative materials can often be used to minimize the risk.

Cosmetic effects: Some medical devices, such as external braces or orthopedic supports, may have temporary cosmetic effects, such as visible marks or indentations on the skin. However, these effects are typically reversible and do not pose a significant health risk.

Device malfunctions: While medical devices are designed to function reliably, there is always a small possibility of device malfunctions, such as battery failure or sensor inaccuracies. However, these malfunctions are often detected through quality control processes, and the risk of harm to patients is generally low.

Non-serious side effects: Certain medical devices, particularly those used for non-invasive procedures or diagnostic tests, may have minor side effects such as mild dizziness, temporary pain, or transient changes in vital signs. These side effects are typically short-lived and resolve without causing significant harm.

It is important to note that the classification of risks as "non-significant" may vary depending on the context and the individual patient's circumstances. Healthcare professionals and regulatory authorities continually assess and monitor the safety of medical devices to ensure that risks are minimizedand the benefits outweigh any potential harm.

1.8 Regulations and Guidelines

Regulations and guidelines for managing significant risks associated with medical devices vary across countries and regions. Here are some examples of regulatory frameworks and guidelines that aim to address these risks:

United States: In the United States, the Food and Drug Administration (FDA) regulates medical devices through the Federal Food, Drug, and Cosmetic Act (FD&C Act) and the Medical Device Amendments of 1976. The FDA provides guidelines and regulations for device classification, premarket approval, post-market surveillance, adverse event reporting, and quality system requirements. The FDA's Center for Devices and Radiological Health (CDRH) focuses specifically on the regulation of medical devices.

European Union: The European Union (EU) regulates medical devices through the Medical Device Regulation (MDR) and the In Vitro Diagnostic Medical Device Regulation (IVDR). These regulations aim to ensure the safety, performance, and effectiveness of medical devices. They outline requirements for device classification, conformity assessment, clinical evaluation, post-market surveillance, and vigilance reporting. The regulations also establish the role of Notified Bodies in assessing the conformity of devices.

International Electrotechnical Commission (IEC): The IEC is an international standardization organization that develops standards for electrical and electronic devices, including medical devices. The IEC 60601 series of standards focuses on the safety and essential performance of medical electrical equipment, addressing areas such as electromagnetic compatibility, risk management, and usability.

International Organization for Standardization (ISO): The ISO has developed several standards relevant to medical devices, including ISO 13485, which outlines requirements for a quality management system specific to the medical device industry. ISO 14971 provides guidance on risk management for medical devices, assisting manufacturers in identifying and addressing potential risks associated with their products.

Medical Device Reporting (MDR) Systems: Many regulatory bodies require the implementation of Medical Device Reporting systems, where manufacturers, healthcare providers, and users are obligated to report adverse events, malfunctions, or incidents related to medical devices. These systems help identify potential risks, track device performance, and enable regulatory bodies to take appropriate actions.

It's important to note that regulations and guidelines for managing risks associated with medical devices are continually evolving to keep pace with technological advancements and emerging challenges. Healthcare professionals and manufacturers should stay updated on these regulations and actively engage in compliance to ensure patient safety and regulatory compliance.

1.9 Pre-market Approval Process for Medical Devices

The pre-market approval process for medical devices involves the regulatory evaluation and approval of devices before they can be marketed and sold. The specific requirements and processes may vary depending on the regulatory authority and the risk classification of the device.

Determine the regulatory pathway: Identify the appropriate regulatory pathway based on the device's risk classification and intended use. This can involve determining whether the device requires pre-market notification (510(k)) or pre-market approval (PMA) in the United States, or a Conformity Assessment procedure in the European Union, among other regulatory pathways.

Compilation of data and documentation: Prepare the necessary data and documentation to support the device's safety, efficacy, and performance. This includes detailed technical specifications, design documentation, risk assessments, clinical data (if applicable), and manufacturing processes. The documentation should align with the regulatory requirements and standards.

Preparing a regulatory submission: Compile all the relevant data and documentation into a comprehensive regulatory submission. This submission serves as an application for regulatory approval and typically includes the device's intended use, indications for use, labeling, instructions for use, and evidence demonstrating compliance with regulatory requirements.

Submission review: The regulatory authority reviews the submission to assess the device's safety, efficacy, and compliance with regulatory standards. The review process can involve multiple rounds of evaluations, requests for additional information, and discussions with the regulatory authority to address any questions or concerns.

Regulatory decision: Based on the review and evaluation, the regulatory authority makes a decision regarding the approval or clearance of the device. If the device meets the regulatory requirements, it may receive marketing authorization, such as clearance or approval to market the device for specific indications and claims.

The regulatory authority may also impose certain conditions, such as post-market surveillance requirements or labeling modifications.

Post-market responsibilities: Once the device is approved or cleared for market, the manufacturer is responsible for post-market activities. This includes monitoring the device's performance, collecting post-market surveillance data, addressing any adverse events or complaints, and complying with ongoing regulatory requirements, such as quality system regulations and adverse event reporting.

It's important to note that the pre-market approval process can be complex and time-consuming, particularly for higher-risk devices. Manufacturers are advised to consult the specific regulations and guidelines provided by the regulatory authority in their target market to understand the detailed requirements and procedures for their device's pre-market approval.

1.10 Post-market Surveillance

Post-market surveillance (PMS) of medical devices refers to the ongoing monitoring and evaluation of devices after they have been approved or cleared for market. It involves collecting and analyzing data on device performance, safety, and effectiveness during real-world use. The primary goal of post-market surveillance is to detect and respond to any adverse events, device malfunctions, or other safety concerns that may arise once the device is in widespread use [7].

Adverse event reporting: Manufacturers, healthcare professionals, and users are typically required to report adverse events associated with medical devices to the regulatory authorities. This includes incidents of device malfunction, patient injuries, or other unexpected adverse effects. These reports help identify potential safety issues and contribute to the ongoing evaluation of device performance.

Post-market studies: Regulatory authorities may require manufacturers to conduct post-market studies to gather additional data on device performance and safety. These studies can be in the form of clinical trials, registries, or post-approval studies. They help assess long-term outcomes, evaluate device effectiveness, and identify any potential risks or benefits that may not have been apparent during pre-market testing.

Quality system requirements: Manufacturers are typically obligated to establish and maintain a quality management system (QMS) that includes processes for post-market surveillance. The QMS ensures that the manufacturer has systems in place to collect, analyze, and act upon post-market data, including complaints, adverse events, and other relevant information.

Post-market surveillance plans: Manufacturers may be required to develop post-market surveillance plans that outline their strategies and methodologies for collecting and analyzing post-market data. These plans specify the activities, timelines, and responsibilities for post-market surveillance, including monitoring of complaints, adverse event reporting, and periodic reporting to regulatory authorities.

Signal detection and risk management: Post-market surveillance aims to identify and analyze signals that may indicate potential safety concerns or emerging risks associated with a device. This involves analyzing data from various sources, such as adverse event reports, complaints, literature reviews, and post-market studies. Manufacturers are responsible for assessing the significance of these signals and implementing appropriate risk management measures if needed.

Post-market communication: Regulatory authorities may require manufacturers to maintain open and transparent communication channels with users, healthcare professionals, and regulatory bodies. This includes providing updates on device performance, safety alerts, and any necessary corrective actions or recalls. It also involves actively engaging with users and healthcare professionals to gather feedback, address concerns, and disseminate important safety information.

Post-market surveillance plays a crucial role in ensuring the ongoing safety and effectiveness of medical devices. It enables regulatory authorities and manufacturers to detect and respond to potential safety issues promptly, enhance device quality, and take appropriate measures to protect patient safety.

1.11 Risk Assessment and Management

Ongoing risk assessment and management in medical devices is crucial for several reasons:

Changing risk landscape: The risk landscape associated with medical devices can evolve over time. New information may emerge regarding potential risks or adverse events associated with a device. Ongoing risk assessment allows for the identification and understanding of these evolving risks, ensuring that appropriate risk management measures are in place to mitigate them.

Real-world performance: Pre-market testing and clinical trials provide valuable insights into a device's safety and effectiveness. However, real-world use may reveal additional risks or challenges that were not apparent during the initial testing phase. Ongoing risk assessment enables the collection and analysis of data on device performance in actual clinical settings, allowing for the identification and mitigation of any new risks that may arise.

Continuous improvement: Ongoing risk assessment and management support a culture of continuous improvement in medical device development and use. By monitoring device performance and identifying areas of improvement, manufacturers can refine their designs, manufacturing processes, and instructions for use to enhance device safety and effectiveness over time.

Feedback from users and healthcare professionals: Regularly engaging with users and healthcare professionals can provide valuable insights into the performance and safety of medical devices. Collecting feedback, including user complaints and reports of adverse events, allows for the identification of potential risks or issues that may have gone unnoticed. This feedback can inform ongoing risk assessment and help improve device design, labeling, and instructions for use.

Regulatory compliance: Regulatory authorities often require manufacturers to demonstrate ongoing compliance with safety and performance standards for medical devices. This includes conducting post-market surveillance, reporting adverse events, and implementing appropriate risk management measures. Ongoing risk assessment and management activities help ensure compliance with these regulatory requirements.

Patient safety: Ultimately, the primary goal of ongoing risk assessment and management in medical devices is to prioritize patient safety. By continuously monitoring and mitigating risks, manufacturers and healthcare professionals can take proactive measures to prevent harm and ensure the safe use of medical devices. This contributes to overall patient safety and enhances public trust in the healthcare system.

In summary, ongoing risk assessment and management in medical devices are essential for adapting to the changing risk landscape, improving device performance, complying with regulatory requirements, and, most importantly, safeguarding patient safety. It enables manufacturers and healthcare professionals to stay vigilant, identify emerging risks, and take proactive measures to mitigate them throughout the lifecycle of a medical device.

References

1. http://www.fda.gov/downloads/RegulatoryInformation/Guidances/UCM126418.pdf
2. Sampath T, Thamizharasan S, Krithaksha V, Timiri Shanmugam PS. Medical devices, a practical guide. Boca Raton: CRC Press; 2022.
3. https://www.fda.gov/medical-devices/overview-device-regulation/classify-your-medical-device
4. Medical device regulations: global overview and guiding principles. WHO. https://www.who.int/medical_devices/publications/en/MD_Regulations.pdf
5. International Organisation for Standardization. ISO 10993-2:2006, Biological evaluation of medical devices. 2006. https://www.iso.org/standard/36405.html
6. https://www.geisinger.org/-/media/OneGeisinger/pdfs/ghs/research/research-at-geisinger/hrpp/guidance/fda_signif_nonsignificant_med_device.pdf
7. https://irb.ucsf.edu/significant-vs-non-significant-risk-devices

Chapter 2
Significant Risks Medical Devices – General Medical Use

Pugazhenthan Thangaraju, B. Aravind Kumar, Hemasri Velmurugan, Sajitha Venkatesan, Ripudaman Arora, and Soumitra Trivedi

2.1 Introduction

The US Food and Drug Administration (FDA) has identified several significant risks associated with medical devices in general medical use. These risks include:

- *Mechanical failure:* Malfunctions or defects in the design, materials, or manufacturing of medical devices can result in mechanical failure and potentially harm patients.
- *Software malfunction:* Medical devices that rely on software can experience malfunctions, errors, or failures that can affect their performance and safety.
- *Inadequate labeling or instructions for use*: Poor labeling, unclear instructions for use, or inadequate training can result in incorrect usage of medical devices and potentially harm patients.
- *Adverse reactions:* Medical devices that use materials or chemicals that cause allergic reactions, infections, or other adverse events can harm patients.

P. Thangaraju (✉) · H. Velmurugan
Department of Pharmacology, All India Institute of Medical Sciences, Raipur, Chhattisgarh, India

B. Aravind Kumar
Department of Pharmacology, Pondicherry Institute of Medical Sciences, Pondicherry, India

S. Venkatesan
Department of Microbiology, All India Institute of Medical Sciences, Raipur, Chhattisgarh, India

R. Arora
Department of ENT, All India Institute of Medical Sciences, Raipur, Chhattisgarh, India

S. Trivedi
Department of Anatomy, All India Institute of Medical Sciences, Raipur, Chhattisgarh, India

P. S. Timiri Shanmugam et al. (eds.), *Significant and Nonsignificant Risk Medical Devices*, https://doi.org/10.1007/978-3-031-52838-5_2

- *Insufficient quality control measures:* Inadequate quality control measures during the manufacturing, distribution, or post-market surveillance of medical devices can result in safety hazards or inadequate performance.
- *Human error:* The use of medical devices requires human interaction, and errors or mistakes can occur that can potentially harm patients.
- *Cybersecurity risks*: Medical devices that are connected to networks or the internet can be vulnerable to cybersecurity risks, such as hacking or unauthorized access.

The FDA requires manufacturers to assess and manage these risks associated with medical devices throughout their life cycle to ensure their safety and effectiveness for patients and healthcare providers.

The US Food and Drug Administration (FDA) also categorizes medical devices for general medical use into three classes based on their potential risks to patients. These classes are as follows:

(a) *Class I:* Low-risk devices that are subject to general controls, such as labeling requirements and adherence to good manufacturing practices. Examples include tongue depressors and bandages.
(b) *Class II:* Moderate-risk devices that require special controls in addition to general controls, such as performance standards and post-market surveillance. Examples include powered wheelchairs and infusion pumps.
(c) *Class III:* High-risk devices that support or sustain human life or have a potential risk of serious injury or illness. These devices require premarket approval (PMA) before they can be marketed in the United States. Examples include implantable pacemakers and heart valves.

"Information Sheet Guidance Document for IRBs, Clinical Investigators and Sponsors on Significant Risk and Nonsignificant Risk Medical Devices" was revised to update the list of examples of significant and nonsignificant risk devices, to clarify the institutional review board (IRB)'s responsibilities when making the risk determination for investigational medical devices, and to make the guidance consistent with the Agency's good guidance practices regulations (21 CFR 10.115). The FDA regulates a wide range of medical devices in general medical use, including diagnostic tests, surgical instruments, monitoring devices, and others. Each device is classified based on its intended use, potential risks, and other factors. The following are examples of medical devices that are categorized under general medical use [1]:

- Catheters for general hospital use—except for conventional long-term percutaneous, implanted, subcutaneous, and intravascular
- Collagen implant material for use in ear, nose, and throat, orthopedics, plastic surgery, urological, and dental applications
- Surgical lasers for use in various medical specialties
- Tissue adhesives for use in neurosurgery, gastroenterology, ophthalmology, general and plastic surgery, and cardiology

2.2 Catheters

These include catheters for general hospital use except for conventional long-term percutaneous, implanted, subcutaneous, and intravascular. The U.S. Food and Drug Administration (FDA) categorizes medical devices into three classes based on the level of control necessary to ensure their safety and effectiveness. The general hospital use (GHU) category is a subset of Class II devices that require special control to ensure their safety and effectiveness. Catheters are medical devices inserted into the body to allow for the drainage of bodily fluids, delivery of medication or nutrients, or measurement of pressure, temperature, or other bodily functions.

Catheters for general hospital use can be classified into several types based on their intended use, such as:

- Urinary catheters: used for draining the bladder
- Central venous catheters: used for delivering medication and nutrients or for drawing blood
- Arterial catheters: used for measuring blood pressure in arteries
- Peripherally inserted central catheters (PICC): used for long-term access to veins
- Epidural catheters: used for pain management during childbirth, surgery, or cancer treatment

It is important to use sterile techniques and proper care when handling and inserting catheters. The risks associated with urinary catheters include urinary tract infection, bladder spasms, urethral damage, bloodstream infection, blockage or dislodgement of the catheter, urinary incontinence, and bladder stone formation.

The risk determination of urinary catheters involves evaluating the potential harm to patients that may result from the use of this devices [2]. The process typically includes the following steps:

1. Identifying the hazards associated with urinary catheter use, such as urinary tract infections, blood clots, and bladder damage.
2. Analyzing the likelihood and severity of harm, taking into account factors such as the type of catheter, the patient's health status, and the duration of catheterization.
3. Evaluating existing controls and their effectiveness, including infection prevention and control measures, appropriate catheter selection, and proper catheter insertion techniques.
4. Developing and implementing risk management plans, including regular monitoring of patients for adverse events and establishing protocols for prompt catheter removal.
5. Monitoring and updating risk management plans as needed, taking into account new information and changes in patient populations or medical practices.

It is important to minimize the risk of harm associated with urinary catheter use by applying evidence-based best practices and following appropriate protocols for catheter selection, insertion, maintenance, and removal.

Prior Investigations of Catheters for General Hospital Use

Catheters used in general hospitals have undergone numerous prior investigations to ensure their safety and effectiveness. The FDA requires medical device manufacturers to conduct extensive testing and clinical trials before their products can be approved for clinical use. Some examples of investigations that have been conducted on general hospital catheter use are as follows:

1. *Clinical trials:*

   Before a catheter can be approved for use in patients, clinical trials must be performed to evaluate its safety and effectiveness. These trials typically involve testing the catheter in a group of patients with the condition for which the catheter is intended and comparing the results with those of a control group.
2. *Biocompatibility testing:*

   The materials used in catheters must be biocompatible, meaning that they do not cause adverse reactions when they come into contact with living tissues. Manufacturers must conduct biocompatibility testing to ensure that their catheters meet these requirements.
3. *Mechanical testing:*

   Catheters must be able to withstand the mechanical stresses they are subjected to during use, such as bending and twisting. Manufacturers must conduct mechanical testing to ensure that their catheters are sufficiently strong to withstand stress.
4. *Sterility testing:*

   Catheters must be sterile to prevent the spread of the infection. Manufacturers must conduct sterility tests to ensure that their catheters are free of bacteria, viruses, and other pathogens.
5. *Shelf-life testing:*

   Catheters must have a shelf life that is long enough to allow for distribution and use, but not so long that they lose their effectiveness or become unsafe. Manufacturers must conduct shelf-life testing to determine the appropriate expiration date of their catheters.

2.3 Collagen Implant Material [3]

Collagen implants are biomaterials used to augment or repair soft tissues. The FDA categorizes certain collagen implants for general hospital use. Collagen is a biocompatible and biodegradable material that has been used in various medical applications, including ear, nose, and throat (ENT), orthopedics, plastic surgery, urology, and dentistry.

- In ENT, collagen implants are used for reconstructive surgeries of the larynx, pharynx, and nasal septum.
- In orthopedics, collagen is used as a scaffold for tissue regeneration in sports injuries, osteoarthritis, and cartilage repairs.

- In plastic surgery, collagen is used as an injectable filler to reduce wrinkles, enhance the lips, and improve facial contours.
- In urology, collagen is used to treat stress urinary incontinence and improve bladder function.
- Collagen is used as a membrane in dentistry to protect and promote healing during periodontal and endodontic procedures.

Collagen implants are well-tolerated by the body and often do not elicit an immune response. However, it is important to consider the potential risks and benefits of using collagen implants and select a reputable supplier to ensure product quality.

Risks associated with collagen implant material

- Infection
- Rejection
- Migration or displacement
- Impaired wound healing
- Tissue damage
- Allergic reaction
- Scarring [4]

Risk determination for collagen implants involves evaluating the potential harm to patients that may result from the use of these devices. The process typically includes the following steps:

1. Identifying the hazards associated with collagen implants, such as implant-related infections, tissue reactions, and implant migration or extrusion.
2. Analyzing the likelihood and severity of harm, taking into account factors such as the type of implant, the patient's health status, and the intended use of the implant.
3. Evaluating existing controls and their effectiveness, including the quality of the manufacturing process, the sterility of the implant at the time of use, and the surgeon's technique for implant placement.
4. Developing and implementing risk management plans, including close monitoring of patients for adverse events and establishing protocols for the prompt recognition and management of implant-related complications.
5. Monitoring and updating risk management plans as needed, taking into account new information and changes in medical practices or patient populations.

It is important to minimize the risk of harm associated with collagen implants using high-quality products, following evidence-based best practices, and closely monitoring patients for adverse events.

Prior Investigations of Collagen Implant Materials for General Hospital Use
Collagen implant materials have been extensively investigated to ensure safety and effectiveness. The FDA requires medical device manufacturers to conduct extensive testing and clinical trials before their products can be approved for use in patients.

Some examples of investigations conducted using collagen implant materials are as follows.

1. *Biocompatibility testing:*
   Collagen implant materials must be biocompatible, meaning that they do not cause adverse reactions when they come into contact with living tissues. Manufacturers must conduct biocompatibility testing to ensure that their collagen implant materials meet these requirements.
2. *Animal testing:*
   Before collagen implant materials can be tested in human clinical trials, they must be tested in animals to evaluate their safety and efficacy. This can involve testing materials in rats, rabbits, or other animals.
3. *Clinical trials:*
   Collagen implant materials must undergo clinical trials to evaluate their safety and efficacy in humans. These trials typically involve testing the materials in a group of patients with the condition for which the implant is intended, and comparing the results to those of a control group.
4. *Manufacturing process validation:*
   Collagen implant materials must be manufactured to ensure safety and consistency. Manufacturers must validate their manufacturing processes to ensure that the materials are of consistent quality.
5. *Shelf-life testing:*
   Collagen implant materials must have a shelf life that is long enough to allow for distribution and use, but not so long that they lose their effectiveness or become unsafe. Manufacturers must conduct shelf-life testing to determine the appropriate expiration date of collagen implant materials.

2.4 Surgical Lasers

Surgical lasers are a powerful tool for medical professionals, allowing for precise and efficient treatment of a variety of conditions. They are used in a variety of medical procedures, ranging from eye surgery to cosmetic procedures.

Surgical lasers use a concentrated beam of light to produce a precise cut, making them ideal for delicate operations. They are also used to treat a wide range of conditions, from cancers to vascular diseases. The LASER device is derived from "light amplification by the stimulated emission of radiation, which creates, amplifies, and emits a beam of light with coherent photons.

The surgeon will have control over three variables in most surgical lasers used: power (measured in watts), spot size (measured in square millimeters or square centimeters), and exposure time (measured in seconds). The most appropriate laser system for appropriate application requires sound knowledge from the surgeon regarding the characteristics of the interaction of laser light with biological tissue. The lasers used in surgery are of the ultraviolet type, which means that interactions are a complex mix of heat and photodissociation of chemical bonds. More commonly used lasers emit light in the visible or infrared region of the electromagnetic spectrum and primarily heat interacting tissues. To exert its effect, radiation must be absorbed by the tissue and converted to heat. Scattering spreads radiation to a large surface target tissue and limits the penetration depth. The shorter the wavelength of the radiation, the more scattered is the tissue. If radiation is reflected or transmitted through the target tissue, no effect is observed. Figure 2.1 shows laser machine.

Types

There are various types of surgical lasers, each of which is designed for a specific procedure. The most common types are CO_2 lasers, which are used for cutting and reshaping tissues, and erbium lasers, which are used for resurfacing the skin. Other types of lasers include diode lasers, which are used for hair and tattoo removal, and YAG (yttrium aluminum garnet) lasers, which are used for eye surgery. Each type of laser has its own specific set of benefits and risks and should be used only by a trained professional.

Benefits

Surgical lasers offer several benefits over the traditional surgical methods. They are more precise and accurate, resulting in less damage to surrounding tissues. They also cause fewer traumas to patients, resulting in shorter recovery times. Surgical lasers also reduce the risk of infection because they do not require the use of scalpels or other instruments that can carry bacteria. This reduces the risk of

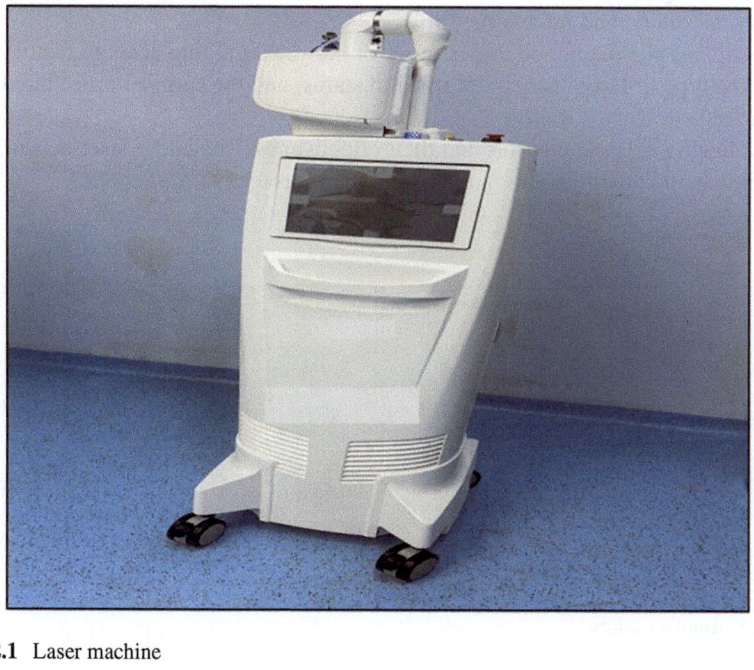

Fig. 2.1 Laser machine

postoperative complications as well as the amount of time a patient needs to spend in the hospital.

Risks Associated with Surgical Laser Therapy
Surgical lasers are powerful tools, but they do have some risks associated with their use. The most common risks include burns, scarring, and infections. These risks can be minimized by carefully following the instructions of medical professionals, and it is important to remember that surgical lasers can damage the eyes; therefore, it is important to wear protective eyewear when operating them. Additionally, it is important to be aware of the potential for laser-induced fires, which can occur if the laser beam comes into contact with combustible materials [4]. Figure 2.2 shows bipolar cautery used in surgery.

Use of Surgical Lasers
Surgical lasers are used in a variety of medical procedures, ranging from eye surgery to cosmetic procedures. They are also used to treat a wide range of conditions, from cancers to vascular diseases. Additionally, they are used in a variety of aesthetic procedures, such as hair removal and tattoo removal [5]. Surgical lasers are also used in veterinary medicine, as they can be used to treat a variety of conditions in animals. They are also used in industrial applications such as cutting and welding metals.

Risk determination of surgical lasers involves evaluating the potential harm to patients and healthcare providers that may result from the use of these devices. The process typically includes the following steps:

1. Identifying the hazards associated with surgical lasers, such as tissue damage, thermal injury, and infection.
2. Analyzing the likelihood and severity of harm, taking into account factors such as the type of laser, the patient's health status, and the surgical procedure being performed.
3. Evaluating existing controls and their effectiveness, including laser design, user training and qualifications, and proper use of protective equipment.

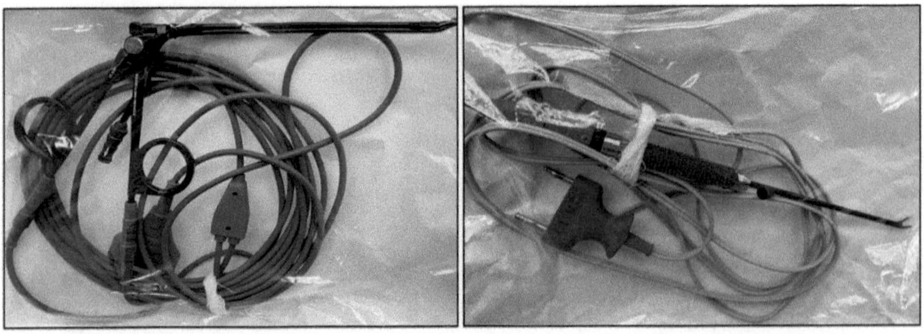

Fig. 2.2 Bipolar cautery

4. Developing and implementing risk management plans, including proper maintenance and calibration of the laser, regular monitoring of patients for adverse events, and establishing protocols for the emergency management of laser-related complications.
5. Monitoring and updating risk management plans as needed, taking into account new information and changes in medical practices or patient populations.

It is important to minimize the risk of harm associated with surgical lasers by following established guidelines and protocols, ensuring proper training and qualification of users, and maintaining and properly using protective equipment.

Prior Investigations of Surgical Lasers

Surgical lasers have been extensively investigated to ensure their safety and effectiveness. The FDA requires medical device manufacturers to conduct extensive testing and clinical trials before their products can be approved for use in patients. Here are some examples of the investigations that have been conducted on general hospital use surgical lasers:

1. *Biocompatibility testing:*
   Lasers must be biocompatible, meaning that they do not cause adverse reactions when they come into contact with living tissue. Manufacturers must conduct biocompatibility tests to ensure that their lasers meet this requirement.
2. *Mechanical testing:*
   Lasers must be able to withstand the mechanical stresses that they will be subjected to during use, such as vibration and shock. Manufacturers must conduct mechanical tests to ensure that their lasers are sufficiently strong to withstand these stresses.
3. *Electrical safety testing:*
   Lasers must be designed and tested to ensure that they meet electrical safety standards such as preventing electrical shock and reducing the risk of fire.
4. *Thermal safety testing:*
   Lasers must be designed and tested to ensure that they do not cause thermal injury to the surrounding tissues. Manufacturers must conduct thermal safety tests to ensure that their lasers are safe for use in patients.
5. *Clinical trials:*
   Lasers must undergo clinical trials to evaluate their safety and effectiveness in humans. These trials typically involve testing the laser in a group of patients with the condition for which the laser is intended and comparing the results to those of a control group.
6. *Laser output measurements:*
   Lasers must be measured to ensure that they produce the correct amount of energy and are operating within a safe range. Manufacturers must conduct laser output measurements to ensure that the lasers operate as intended.

2.5 Tissue Adhesives

Tissue adhesives are medical products used to bond tissues together, reducing the need for sutures or staples in certain surgical procedures. They can be used in a variety of medical specialties, including neurosurgery, gastroenterology, ophthalmology, general and plastic surgery, and cardiology [6]. Tissue adhesives are medical devices used to close wounds or incisions by bonding the edges of tissues together. The FDA categorizes certain tissue adhesives under general hospital use. These include:

(a) *Cyanoacrylate tissue adhesives:*
These are synthetic adhesives that polymerize upon contact with moisture to form strong bonds between tissues. They are commonly used to close incisions or lacerations on the skin and seal punctures in the blood vessels.

(b) *Fibrin sealants:*
These are biological adhesives made from human or animal plasma proteins that function by forming a clot at the site of tissue injury. They are commonly used in surgery to control bleeding and promote tissue healing.

(c) *Collagen-based tissue adhesives:*
These are adhesives made from purified collagen that bond the edges of the tissue together. They are commonly used in surgeries to seal incisions and wounds and to promote tissue healing.

(d) *Albumin-based tissue adhesives:*
These are adhesives made from purified albumin that work by cross-linking proteins in the tissue to form a strong bond. They are commonly used to close small incisions and wounds.

Use of Tissue Adhesives in Various Specialties

- In neurosurgery, tissue adhesives can be used to seal dural tears and prevent cerebrospinal fluid leakage.
- In gastroenterology, tissue adhesives can be used to stop bleeding in the digestive tract, particularly in the esophagus, stomach, and rectum.
- In ophthalmology, tissue adhesives can be used to secure conjunctival or scleral flaps during corneal transplantation procedures.
- In general and in plastic surgery, tissue adhesives can be used to close incisions and wounds, particularly in areas with limited blood flow or high tension.
- In cardiology, tissue adhesives can be used to seal punctures in the heart during catheterization.

Tissue adhesives are considered safe for use and have several advantages such as quick and easy application, reduced wound size, and improved cosmetic outcomes. However, it is important to carefully evaluate the risks and benefits of using tissue adhesives in each individual case, and to follow proper application techniques to ensure optimal results.

Risks Associated with Tissue Adhesives

- Allergic reaction
- Infection
- Tissue damage
- Over-application leading to tissue swelling
- Inadequate wound healing
- Toxicity [7]

Risk determination of tissue adhesives involves evaluating the potential harm to patients that may result from the use of these devices. The process typically includes the following steps:

1. Identifying the hazards associated with tissue adhesives, such as allergic reactions, wound infections, and implant failure.
2. Analyzing the likelihood and severity of harm, taking into account factors such as the type of adhesive, the patient's health status, and the surgical procedure being performed.
3. Evaluating existing controls and their effectiveness, including the quality of the manufacturing process, sterility of the adhesive at the time of use, and the surgeon's technique for adhesive application.
4. Developing and implementing risk management plans, including close monitoring of patients for adverse events and establishing protocols for the prompt recognition and management of adhesive-related complications.
5. Monitoring and updating risk management plans as needed, taking into account new information and changes in medical practices or patient populations.

It is important to minimize the risk of harm associated with tissue adhesives by using high-quality products, following evidence-based best practices, and by closely monitoring patients for adverse events.

Prior Investigations of Tissue Adhesives

Tissue adhesives have undergone numerous investigations to ensure their safety and effectiveness. The FDA requires medical device manufacturers to conduct extensive testing and clinical trials before their products can be approved for use in patients. Here are some examples of the investigations that have been conducted on general hospital use tissue adhesives:

1. *Biocompatibility testing:*

   Tissue adhesives must be biocompatible, meaning that they do not cause adverse reactions when they come into contact with living tissue. Manufacturers must conduct biocompatibility testing to ensure that their tissue adhesives meet these requirements.
2. *Tensile strength testing:*

   Tissue adhesives must be able to withstand forces that will be applied during use, such as tension or bending. Manufacturers must conduct tensile strength testing to ensure that their tissue adhesives are sufficiently strong to withstand these forces.

3. *Setting-time testing:*

   Tissue adhesives must be set within a certain amount of time to be effective. Manufacturers must conduct setting-time testing to ensure that their tissue adhesives are set within an appropriate amount of time.

4. In vitro *studies:*

   In vitro studies were conducted to evaluate the properties of tissue adhesives, such as their adhesive strength and ability to bond to different types of tissues.

5. *Animal studies:*

   Animal studies were conducted to evaluate the safety and effectiveness of tissue adhesives before they were tested in humans. These studies typically involve testing tissue adhesion in rats, rabbits, or other animals.

6. *Clinical trials:*

   Tissue adhesives must undergo clinical trials to evaluate their safety and effectiveness in humans. These trials typically involve testing the tissue adhesive in a group of patients with the condition for which the adhesive is intended and comparing the results to those of a control group.

References

1. US Food and Drug Administration. Information Sheet Guidance for IRBs, Clinical Investigators, and Sponsors. Significant Risk and Nonsignificant Risk Medical Device Studies. http://www.fda.gov/downloads/RegulatoryInformation/Guidances/UCM126418.pdf.
2. US Food and Drug Administration. Information Sheet Guidance for Clinical Investigations of Devices Indicated for the Treatment of Urinary Incontinence - Guidance for Industry and FDA Staff. https://www.fda.gov/regulatory-information/search-fda-guidance-documents/clinical-investigations-devices-indicated-treatment-urinary-incontinence-guidance-industry-and-fda.
3. Wang H. A review of the effects of collagen treatment in clinical studies. Polymers (Basel). 2021;13(22):3868.
4. Smalley PJ. Laser safety: risks, hazards, and control measures. Laser Ther. 2011;20(2):95–106.
5. Woo SH, Chung P, Lee SJ. Safe use of medical lasers. Med Lasers. 2021;10:68–75.
6. Ge L, Chen S. Recent advances in tissue adhesives for clinical medicine. Polymers. 2020;12(4):939.
7. Bal-Ozturk A, Cecen B, Avci-Adali M, Topkaya SN, Alarcin E, Yasayan G, Ethan YC, Bulkurcuoglu B, Akpek A, Avci H, Shi K, Shin SR, Hassan S. Tissue adhesives: from research to clinical translation. Nano Today. 2021;36:101049.

Chapter 3
Significant Risk Medical Devices – Anaesthesiology

Hemasri Velmurugan, Habib Md Reazaul Karim, Krishnapriya Neelambaran, and Pugazhenthan Thangaraju

3.1 Introduction

Nowadays, technology is grown and dependable. Several technologies and device classes are available, depending on the requirements, ranging from mechanical mixing valves to mass flow controllers and from small mobile gas mixers to complicated large-scale gas mixing systems. This chapter provides an overview of the different anaesthetic devices currently in use as well as details on the most prevalent device mechanism, types, applications, benefits, drawbacks, and limitations. In this section, we will look at some FDA-approved anaesthetic devices.

3.2 Breathing Gas Mixers

Lundsgaard and Degn (1973) describe a simple, dependable, and versatile mechanism for mixing gases. It distributes the blending of two commercially available gases from cylinders using a mixture of stainless steel capillary tubes functioning as flow resistors. Mixed gases are employed in a wide range of industrial applications. Many common mixed gases are currently available in premixed form. However, it is often preferable to mix the required gases on-site. It is used particularly when excessive consumption, odd combinations, or frequent changes in the gas mixture are evident. It is an odd method for anaesthetists who are trained to setting a process

H. Velmurugan · K. Neelambaran · P. Thangaraju (✉)
Department of Pharmacology, All India Institute of Medical Sciences (AIIMS), Raipur, Chhattisgarh, India

H. M. R. Karim
Department of Anaesthesiology Critical Care and Pain Medicine, All India Institute of Medical Sciences (AIIMS), Raipur, Chhattisgarh, India

P. S. Timiri Shanmugam et al. (eds.), *Significant and Nonsignificant Risk Medical Devices*, https://doi.org/10.1007/978-3-031-52838-5_3

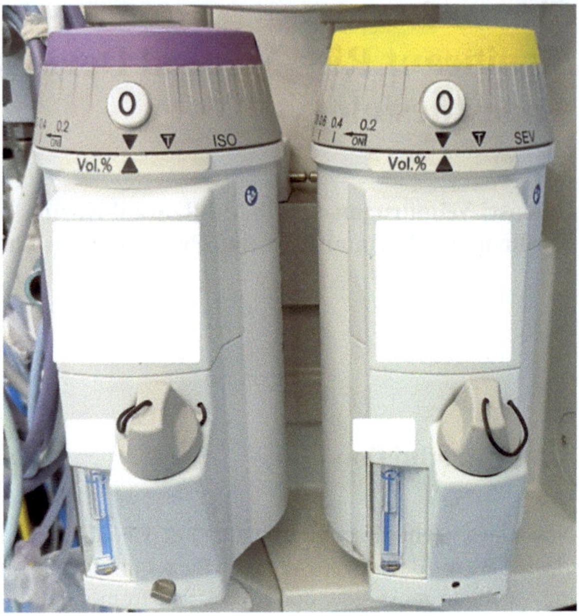

Fig. 3.1 Breathing gas

variable rather than specifying the intended output and allowing the machine handle the process [1]. Figure 3.1 shows breathing gas container.

Types

They are classified as manual or electronic flowmeters as shown in Table 3.1.

How It Works?

- *Traditional flowmeters:* They use a needle valve to mechanically control and display flow with a glass tube. The parts are as follows: needle valve, indicator float, knobs, and valve stops. When the knob is turned anticlockwise, the flow increases. The annular-shaped orifice around the float is comparably tubular at low flows allowing viscosity to control flow. At high flow rates, the annular hole resembles an aperture, and density determines flow.
- *Transitional flowmeters:* Flow is mechanically operated (needle valve) and electronically displayed in a transitional flowmeter. Glass tubes are not used in transitional flowmeters. There is a needle valve, so flow can be generated even when no electricity is present. Flows are electronically recorded and displayed.
- *Electronic flowmeters:* The gas mixer regulates the flow of all gas and vapour to the patient as shown on a monitor screen. The user sets the required carrier gas (nitrous oxide, air), FIO2 percentage, and total fresh gas flow (FGF). Thus, flowmeters (both transitional and electronic) enable automated anaesthesia record-keepers to document fresh gas flows. Figure 3.2 shows the blood gas analyser.

Table 3.1 Types of breathing gas mixers

S.No	Types	Control	Display
1	Traditional	Needle tube	Glass tube
2	Transitional (Hybrid)	Needle tube	Electronic
3	Electronic	Electronic	Electronic

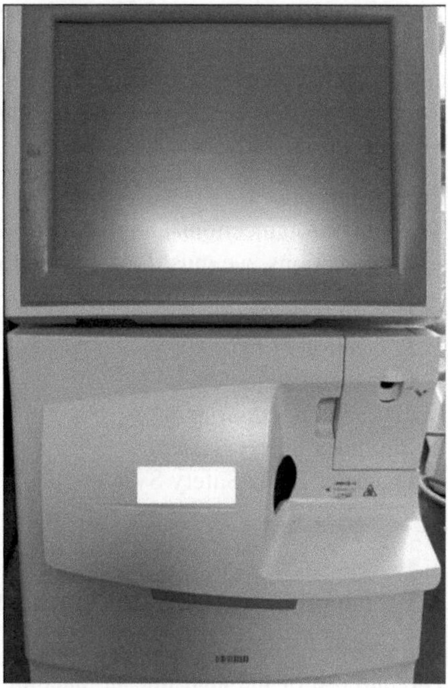

Fig. 3.2 Blood gas analyser

Applications

- This consistency ensures maximum process reliability.
- Gas mixers provide the greatest amount of versatility in terms of needed mixing ratios, gas quantities, and application location.
- Operating a system with changeable gas compositions is not an issue. The gas mixture parameters can be modified at any moment to obtain the ideal gas combination in seconds. With the correct gas mixers, even mobile use is conceivable.
- It eliminates the time-consuming storage of various essential combinations which is especially important with frequent mixture changes [2].

Advantages

- Homogeneity
- Quality
- Flexibility
- Profitability from lower purchasing prices for standard products

Disadvantages

Even when oxygen (O_2) is in the downstream position, a leak in the O_2 flow tube may result in the formation of a hypoxic mixture. O_2 escapes through the leak-, whereas nitrous oxide (N_2O) continues to flow towards the common outlet especially at high N_2O to O_2 flow ratios [3].

Limitation

There have been instances of hypoxic gas mixes when using nitrous oxide without an enough complementing oxygen concentration. But in modern machines, a mechanical or electronic linkage exists between the two flow control valves to allow for correct proportioning of the nitrous oxide: oxygen ratios to guard against hypoxic gas mixes with nitrous oxide [3].

Precaution

To avoid misconnections and the distribution of incorrect gases to patients, as well as complications, we must look for the following.

- Pressure regulator
- Prevention of hypoxia
- Flowmeter sequence
- Gas scavenging system
- Pin Index Safety System
- Diameter Index Safety System

3.3 Electroanaesthesia Apparatus

The purpose of the anaesthesia machine has gradually changed from being just a way to put a patient to sleep and give them oxygen to a workstation that includes increasingly sophisticated ventilator modes, end-tidal CO2 monitors, end-tidal anaesthetic concentrations, limited alveolar concentration estimators and a way for maintaining records of vital signs [4].

Types

There are various kinds of anaesthesia circuits.

1. *The circle system*

   It is the most frequently used system in modern anaesthesia machines and it is based on the fresh gas inflow settings, circle systems can be closed (fresh gas flow [FGF] = oxygen and anaesthetic update), semi-closed (high FGF, gas exits through the expiratory valve) or semi-open.
2. *The Mapleson system*

   It includes the Mapleson A, B, C, D (Bain modification) and E and F (Jackson-Rees) systems.

   - For Mapleson D, E, and F systems, there is a T piece near to the patient. They have the advantage of preserving airway moisture and heat while limiting anaesthetic gas leakage out of the system and into the atmosphere.

- The Mapleson D (Bain) system is a variation of the Mapleson D that includes a tube inside the corrugated expiratory tube that supplies fresh gas to the patient (coaxial formation). This has the advantage of preserving moisture and warming the fresh gas as it travels to the patient.
- The Mapleson F (Jackson-Rees) scheme is a variant of the Mapleson E. It is equipped with a reservoir bag connected to the expiratory limb as well as an adjustable overflow bag. It has little dead area and little resistance to spontaneous ventilation, making it suitable for paediatric patients. However, this is an inefficient system, because it needs a high FGF level to prevent rebreathing [4, 5].

Components

1. *Vaporisers*

   There are two types of vaporisers: measured flow vaporisers and changeable bypass vaporisers.

   Variable bypass vaporisers: It operates by adjusting a splitting ratio on a dial that controls the vaporisers. It automatically compensates for a broad variety of operating room temperatures in order to maintain a consistent anaesthetic output at a given atmospheric pressure. Each vaporiser is designed to be used with a particular volatile anaesthetic, such as enflurane, halothane, isoflurane, or sevoflurane.

   Measured flow vaporised: The desflurane vaporiser is the most prevalent type of measured flow vaporised. Desflurane is heated to a constant temperature of 39 degrees Celsius because of its low boiling point and propensity to volatility. As a result, the circuit of the vaporiser starts in the vaporiser itself rather than with fresh gas flowing over the volatile anaesthetic.

2. *Oxygen Flush Button*

   - It is most frequently utilised during mask ventilation when an inadequate mask seal cannot be achieved for a variety of causes, including a patient's beard, operator error, and patients with challenging airways.
   - Even if the volatile anaesthetic or nitrous oxide is switched on, the anaesthesia provider must be aware that only oxygen is being administered to the patient when the oxygen flush button is pressed.
   - Due to the flow of gas at higher pressures than the usual low-pressure system of the anaesthesia machine, the use of the oxygen flush may result in periods of awareness during anaesthesia and barotrauma to the patient's lungs.

3. *Adjustable Pressure Limiting Valve (APL Valve)*

   - Between the expiratory unidirectional valve and the carbon dioxide collector is a valve known as the APL valve, also referred to as the pop-off valve.
   - When tubing may become blocked, the APL valve acts as a pressure relief valve to avoid excessive pressures in the breathing circuit.
   - Excessive pressures can harm flowmeters and vaporisers as well as cause barotrauma to the patient. The valve stays open during spontaneous ventilation to enable easier breathing.

- When positive pressure ventilation is needed following induction, the valve can be partially closed by squeezing the reservoir bag.
- To reduce operating room pollution, any gas released from the APL valve to restrict pressure is routed to the scavenger system.

4. *Carbon Dioxide Absorbent*

- CO2 absorbent is a combination of calcium hydroxide, sodium hydroxide, potassium hydroxide, and barium hydroxide that prevents carbon dioxide from entering the anaesthesia machine's inspiratory limb.
- Chemical indicators in CO2 absorbents used in anaesthesia machines usually change colour as the filter becomes saturated.
- When the filter is two-thirds saturated, it should be changed to avoid carbon dioxide re-breathing [6].

Functions

The modern anaesthesia machine has four primary functions as shown in Fig. 3.3.

Advantages

- Closed-circuit systems, in particular, enable for the recycling of inhaled gases, resulting in a significant reduction in environmental pollution.
- Carbon dioxide absorbent advancements have enabled carbon dioxide to be absorbed and filtered out of exhaled gases, allowing oxygen and volatile anaesthetics to be reused.
- The efficacy of recycling gases can be increased by employing low-flow anaesthesia [7, 8].

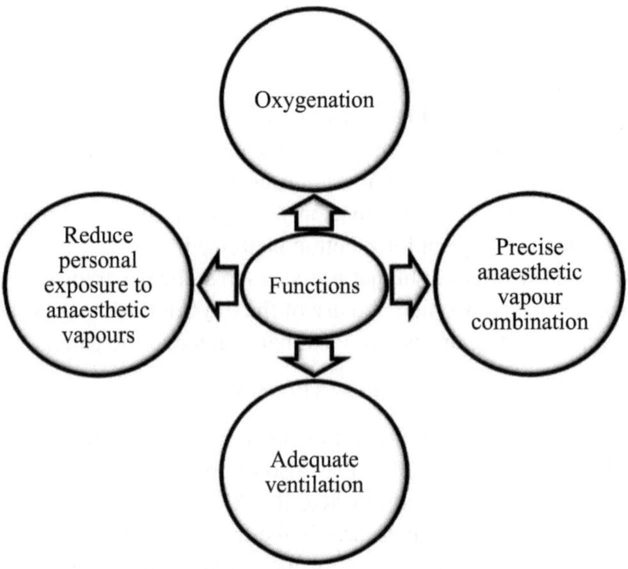

Fig. 3.3 Function of electroanaesthesia apparatus

Disadvantages

- Mask closure is inadequate.
- Endotracheal catheter dislodged.
- Inspiratory or expiratory limb disconnections, open filling reservoir on a variable bypass vaporiser, etc.
- A carbon dioxide absorbent canister that is not properly connected or detached.
- Anaesthesia circuits without a closed-circuit system can be used to administer anaesthesia; however, this comes at a significant financial cost and has a significant environmental effect when widely implemented [7].

3.4 Epidural Anaesthesia and Equipment

Epidural anaesthesia is a method that can be used as either a primary surgical anaesthetic or to manage postoperative pain. In surgical, obstetric, and chronic pain situations, it is used as an adjunct or substitute to general anaesthesia. Local anaesthetic and other adjuvants can be continuously infused or administered on an as-needed basis via an epidural catheter, inhibiting pain impulses at the nerve root. This administration method may decrease adverse side effects and hospital length of stay while maintaining or improving patient safety [9].

Types
Epidural can be given at any level of spinal cord if selection blockage is possible.

1. Cervical epidural
2. Thoracic epidural
3. Lumbar epidural
4. Caudal epidural

Equipment

- *Epidural needles*: Several kinds have been developed, namely, Tuohy, Hustead, Crawford, Weiss, etc., with Tuohy being the most commonly used (Fig. 3.4).
- *Syringe for resistance loss*: Constructed of glass or plastic. It has very little friction between the plunger and the barrel, allowing it to sense changes in resistance at the epidural region. The syringes can be filled with air, saline, or both; this does not appear to impact the effectiveness of identifying the epidural space or the complication rate.
- *Epidural catheters*: These are used to provide constant epidural anaesthesia or analgesia. They can be either flexible or rigid. They could also have a solitary or multiple holes [10]. Figure 3.5 shows epidural kit and drugs.

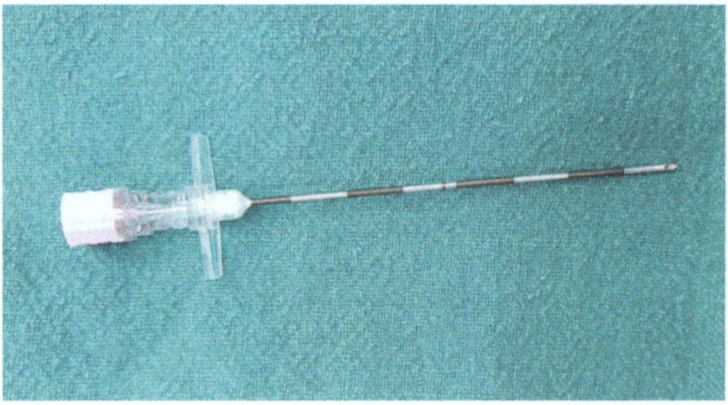

Fig. 3.4 Epidural needle

EPIDURAL KIT

- Skin Preparation: betadine, chlorhexidine
- Topical Local Anaethetic Syringe and Needle: 25 gauge-needle, lidocaine 1% to 2%
- Toughy or Housted needle: 17 gauge to 18 gauge 3.5 inch (up to 6 inches for obese patients).
- Loss of Resistance Syringe (glass or plastic)
- Catheter single or multi-orifice catheter
- Test Dose: commonly 3 ml 1.5% lidocaine with epinephrine (1:200,000)
- Tape and Tegaderm: to secure the catheter

EPIDURAL DRUGS

- Long-acting: commonly lidocaine, bupivacaine, levobupivacaine, ropivacaine, or 2-chloroprocaine.
- Adjuvants
- Opiate Medications: fentanyl, sufentanil, morphine, hydromorphone
- Alpha-adrenergic Agonists: epinephrine, clonidine
- Sodium bicarbonate

Fig. 3.5 Epidural kit and drugs

Technique

- Single dosage injection
- Fractional–continuous epidural—repeated injections of Local Anaesthesia through catheter placed into epidural space
- Positions: Cervical epidural (C7) in sitting posture. For thoracic epidural (T7) and lumbar epidural (L1-L2, L2-L3, L3-L4, L4-L5) in decubitus lateralis and full extension posture [10]

Method for Determining Epidural Space
Principle: Negative pressure in space

- Loss of resistance: Plunger of syringe pressed without resistance once epidural needle is in
- Hanging Drop: Saline drop at hub of epidural needle is sucked in once it enters space [11].

Indications

- For individuals with difficult airways or other respiratory issues related to general anaesthesia's side effects, epidural anaesthesia can also be taken into consideration.
- Obstetric, abdominal, urogenital, and lower extremity procedures are all common applications for epidural anaesthesia.
- While upper epidural abdominal procedures are possible with epidural anaesthesia, they are less common due to the difficulty in safely obtaining the desired sensory level of anaesthesia.
- If muscular relaxation is not required, epidurals can be used for thoracic surgery, major intra-abdominal surgery, or spine surgery.
- This method may also be used to manage pain during or after surgery.
- It may reduce surgical risk and morbidity in certain patient groups, such as those with ischemic cardiac disease.
- It has also been shown to reduce post-operative pulmonary complications [9, 10].

Contraindications
There are absolute and relative contraindications of epidural anaesthesia (Fig. 3.6).

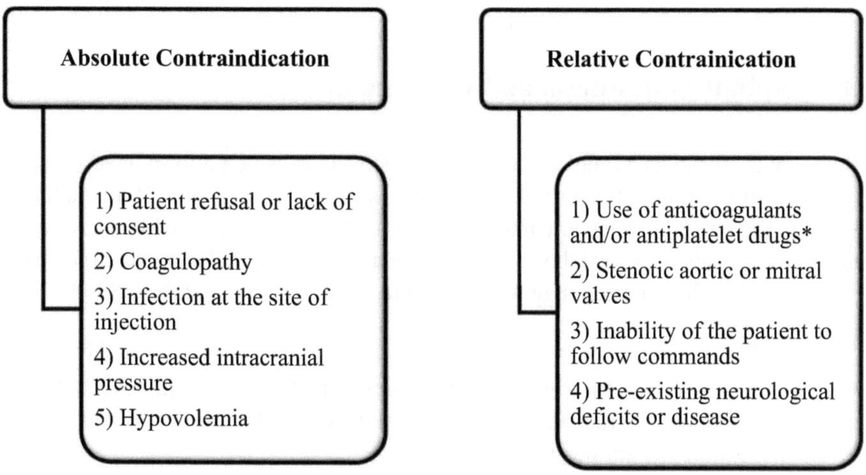

Fig. 3.6 The absolute and relative contraindications of epidural anaesthesia

Complications

- Hypotension
- Nausea and vomiting
- Bronchoconstriction
- Dural perforation post-puncture pain
- Temporary neurological condition (symmetrical back pain, radiated to the buttocks and legs, without sensitive or motor component)
- Nerve injury with neuropathy—paresis is exceedingly uncommon.
- Hematoma in the epidural space
- Abscess in the epidural space
- Meningitis
- Intrathecal injection by accident with complete spinal anaesthesia
- Osteomyelitis [9]

Advantages

- Well-defined anaesthesia region
- Longer duration
- Less severe spinal anaesthesia disturbances minimised
- GI complaints minimised
- Catheterisation minimised less respiratory affects

Disadvantages

- Technically more challenging
- Incomplete muscle relaxation
- Large amount required
- Danger of dural puncture
- Incomplete/patchy block physiological consequences [12]

3.5 Spinal Anaesthesia and Equipment

Spinal anaesthesia is a type of neuraxial anaesthesia in which a local anaesthetic is injected straight into the intrathecal space (subarachnoid space). Subarachnoid space is the space between arachnoid and pia matter that includes CSF, nerve roots, and blood vessels that feed the spinal cord. In adults, the spinal subarachnoid space stretches from the foramen magnum to S2 and in children it extends to S3 [13, 14].

Equipment

- *Spinal needles*: Several kinds have been developed, namely, Quincke Babcock needle, Whitacre needle, Sprotte needle, Pitkin needle, Tuohy needle, and Greene needle. Figure 3.7 shows spinal needle.

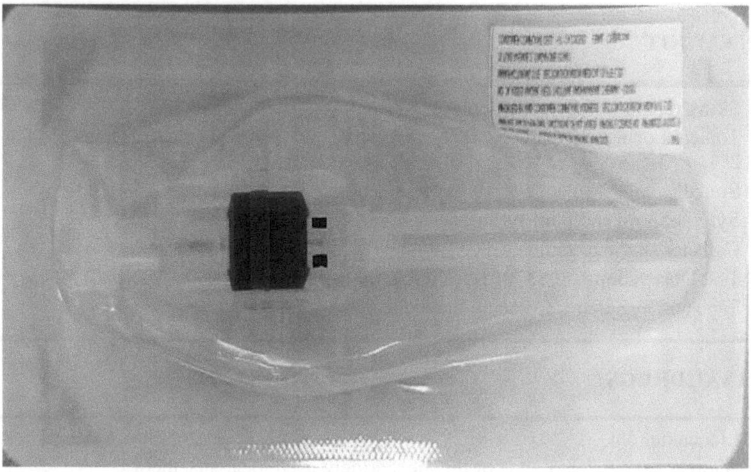

Fig. 3.7 Spinal needle

Technique

- For median approach, the physician inserts the needle until the dura-subarachnoid membrane are penetrated, which is indicated by free-flowing CSF.
- For paramedian approach, the skin wheal from the local anaesthetic is positioned about 2 cm from the midline and the spinal needle is advanced at an angle towards the midline. The supraspinous and interspinous ligaments are rarely met in this approach. As a result, there is little opposition until reaching the ligamentum flavum.
- Positions: Lateral (left lateral), sitting and prone. Figure 3.8 shows spinal kit and drugs.

Indications

- Spinal anaesthesia is commonly used for surgical operations affecting the lower abdomen, pelvis, perineum, and lower extremities
- Useful in hernia (ingunial or epigastric), haemorrhoidectomy, fistula, fissure, nephrectomy, and cystectomy in combination with general anaesthesia, transurethral resection of the prostate and bladder tumours, abdominal and vaginal hysterectomies, laproscopic assisted vaginal hysterectomies combined with general anaesthesia and caesarean sections.
- It is especially useful for procedures performed below the umbilicus.
- Short operations benefit from spinal anaesthesia. For longer procedures or processes that would jeopardise respiration, general anaesthesia should be used [13].

Contraindications

There are absolute and relative contraindications of spinal anaesthesia (Fig. 3.9).

SPINAL KIT

- Skin Preparation: betadine, chlorhexidine
- Topical Local Anaethetic Syringe and Needle: 25 gauge-needle, lidocaine 1% to 2%
- Spinal needle: 18 gauge to 30 gauge • Length : 3.5 to 4 inches
- Syringe: 3 ml and 5 ml syringes
- Catheter single or multi-orifice catheter
- Test Dose: commonly 3 ml 1.5% lidocaine with epinephrine (1:200,000)
- Tape and Tegaderm: to secure the catheter

SPINAL DRUGS

- Lidocaine (5%)
- Bupivacaine (0.75%): One of the most widely used local anaethetics
- Lidocaine 5%
- Tetracaine 0.5%
- Mepivacaine 2%
- Ropivacaine 0.75%
- Levobupivacaine 0.5%
- Chloroprocaine 3%

Fig. 3.8 Spinal kit and drugs

Absolute Contraindication

1) Patient refusal
2) Infection at the site of injection
3) Coagulopathy or other bleeding diathesis
4) Severehypovolemia
5) Increased intracranial pressure
6) Severe aortic stenosis or mitral stenosis
7) Post traumatic vertibral injuries
8) Myocardial infract

Relative Contrainication

1) Sepsis
2) Un co-operative patient
3) Preexisting neurological deficit
4) Stenotic valvular heart disease
5) Severespinal deformity
6) Spinal congenital anomalies
7) Hypotension and Hyper tension
8) Severeanaemia
9) Shock(haemorragic,septic)
10) Brain tumor

Fig. 3.9 The absolute and relative contraindications of spinal anaesthesia

Complications

- Backache
- Headache caused by a spinal puncture (as high as 25% in some cases)
- Nausea and vomiting
- Hypotension
- Low-frequency hearing impairment
- Complete spine anaesthesia
- Injuries to the nervous system
- Hematoma of the spine
- Arachnoiditis
- Transient neurological syndrome (especially with lidocaine) [15]

Advantages

- Cost is extremely low.
- Patient contentment.
- There are few side effects on the respiratory system as long as excessively elevated blocks are avoided.
- Patent airway—the risk of airway obstruction or aspiration of gastric contents is decreased.
- Diabetic patients—there is little danger of unrecognised hypoglycaemia in awake patient.
- Excellent for lower abdominal and lower leg surgery muscular relaxation.
- Blood loss during surgery is less than when the same procedure is performed under general anaesthetic.
- Because it increases blood flow to the gut, spinal anaesthesia lowers the likelihood of anastomotic dehiscence.
- The bowel is contracted and the sphincters are relaxed, but peristalsis persists. Following surgery, normal gut function resumes quickly.
- Deep venous thrombosis and pulmonary emboli are less common after surgery [14–16].

Disadvantages

- Difficult needle placement
- Inability to obtain CSF
- Failed spinal
- Dural rupture
- Urinary retention [15]

3.6 Esophageal Obturators

The esophageal obturators (EOA) insertion was faster (mean 6 sec vs 20 sec), more accurate (98 percent vs 48 percent), and simpler to teach paramedics. When optimal conditions for endotracheal intubation are not possible, the EOA should be used first in the management of the airway in aneflexic, apneic patients. It is placed in the esophagus to keep stomach contents from accessing the lungs while the patient is being ventilated artificially [17].

Technique

- Ventilation is provided by a mask and a cuffed esophageal conduit with a sealed distal end.
- Before inflating, the cuff must be moved beyond the carina.
- Cuff was inflated with 35 cc of air.
- For ventilation, the mask needs to suit tightly.

Indications

- Deeply unconscious patients.
- Aneflexic or apneic patients.
- When airway management was required but intubation was not possible, this method was recommended [18].

Contraindications
Patients younger than 16 years of age.

Advantages

- Fast and accurate
- Required less technical skill

Disadvantages
Mask seal requires ventilating.

Limitation
It was once widely used, but due to difficulties and EndoTracheal training, its popularity has waned.

3.7 Gas Machines for Anaesthesia

Anaesthetic machines and ventilators are used to improve a patient's breathing. Anaesthetic gas machine is shown in Fig. 3.10.

It is a continuous flow gas anaesthetic device that delivers anaesthetic vapours. It has automated and manual ventilation modes as well as a ventilation monitoring system for inspired and expired gas monitoring and agent identification for system monitoring [19, 20]. Table 3.2 shows the classification of Anaesthesia breathing systems based on gas flow. The parts of machine components are shown in Fig. 3.11.

Functions
The machine performs four essential functions:

- Provides oxygen
- Correctly mixes anaesthetic gases and vapours

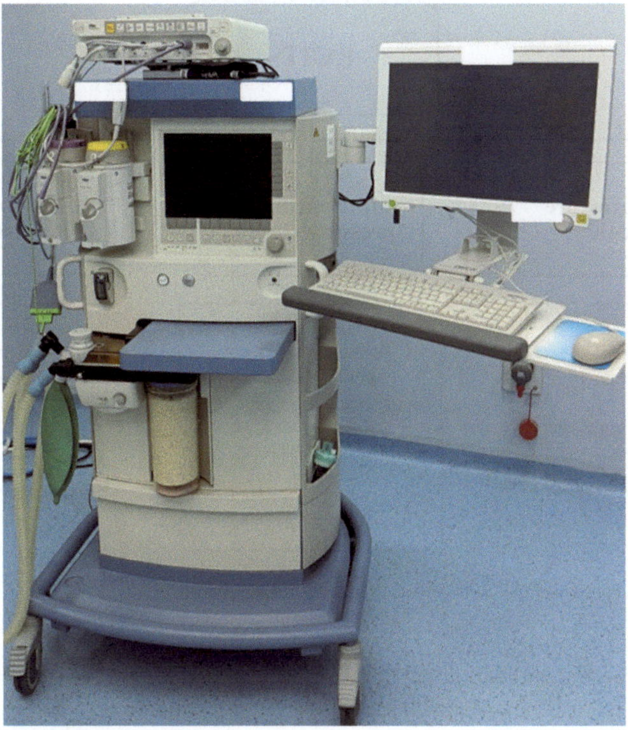

Fig. 3.10 Anaesthetic gas machine

Table 3.2 Classification of anaesthesia breathing systems based on gas flow

Features	Open breathing systems	Semi-open breathing systems	Semi-closed breathing systems	Closed breathing systems
Reservoir bag	Absent	Present	Present	Present
Functions	No functional rebreathing of exhaled gases, no tubing, and no valves	No functional re-breathing of exhaled gases and have high fresh gas flows	Partial re-breathing of exhaled gases, unidirectional valves, neutralising carbon dioxide, and low fresh gas flows	Total rebreathing of exhaled gases, unidirectional valves, and neutralise carbon dioxide
System involved	Insufflation and open-drop anaesthesia	Include Mapleson breathing systems	Circle systems with an adjustable pressure-limiting valve (APL valve) that is at least partially open	Circle systems with the APL valve closed

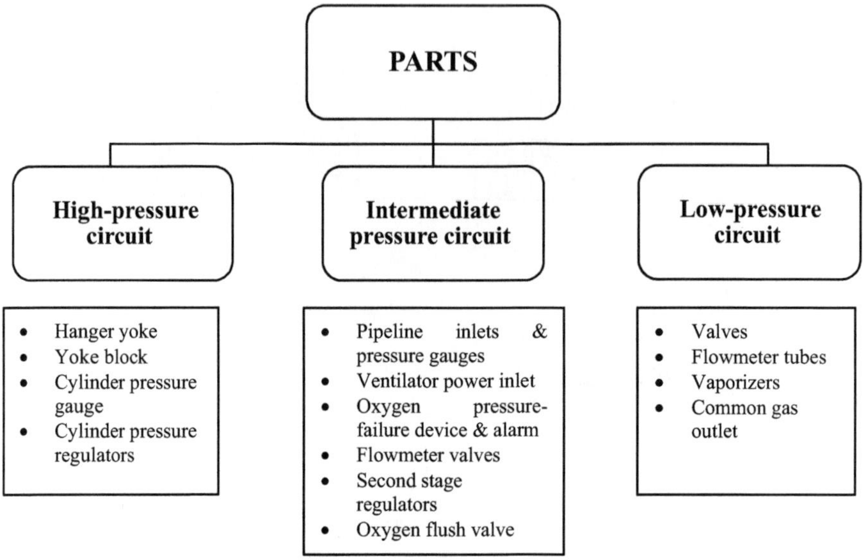

Fig. 3.11 Parts of machine components

- Allows for patient ventilation
- Reduces anaesthesia-related hazards to staff and patients

Advantages

- Smaller and lighter than traditional designs
- Most are dependent on pressure-controlled breathing.
- Enable for more readily automated record keeping.
- Improve patient safety by using more reliable and capable fundamental components such as ventilators, vaporisers, and flowmeters.
- Carry out compliancy and leak testing on the breathing circuit, promoting low flow anaesthesia.
- Include improved monitors and innovative new monitoring capabilities

Disadvantages

- Kuhn bag can twist and obstruct breathing.
- Extensive gas flow demands.
- A lack of humidity.
- Knobs may respond to light touch or accidental brushing Leakage through open flow control valves Inability to turn the knob Failure to allow adequate gas flow.

Precaution

- Backup battery.
- Alarms.
- Examination of the breathing circuit pressure, oxygen content, exhaled volume, or carbon dioxide levels (or both).

- Monitoring devices are required.
- The concentration of anaesthetic vapour must be watched.
- EKG, pulse oximetry, and blood pressure monitoring are all required.
- Check flowmeters, vaporiser, and pipeline gas source [21, 22].

3.8 High Frequency Ventilators Greater Than 150 bpm

High Frequency Oscillatory Ventilator (HFOV) gives small tidal volumes that are typically equal to or less than the dead space, 150 millilitres, at a very fast rate (Hertz-Hz) of between 4–5 breaths per second. Delivering tidal volumes of empty space or less at very high frequencies allows for the preservation of a minute volume. A mean pressure adjust device keeps the lungs open to a constant airway pressure. HFOV provides for the decoupling of oxygenation and ventilation, allowing the clinician to adjust either oxygenation or ventilation independently [23].

Types
They are mainly four types [24].

1. High-frequency oscillatory ventilation (HFOV)
2. High-frequency positive pressure ventilation (HPPV)
3. High-frequency jet ventilation (HJV)
4. High-frequency percussive ventilation (HFPV)

How It Works?

- It operates within the safe window, enabling the recruitment of collapsed alveoli and limiting atelectasis.
- This is accomplished by using rapid, tiny oscillations generated by a reciprocating piston. Both inspiration and expiration are active processes caused by the piston striking forward and producing positive pressure in the airway before recoiling and creating negative pressure.
- The amplitude of the wave, as determined by the power control, decides the forward and backward excursion of the piston and contributes to the tidal volume.
- Tiny tidal volumes are forced in and out of the lungs, oscillating around a fixed pressure, hence the oscillator's name.
- In HFOV, a rigid ventilator circuit with minimal compliance is used [25, 26].

Indications
Figure 3.12 depicts indications of high frequency ventilators in adults and neonates.

Advantages

- Safe—prevent over distension
- It also helps to avoid derecruitment and atelectrauma and barotrauma.
- It also improves the ventilation/perfusion (V/Q) ratio by providing for uniform aeration of the lungs.

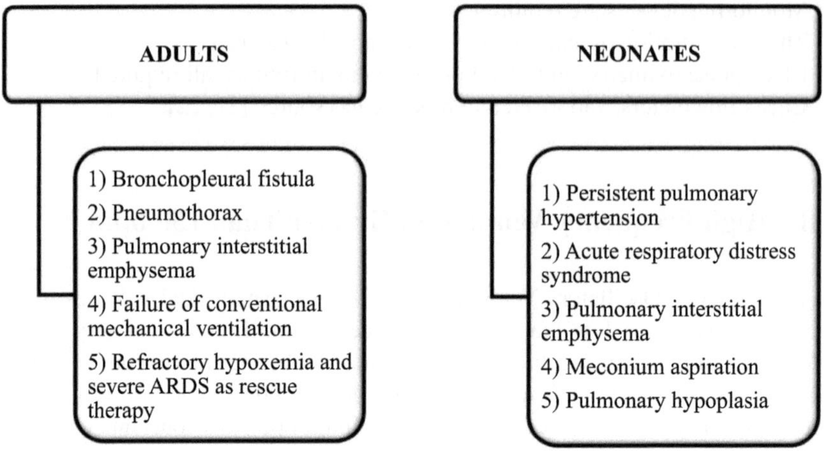

Fig. 3.12 Indications of high frequency ventilators in adults and neonates

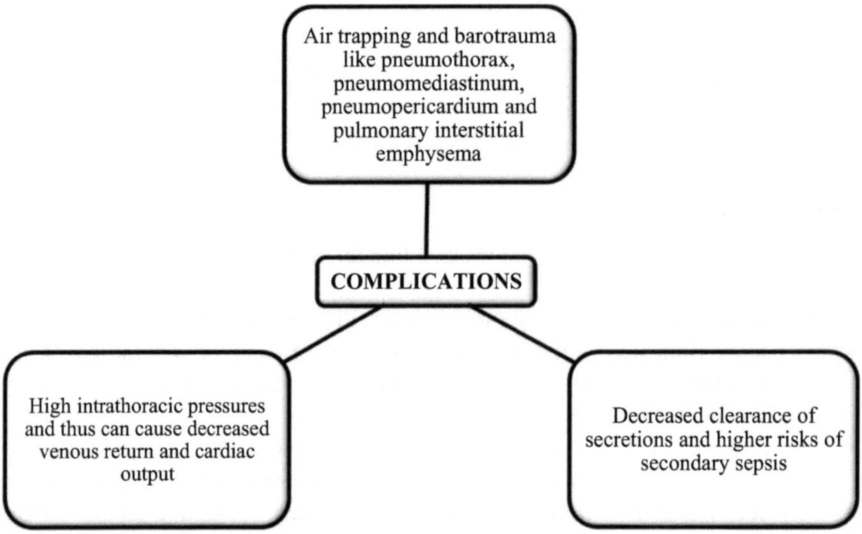

Fig. 3.13 Complications of high frequency ventilators

- The tidal volume used in HFOV is less than the dead space volume, stopping the alveoli from cyclically opening and closing.
- It maintains a consistent Mean Arterial Pressure.
- It also aids in employment by utilising a high PEEP [25, 26].

Complications

The complications of high frequency ventilators are listed in Fig. 3.13.

Limitations

- Transportation issues
- Noisy machines disrupting clinical exams
- Delay in identifying problems [27]

Conclusion

The effective implementation of high-frequency ventilation (HFV) emphasises the importance of the multidisciplinary team in reviewing and enhancing care for patients who require direct interaction and collaboration between various health care providers and specialised teams.

3.9 Rebreathing Devices

A rebreather is a breathing apparatus that absorbs carbon dioxide from a user's exhaled breath, enabling for the rebreathing (recycling) of each breath's substantially unused oxygen content and, if present the unused inert content. Oxygen is supplied to replenish the amount metabolised by the patient. Anaesthetic machines can be set as rebreathers, which contain an absorbent to eliminate the carbon dioxide from the loop [28].

Types

There are three types of rebreathing devices namely oxygen rebreather, semi-closed circuit rebreather and closed-circuit rebreather. Table 3.3 depicts the types of rebreathing devices and their differences.

How It Works?

Anaesthetic machines can use both semi-closed and completely closed circuit systems, as well as push-pull (pendulum) two-directional flow and one-directional loop systems. The pendulum and loop systems are two basic components that regulate the flow of breathing gas inside the rebreather. In the pendulum part, the user inhales gas from the counter lung via a breathing hose and exhaled gas flows back to the counter lung via the same hose. The user inhales gas through one hose and exhales through another in the loop system. Oxygen delivery devices are shown in Fig. 3.14.

Table 3.3 Types of rebreathing devices and their differences

S.no	Features	Oxygen rebreathers	Semi-closed circuit rebreathers	Closed-circuit rebreathers
1	Gas supply	Pure oxygen	Gas mixtures	Both pure oxygen and mixed gases
2	Salient points	No decompression depths and carry a danger of oxygen toxicity	Can be used for greater depths without risking oxygen toxicity	They differ others by the way that they maintain the oxygen concentration

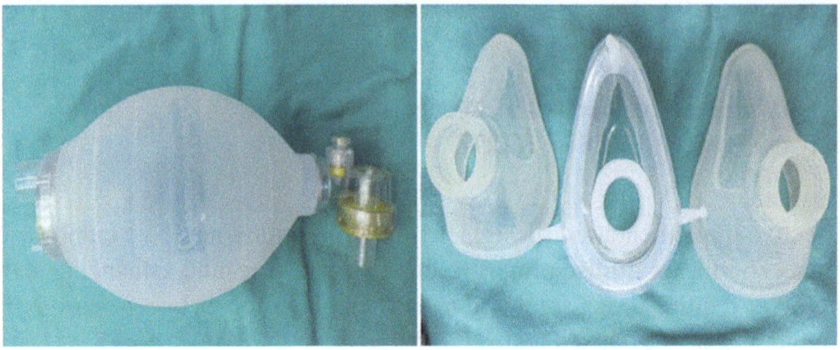

Fig. 3.14 Oxygen delivery devices

Fig. 3.15 Functions of rebreathing devices

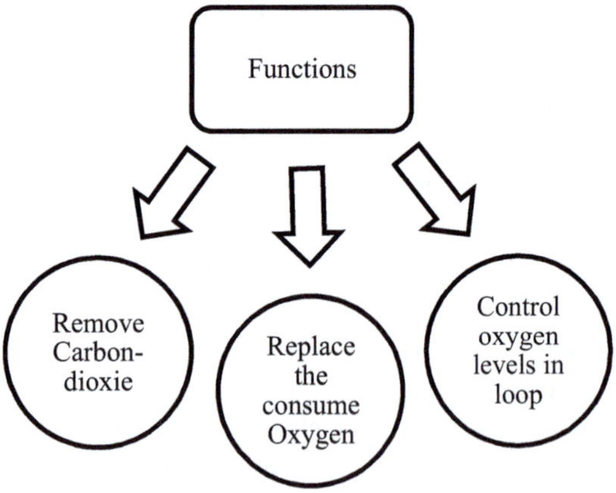

For rebreathing, the device must do the following:

1. Use a cannister of sodium hydroxide to remove the expelled carbon dioxide. Calcium carbonate is formed when carbon dioxide interacts with sodium hydroxide and calcium hydroxide.
2. Replace the oxygen that have ingested by injecting new oxygen into the breathing loop using small tanks of pure oxygen or mixed gases (nitrogen-oxygen or helium-oxygen).
3. Solid-state oxygen sensors monitor the partial pressure of oxygen in the breathing loop and transmit this information to a microprocessor that controls the oxygen-delivery system. Figure 3.15 shows the functions of rebreathing devices [28, 29].

Applications

- Supplying oxygen and anaesthetic gases to a patient during surgery or other sedation treatments.
- The anaesthetic generator can also supply gas to ventilated patients who are unable to breathe on their own.

Advantages

- The device is used to administer controlled concentrations of anaesthetic gases to patients while avoiding contaminating the air that the staffs inhale and conserving anaesthetic gas.
- Create few or no bubbles, causing no disturbance to marine life or revealing the diver's presence.
- Less decompression.
- Better gas efficiency.
- Lighter weight.
- To prevent contamination, a waste gas scavenging device removes any gases from the operating room [30].

Disadvantages

- Leakage of gas.
- Failing to monitor oxygen levels and failure of the gas injection system is primarily an issue with mixed gas rebreather.
- Flooding of the ambient pressure space can occur.
- Oxygen rebreathers are typically robust and reliable, and they can be overridden directly if they fail, which is signalled by an insufficient volume of gas in the rebreather's ambient pressure volume [31].

3.10 New Modes of Ventilation

A ventilator setting that was adequate at one moment may no longer be appropriate as the patient deteriorates or improves. Figure 3.16 shows ventilator machine.

Conventional ventilators only provide the programmed settings and do not account for patient variables. As a consequence, all traditional volume and pressure controls are Open Loop. The newer closed loop modes seek to change with the changing lung and take input from patient parameters, completing the feedback loop. The variables self-adjust and are no longer limited to a single parameter, but when one component's threshold is reached, they move to the other set parameter. The term "dual control" airflow emerged as a result of this [32]. The ventilators on dual control within a breath and in subsequent breath are shown in Fig. 3.17.

Newer Closed Loop Ventilators

1. *Mandatory minute ventilation (MMV)*
    - MMV establishes safety in patients with apneic episodes or central drive pathologies by giving set value ventilation as mandatory ventilation.
    - It occurs when the ventilator uses feedback to adjust both the respiratory rate and the amount of pressure support to achieve the specified minimum minute ventilation.

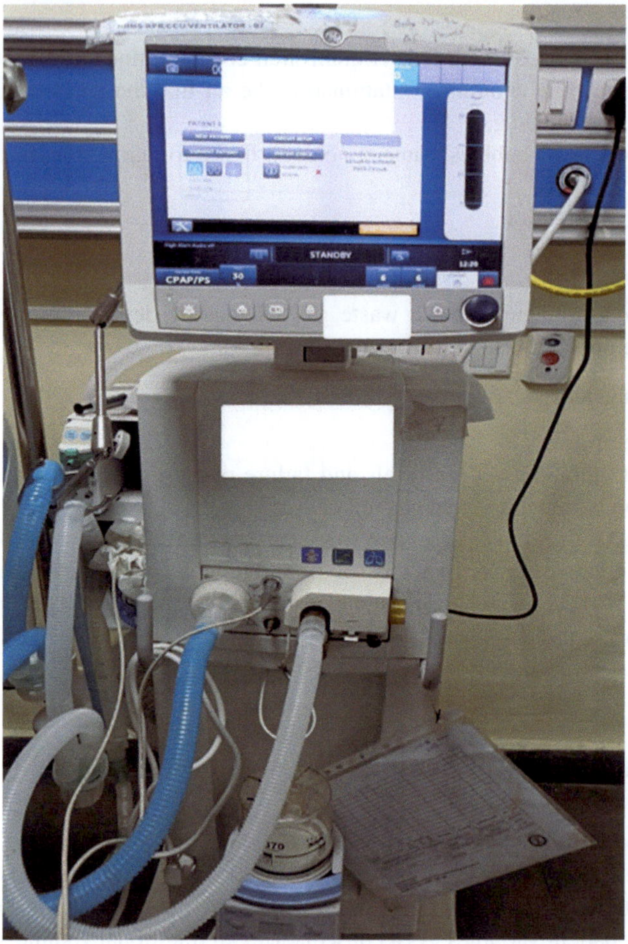

Fig. 3.16 Ventilator machine

- The operator determines the minimum minute ventilation needed based on the current clinical state of the patient.
- If the patient falls short of the target number, the ventilator compensates.
- A too-high maximum respiratory rate can result in a substantial rise in ventilator frequency during ventilator automatic adjustment, resulting in "rapid shallow" breathing [33].

2. *Proportional Assist Ventilation (PAV)*

- PAV relieves respiratory muscles and maintains synchronisation between the ventilator and the patient's neural ventilator drive.
- PAV is a potential newer mode that improves ventilator-patient synchrony.
- The therapist determines the percentage of help to be provided and other parameters are automatically adjusted by patients' air hunger.

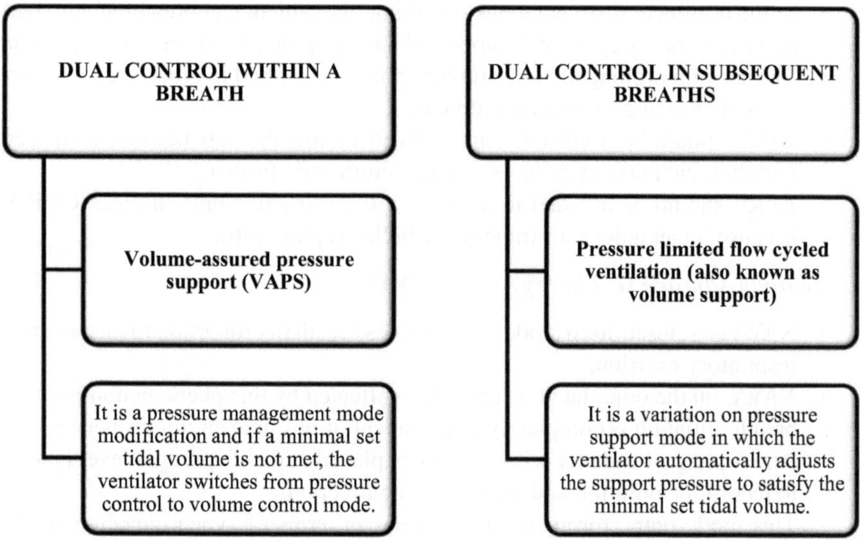

Fig. 3.17 Ventilators on dual control within a breath and in subsequent breath

- Other benefits of PAV include lower airway pressures, easier weaning, and less effort of breathing.
- The possible disadvantage is that if the patient's condition worsens or improves, the proportion of assistance may need to be adjusted to reflect the newer clinical scenario [34].

3. *Adaptive Support Ventilation*

   - This is a one-of-a-kind mode that uses minimal effort of breathing as its goal to achieve the desired minute ventilation.
   - The working principle is based on pressure controlled synchronised intermittent mandatory ventilation with automatic adjustments
   - To establish pressure level and respiratory rate based on previously measured lung mechanics.
   - It can be used in various modes ranging from partial to complete support.
   - Patients with rapid changes in lung physical parameters such as compliance (ARDS) and resistance profit from this treatment [35].

Bi-Level Ventilation Modes

Bi-levels are innovative ventilator modes that allow patients spontaneous breathing during any phase of ventilator cycle.

1. *Airway pressure release ventilation (APRV)*

   - The bi-level pressure functions as time-cycled inverse ratio ventilation in the absence of spontaneous breathing activity.
   - APRV gives two continuous positive airway pressure levels (CPAP).

- In the presence of spontaneous breathing, the patient can breathe in any part of the respiratory cycle with supported breaths, reducing the need for sedation.
- The use of a prolonged high-pressure phase is justified in order to avoid alveolar collapse and sustain recruitment.
- APRV should be used with care in hypovolemic patients because increased intrathoracic pressure reduces venous return even further.
- APRV should be avoided in patients with obstructive lung disease (COPD) because it can induce air trapping or bullae rupture [36].

2. *Neurally Adjusted Ventilatory Assist (NAVA)*

   - NAVA is a closed-loop mode that provides breath in proportion to the patient's inspiratory exertion.
   - NAVA, on the other hand, is entirely unaffected by this phenomenon because breath initiation is completely independent of physical circuit properties.
   - An esophageal catheter with electrodes placed at the diaphragm level is used to record the duration and strength of contraction.
   - This mode outperforms all other modes in terms of synchronisation of the patient's attempts to the ventilator by having the shortest possible delay between the two.
   - Setting up a NAVA ventilator is simple, and the only single input needed is gain over diaphragmatic electrographic potential.
   - The placement of an esophageal catheter with electrodes and the validity of extended durations of ventilation are two practical limitations of NAVA [37].

3. *NeoGanaeh (Smartcare)*

   - It is a closed loop modification of pressure support ventilation with incorporated artificial intelligence.
   - By adjusting ventilator assistance in accordance with the patient's respiratory pattern and research-based weaning protocols, the system replicates clinical practise.
   - This method is founded on basic principles mainly to adapt pressure support to the patient's present clinical condition, wean the ventilator from pressure support if the patient is stable, and begin spontaneous breathing trials according to pre-recorded clinical guidelines.
   - Once the patient is stable on a setting, the attempt is made to gradually wean the patient off in order to keep the ideal respiratory variables, which are already known as the patient comfort zone [38].

3.11 Tracheal Tubes

Endotracheal (ET) tubes are curved tubes used for intubation. It is a polyvinylchloride (PVC) tube that is inserted between the vocal cords and travels through the trachea to deliver oxygen and inhaled gases to the lungs. It also protects the lungs

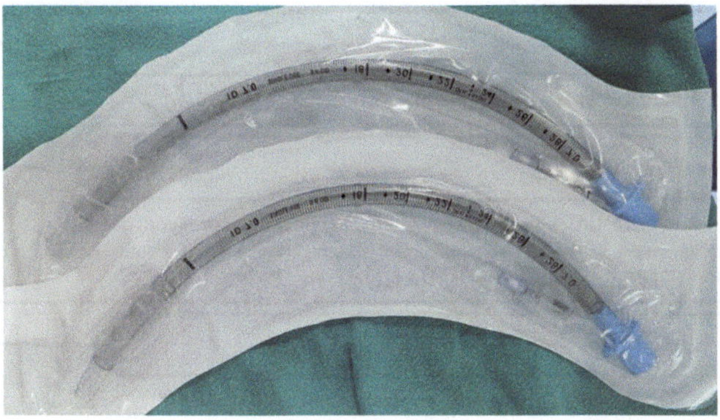

Fig. 3.18 Endotracheal tube

from contaminants like gastric fluids and blood. Endotracheal tube is shown in Fig. 3.18.

Types

There are two types of ET tubes: cuffed and uncuffed tubes.

- Cuffed ET tubes: Used in children over the age of 8 years. On inflation, the cuff keeps the ET tube in place and prevents aspiration of contents from the GI tract into the respiratory tract.
- Uncuffed ET tubes: Used in children under the age of 8 as the narrow subglottic area acts as a cuff and keeps the ET tube from slipping.

Components

The endotracheal tube has four parts which are shown in Fig. 3.19.

Applications

The use of endotracheal tube is listed in Fig. 3.20.

Advantages

While latex tubes are still available, plastic tubes (PVC) are preferred:

- Disposable
- Less allergic reactions compared to latex
- Transparent (allows simple viewing of ETT blockage caused by blood, pus, or secretions)
- Less chances of infection

Complications

- Prolonged intubation may result in pressure necrosis of the laryngeal structures resulting in persistent hoarseness.
- Mechanical trauma to the tongue, teeth, palate, pharynx, and larynx.
- Stimulation of the posterior pharyngeal wall, resulting in coughing, regurgitation, or a vasovagal episode with hypoxia and bradycardia.
- Pneumothorax.

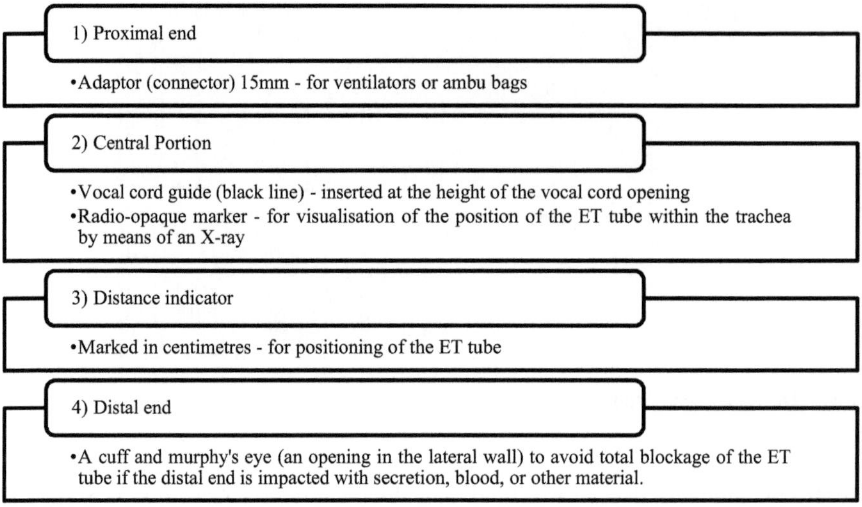

Fig. 3.19 Components of endotracheal tube

Fig. 3.20 Applications of endotracheal tube

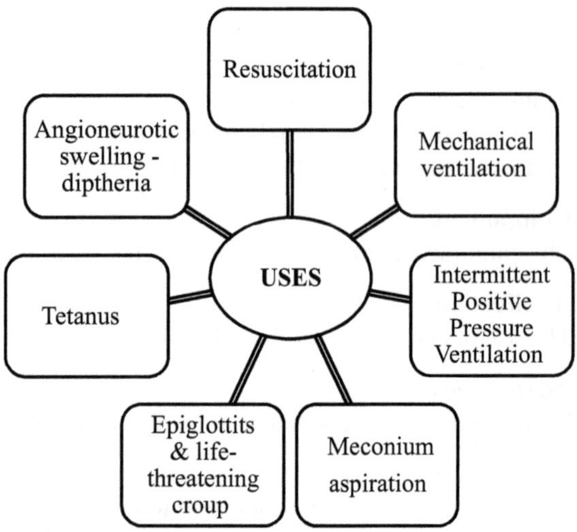

References

1. Ward CS. Anaesthetic equipment physical principles and maintenance. 2nd ed. London: Bailliere Tindall; 1985. p. 104–21.
2. Dorsch JA, Dorsch SE. Understanding Anaethesia equipment, construction, care and complications. 2nd ed. Baltimore: Williams and Wilkins; 1984. p. 38–76.
3. Brockwell RC, Andrews JJ. Inhaled anaethetic delivery systems. In: Miller RD, editor. Miller's Anaethesia. 7th ed. Philadelphia: Churchill Livingstone; 2010. p. 667–716.

4. Gurudatt C. The basic anaesthesia machine. Indian J Anaesth. 2013;57(5):438–45.
5. Subrahmanyam M, Mohan S. Safety features in anaesthesia machine. Indian J Anaesth. 2013;57(5):472–80.
6. Kaul TK, Mittal G. Mapleson's breathing systems. Indian J Anaesth. 2013;57(5):507–15.
7. Upadya M, Saneesh PJ. Low-flow anaesthesia - underused mode towards "sustainable anaesthesia". Indian J Anaesth. 2018;62(3):166–72.
8. Laster AM, Pressman RS. An evaluation of an electroanaethetic device. J Am Dent Assoc. 1975;90(4):816–21. https://doi.org/10.14219/jada.archive.1975.0153.
9. Gerheuser F, Roth A. Epidural anaethesia. Anaesthesist. 2007;56(5):499–523; quiz 524–6.
10. Harrison GR, Clowes NW. The depth of the lumbar epidural space from the skin. Anaesthesia. 1985;40(7):685–7.
11. Moriarty A. Pediatric epidural analgesia (PEA). Paediatr Anaesth. 2012 Jan;22(1):51–5.
12. Turner A, Lee J, Mitchell R, Berman J, Edge G, Fennelly M. The efficacy of surgically placed epidural catheters for analgesia after posterior spinal surgery. Anaesthesia. 2000;55(4):370–3. https://doi.org/10.1046/j.1365-2044.2000.01117.x.
13. Hartung HJ, Luiz T. Die totale Spinalanästhesie. Eine Komplikation der lumbalen Katheterperiduralanästhesie zur postoperativen Schmerztherapie [Total spinal anaethesia. A complication of lumbar catheter peridural anaethesia for postoperative analgesia]. Anaesthesist. 1992;41(5):285–7.
14. McKenzie CP, Carvalho B, Riley ET. The Wiley Spinal Catheter-Over-Needle system for continuous spinal Anaethesia: a case series of 5 Cesarean deliveries complicated by Paresthesias and headaches. Reg Anaeth Pain Med. 2016;41(3):405–10. https://doi.org/10.1097/AAP.0000000000000036.
15. Hyderally H. Complications of spinal anaethesia. Mt Sinai J Med. 2002;69(1–2):55–6.
16. Concepcion MA. Spinal anaethetic agents. Int Anaethesiol Clin. 1989;27(1):21–5. https://doi.org/10.1097/00004311-198902710-00005.
17. Don Michael TA. Esophageal obturator airway. Med Instrum. 1977;11(6):331–3.
18. Johnson KR Jr, Genovesi MG, Lassar KH. Esophageal obturator airway: use and complications. JACEP. 1976;5(1):36–9. https://doi.org/10.1016/s0361-1124(76)80165-7.
19. Struys MM, Kalmar AF, De Baerdemaeker LE, et al. Time course of inhaled anaesthetic drug delivery using a new multifunctional closed-circuit anaesthesia ventilator. In vitro comparison with a classical anaesthesia machine. Br J Anaesth. 2005;94(3):306–17. https://doi.org/10.1093/bja/aei051.
20. Odin I, Feiss P. Low flow and economics of inhalational anaesthesia. Best Pract Res Clin Anaesthesiol. 2005;19(3):399–413. https://doi.org/10.1016/j.bpa.2005.01.006.
21. Szpisjak DF, Lamb CL, Klions KD. Oxygen consumption with mechanical ventilation in a field anaethesia machine. Anaeth Analg. 2005;100(6):1713–7. https://doi.org/10.1213/01.ANE.0000149897.87025.A8.
22. Kennedy R, French R. An audit of anaesthetic fresh-gas flow rates and volatile anaesthetic use in a teaching hospital. N Z Med J. 2003;116(1174):U438. Published 2003 May 16.
23. Young D, Lamb SE, Shah S, MacKenzie I, Tunnicliffe W, Lall R, Rowan K, Cuthbertson BH. High-frequency oscillation for acute respiratory distress syndrome. N Engl J Med. 2013;368(9):806–13. https://doi.org/10.1056/NEJMoa1215716.
24. Mutz N, Baum M, Benzer H, Putz G. Clinical experience with several types of high frequency ventilation. Acta Anaesthesiol Scand Suppl. 1989;90:140–4.
25. Tobin MJ. Principles and practice of mechanical ventilation. Shock. 2006;26(4):426. https://doi.org/10.1097/01.shk.0000245023.16612.dd.
26. Wattwil LM, Sjostrand UH, Borg UR. Comparative studies of IPPV and HFPPV with PEEP in critical care patients. I: a clinical evaluation. Crit Care Med. 1983;11(1):30–7. https://doi.org/10.1097/00003246-198301000-00009.
27. Singh JM, Stewart TE. High-frequency mechanical ventilation principles and practices in the era of lung-protective ventilation strategies. Respir Care Clin N Am. 2002;8(2):247–60.
28. McIntyre JW. Anaesthesia breathing circuits. Can Anaesth Soc J. 1986 Jan;33(1):98–105.

29. Harvey D, Pollock NW, Gant N, Hart J, Mesley P, Mitchell SJ. The duration of two carbon dioxide absorbents in a closed-circuit rebreather diving system. Diving Hyperb Med. 2016;46(2):92–7.
30. Merchant R, Chartrand D, Dain S, Dobson G, Kurrek M, Lagacé A, et al. Guidelines to the practice of anaethesia revised edition 2013. Can J Anaesth. 2013;60:60–84.
31. Brockwell RC, Andrews JJ. Complications of inhaled anaethesia delivery systems. Anaethesiol Clin North Am. 2002;20:539–54.
32. Lucangelo U, Bernabé F, Blanch L. Respiratory mechanics derived from signals in the ventilator circuit. Respir Care. 2005;50:55–65.
33. Burns KE, Lellouche F, Lessard MR. Automating the weaning process with advanced closed-loop systems. Intensive Care Med. 2008;34:1757–65.
34. Ambrosino N, Rossi A. Proportional assist ventilation (PAV): a significant advance or a futile struggle between logic and practice? Thorax. 2002;57:272–6.
35. Dongelmans DA, Veelo DP, Paulus F, de Mol BA, Korevaar JC, Kudoga A, et al. Weaning automation with adaptive support ventilation: a randomized controlled trial in cardiothoracic surgery patients. Anaeth Analg. 2009;108:565–71.
36. Modrykamien A, Chatburn RL, Ashton RW. Airway pressure release ventilation: an alternative mode of mechanical ventilation in acute respiratory distress syndrome. Cleve Clin J Med. 2011;78:101–10.
37. Biban P, Serra A, Polese G, Soffiati M, Santuz P. Neurally adjusted ventilatory assist: a new approach to mechanically ventilated infants. J Matern Fetal Neonatal Med. 2010;23(Suppl 3):38–40.
38. Lellouche F, Brochard L. Advanced closed loops during mechanical ventilation (PAV, NAVA, ASV, SmartCare). Best Pract Res Clin Anaesthesiol. 2009;23:81–93.

Chapter 4
Significant Risk Medical Devices – Cardiovascular

Aswini Saravanan, Abhishek Anil, Surjit Singh, and Shoban Babu Varthya

4.1 Defibrillators

Cardiac defibrillation involves administering an electrical current to individuals experiencing life-threatening ventricular dysrhythmias, namely, ventricular fibrillation (VF) or pulseless ventricular tachycardia (VT). According to Advanced Cardiac Life Support (ACLS) guidelines, pulseless VT and VF are treated similarly. Among adults, VF is the most frequent cause of sudden cardiac arrest. The preferred and definitive treatment for VF is electrical defibrillation. This procedure delivers an electrical shock to the chest using either manual paddles or adhesive "hands-free" pads. Modern defibrillators commonly employ biphasic waveforms, which require lower energy levels for effective defibrillation compared to the older monophasic waveforms [1]. Prompt delivery of defibrillation has been associated with reported survival rates as high as 75% [2, 3]. However, the likelihood of a positive outcome decreases by approximately 10% for each minute that defibrillation is delayed [4, 5].

Majority of defibrillators operate based on energy, where the device charges a capacitor to a specified voltage and then releases a predetermined amount of energy measured in joules. The actual energy delivered to the heart muscle depends on the selected voltage and the transthoracic impedance, which varies among patients. Although most automated external defibrillators (AEDs) currently in use are energy-based, there are two less commonly employed types of defibrillators in clinical practice: impedance-based and current-based defibrillators [6].

Energy-based defibrillators offer different waveform options, broadly categorized as monophasic, biphasic, or triphasic. Biphasic waveforms have a lower

A. Saravanan · A. Anil · S. Singh · S. B. Varthya (✉)
Department of Pharmacology, All India Institute of Medical Sciences, Jodhpur, Rajasthan, India

P. S. Timiri Shanmugam et al. (eds.), *Significant and Nonsignificant Risk Medical Devices*, https://doi.org/10.1007/978-3-031-52838-5_4

defibrillation threshold (DFT), allowing for the use of lower energy levels, which may result in reduced myocardial damage. Furthermore, the utilization of biphasic waveforms enables the development of smaller and lighter AEDs [7].

4.1.1 Automated External Defibrillators (AED)

Since the introduction of direct current defibrillators in 1962, it has been observed that electrical countershock or cardioversion applied externally to the closed chest can successfully terminate various cardiac arrhythmias, not limited to ventricular fibrillation [8]. An automated external defibrillator (AED) is a portable and lightweight computerized device designed to analyze cardiac rhythms and deliver defibrillation when necessary. It employs voice and/or visual prompts to guide both lay rescuers and healthcare providers in safely defibrillating individuals experiencing cardiac arrest due to ventricular fibrillation or pulseless ventricular tachycardia. Extensive in vitro and clinical studies have demonstrated high accuracy, with sensitivity and specificity surpassing 90% [9, 10]. Two types of AEDs exist: semi-automatic, which prompts the operator to deliver the shock by pressing a button, and fully automatic AEDs, which can administer a shock without requiring external intervention.

Intended Use [1]: The intended use of these devices is to treat ventricular fibrillation (VF) or pulseless ventricular tachycardia (VT).

Mechanism: These devices primarily consist of a battery, a capacitor, electrodes, and an electrical circuit designed to analyze the heart rhythm and deliver an electric shock if necessary.

Batteries: Batteries play a crucial role in the AED system. Traditionally, lead batteries and nickel-cadmium batteries were used. However, nonrechargeable lithium batteries have gained popularity due to their smaller size, longer duration without maintenance (up to 5 years), and improved performance. It is important to store defibrillators in controlled environments as extreme temperatures can adversely affect the batteries. Additionally, proper disposal of batteries is essential as they contain corrosive and highly toxic substances, requiring designated containers.

Capacitor: The electrical shock delivered to the patient is generated by high voltage circuits utilizing energy stored in a capacitor. These capacitors can hold up to 7 kV of electricity. The energy delivered by the AED system can range from 30 to 400 joules (J).

Electrodes: Electrodes are vital components of a defibrillator system as they collect information for rhythm analysis and deliver energy to the patient's heart. Various types of electrodes are available, including hand-held paddles, internal paddles, and self-adhesive disposable electrodes. In emergency settings, disposable electrodes are generally preferred due to their faster application and improved defibrillation technique.

Several characteristics of electrodes can impact the outcome of defibrillation. These include electrode position, pad size, and the choice between hand-held and patch electrodes. Studies have shown that four pad positions (antero-lateral, antero-posterior, anterior-left infrascapular, and anterior-right infrascapular) are equally effective. Antero-lateral placement is commonly used as the default electrode position due to its ease of placement and education. The size of the electrode pads plays a crucial role in transthoracic current flow during external shocks. Larger paddles result in lower resistance, allowing more current to reach the heart [11] and potentially reducing myocardial necrosis [12]. Consequently, larger paddles are preferred. Manufacturers offer adult paddles, typically ranging from 8 to 13 cm in diameter, as well as smaller pediatric paddles [13].

Hand-held paddle electrodes may be more effective than self-adhesive patch electrodes because they can be applied with pressure to improve electrode-to-skin contact and reduce transthoracic impedance (TTI). However, they are not used with automated external defibrillators (AEDs) due to the need for training [6].

Electrical Circuit: The electrical circuitry of AEDs is highly sophisticated and microprocessor based. These devices analyze multiple features of the surface ECG signal, such as frequency, amplitude, slope, and waveform morphology. They incorporate various filters to eliminate interference from QRS signals, radio transmission, loose electrodes, poor contact, and patient movement.

Controls on an AED typically include a power button, a display screen for trained rescuers to check the heart rhythm, and a discharge button. Manual defibrillators may have additional controls for energy selection and charging. Certain defibrillators feature specific controls for internal paddles or disposable electrodes [6].

Successful defibrillation is achieved when ventricular fibrillation (VF) is terminated for at least 5 s following the shock [14].

Transthoracic impedance (TTI) refers to the dissipation of energy in the lungs, thoracic cage, and other anatomical structures of the chest. In animal studies, only a small percentage (4%) of the supplied energy reaches the heart [15]. The average TTI in adult humans is approximately 70–80 Ω and is influenced by various factors such as energy level, electrode size, interelectrode distance, skin-electrode interface, electrode pressure, ventilation phase, myocardial tissue, and blood conductivity. When TTI is high, a low-energy shock may not generate sufficient current for successful defibrillation [16]. To reduce TTI, defibrillator operators can use conductive materials such as gel pads or electrode paste [17] with paddles or opt for self-adhesive pads.

Technique

During defibrillation, the placement of the defibrillator paddles or hands-free pads is crucial. The paddles are positioned on the chest, with one paddle placed along the upper right sternal border and the other placed at the cardiac apex. Alternatively, hands-free pads can be used in the same locations or in an anteroposterior configuration.

Once the defibrillator is in the defibrillation mode, the capacitor is charged. The initial energy level for biphasic manual defibrillators is typically preset by the manufacturer. According to the 2015 American Heart Association (AHA) guidelines, it is reasonable to use the manufacturer's recommended dose for the first defibrillation shock, usually ranging between 120 and 200 J for biphasic defibrillators and 360 J for monophasic defibrillators. If the recommended dose is unavailable, the AHA suggests delivering the maximum available dose.

Before delivering the shock, it is important to ensure that no one is touching the patient or any objects in contact with the patient. The "shock" button is then pressed, allowing the stored charge to be delivered across the chest from paddle to paddle or hands-free pad to pad. After the shock is delivered, cardiopulmonary resuscitation (CPR) is immediately resumed for 2 min.

If ventricular fibrillation (VF) or pulseless ventricular tachycardia (VT) persists, CPR is resumed, and subsequent shocks may be administered. It is important to increase the defibrillation dose with each successive shock until the maximum available dose is reached, following the manufacturer's guidelines.

The AHA's VF/pulseless VT algorithm recommends administering 1 mg of epinephrine intravenously every 3–5 min after the first unsuccessful defibrillation. CPR is resumed after the second shock for 2 min. If VF or pulseless VT persists, 300 mg of amiodarone can be administered intravenously during the resuscitation. Additional doses of antiarrhythmic drugs, such as 150 mg of amiodarone or 1–1.5 mg/kg of lidocaine, may be given if needed. Epinephrine and amiodarone are preferred over lidocaine according to the current AHA guidelines [18].

For pediatric patients, the initial energy dose for defibrillation is recommended to be 2 J/kg, with subsequent defibrillations dosed at 4 J/kg or higher, up to a maximum of 10 J/kg. The use of infant pads is necessary for patients under 10 kg or less than 1 year of age, while adult pads are used for patients over 10 kg or 1 year of age. In the absence of small pads, adult pads can be used in the anterior-posterior position, ensuring that the pads do not touch and are not cut to fit the patient. It is not recommended to delay defibrillation due to the unavailability of infant pads or low voltage defibrillators. Delivering a shock with adult pads/defibrillator at an adult dose is preferable to not delivering a shock at all. Ideally, manual defibrillators should be used for patients under 1 year of age, but a pediatric attenuator can be used if unavailable. In children under 8 years of age or less than 25 kg, an automated external defibrillator (AED) with a pediatric dose attenuator is used, while standard adult AED pads and cable systems are used for children 8 years of age and older or 25 kg and greater [19, 20].

In the case of VF or pulseless VT during thoracotomy procedures such as emergency department thoracotomy or cardiac surgery, internal defibrillation is performed. The process follows a similar algorithm to external defibrillation, with some differences in the defibrillation energy dose. In internal defibrillation, an initial dose of 20 J is recommended to prevent burn-like injuries to the myocardium. Care must be taken to avoid damaging coronary vessels. Subsequent doses can be increased up to a maximum of 40 J. Sterile internal pads specifically designed for internal defibrillation must be used and should be readily available during any thoracotomy procedures [21].

It is important to note that these guidelines and techniques may vary based on the specific protocols and recommendations of medical organizations and individual healthcare providers. Always consult the latest guidelines and receive proper training before performing defibrillation procedures.

Complications: One potential complication is the induction of ventricular fibrillation (VF) by the "R-on-T phenomenon." This can occur when defibrillation is performed on a patient who originally had a pulse, leading to cardiac arrest. To avoid this complication, defibrillation is only performed for VF or pulseless VT. If a patient requires electrical cardioversion for an unstable tachycardic rhythm but is not in cardiac arrest, synchronized cardioversion is performed instead of defibrillation [22].

Device Maintenance: Failure to properly maintain the defibrillator or power supply is responsible for the majority of reported malfunctions. However, newer automated external defibrillator (AED) models require minimal maintenance. These devices conduct self-checks to ensure proper operation and indicate when they are "ready to use" [6].

Use Errors: Rare errors have been noted in trials when the device failed to recognize certain varieties of VF or when operators did not follow recommended instructions [10, 23]. To diagnose VF, the device must identify an ECG waveform with an amplitude of at least 0.8 mV at a rate faster than a preprogrammed threshold. For VT, the criteria include a frequency of at least 120 beats/min, QRS duration of more than 160 ms, and absence of P wave. ECG analysis is performed in consecutive segments of 2.7 s, and the diagnosis must coincide in two out of three segments to provide a decision [6].

Drawback: One drawback is that defibrillation requires interruptions in cardiopulmonary resuscitation (CPR) to analyze the rhythm and deliver the electric shock. Ongoing efforts are focused on minimizing this interruption time, and technical advancements may eventually enable accurate rhythm interpretation even while CPR is ongoing [23].

It is important to note that these complications and drawbacks should be considered when using defibrillators, and healthcare providers should be adequately trained to minimize the occurrence of errors and ensure patient safety.

4.1.2 Implantable Cardioverter Defibrillator [24]

The implantable cardioverter-defibrillator (ICD), also known as the automated internal cardiac defibrillator or shock box, is a device similar to a pacemaker that can detect abnormal fast heart rhythms and provide immediate treatment. This treatment can include overdrive pacing called anti-tachycardia pacing (ATP) or shock therapy, which can be delivered in a synchronized or asynchronized manner depending on the detected rhythm and preprogrammed algorithm.

Uses: The ICD has become the primary treatment option for patients at high risk of sudden cardiac death (SCD). It has consistently shown a survival benefit in patients who have experienced cardiac arrest (SCA) due to ventricular fibrillation (VF) or ventricular tachycardia (VT). It is also used in patients with heart failure and severe systolic dysfunction (left ventricular ejection fraction (LVEF) less than or equal to 35%), as well as in patients with hypertrophic cardiomyopathy (HCM).

Indications for ICD implantation are generally categorized as secondary or primary. Secondary prevention involves implanting the device in patients who have already experienced and survived cardiac arrest due to VF/VT. Primary prevention, on the other hand, involves implanting the device in patients who are at high risk of sudden cardiac death due to VF/VT but have not experienced such an event before. Primary prevention has become the most common reason for ICD implantation in recent times.

It is important to note that the ICD is a specialized device that requires careful evaluation and consideration of the patient's individual risk factors and medical history. The decision to implant an ICD should be made in consultation with a cardiologist or electrophysiologist who has expertise in managing cardiac arrhythmias and sudden cardiac death.

Class (I) recommendations for ICD implantation include [25]:

- Left ventricular (LV) dysfunction with an ejection fraction (EF) less than or equal to 35% and NYHA (New York Heart Association) class II/III symptoms
- LV dysfunction with an EF less than or equal to 35% due to a previous myocardial infarction (MI) or at least 40 days post-MI, along with NYHA class II/III symptoms
- LV dysfunction with an EF less than or equal to 30% due to a previous myocardial infarction (MI) or at least 40 days post-MI
- LV dysfunction with an EF less than or equal to 40% due to a previous myocardial infarction (MI) or at least 40 days post-MI, and inducible ventricular tachycardia (VT) or ventricular fibrillation (VF) on electrophysiological study (EPS)
- Syncope (fainting) of unknown etiology (cause) and inducible VT/VF on EPS
- Sustained VT in the presence of structural heart disease

Contraindications

The use of an Implantable Cardioverter Defibrillator (ICD) is not recommended in the following situations where reversible causes, such as myocardial ischemia, sepsis, hypoxia, electrolyte imbalance, or electrocution, are responsible for ventricular tachycardia/ventricular fibrillation (VT/VF). It is also contraindicated for patients with atrial arrhythmias without concomitant VT/VF or those with incessant VT/VF.

Equipment

The ICD system comprises three main components:

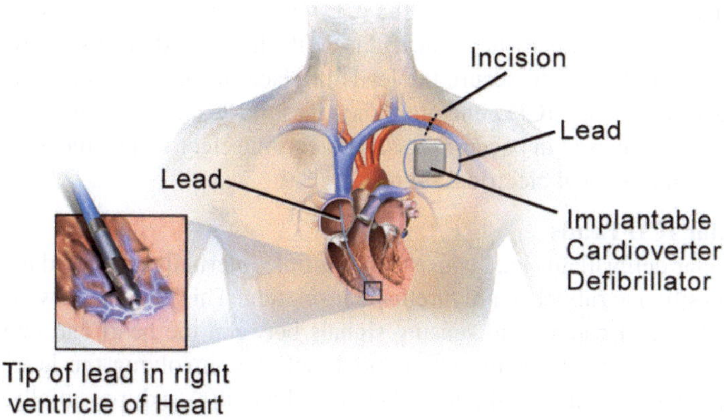

Fig. 4.1 Diagram of implantable cardioverter defibrillator [26]The depicted image presents a single chamber Implantable Cardioverter Defibrillator (ICD). Pulse generator is positioned within a subcutaneous pocket in the pectoral region and has a header featuring ports for leads, battery and capacitors, memory chips, integrated circuits, and telemetry module [27]

1. Pulse Generator: The pulse generator, or generator, is a rounded or ovoid pod that houses sophisticated machinery capable of reading, recognizing, and storing cardiac rhythm tracings. It determines the need for and delivers appropriate therapy, such as antitachycardia pacing (ATP) or defibrillation, based on a preprogrammed algorithm. Typically implanted in a subcutaneous pocket, usually in the pectoral region as shown in Fig. 4.1, it contains a battery and a defibrillator. The battery has a finite charge and capacity to defibrillate and pace the heart as required. The manufacturer code, readable on a magnified plain radiogram, is also included. A device programmer magnet can be placed over the pulse generator for magnet inhibition, interrogation, external programming, or therapy. Additionally, these generators can communicate wirelessly with compatible programmers within a limited radius. The pulse generator incorporates the necessary elements for pacing, sensing, and defibrillation functions, including the battery and electronic circuitry.
2. Battery: Each ICD is equipped with a battery, and its lifespan depends on the frequency of pacing and defibrillation performed by the device. Typically, the battery lasts between 4 and 7 years.
3. Circuitry: The ICD circuitry determines the timing and delivery of bradycardia pacing and anti-tachycardia therapies. It offers a wide range of programming options to tailor the device's function to suit individual needs. The ICD capacitor(s) play a critical role in defibrillation. Capacitors store energy (charge) that can be released across the heart's pathway, allowing for defibrillation or cardioversion.

The ICD Header and Can

The ICD generator communicates with the heart through a ventricular defibrillator lead, with the option of an atrial and left ventricular (LV) lead in specific ICD models. These leads are connected to the ICD via a header, which provides insertion holes for all components of the ventricular defibrillator lead. The set screws in the header can be tightened to secure the leads in place or loosened for their removal. The metal casing of the ICD generator, known as the "can," is composed of materials similar to those used in pacemakers. It protects the electrical elements of the ICD from fluids and external electrical sources.

Ventricular ICD Leads

A ventricular defibrillator lead consists of multiple internally separated metal wires encased in silicone rubber or polyurethane insulation. This design allows the lead to transmit electrical pacing and sensing signals between the heart and generator. It also provides a separate pathway for shock delivery, involving the lead's coil(s). Shocks are delivered across the heart between the coil(s) and, potentially, the ICD can. Each lead element has its own pin that connects to the ICD header. Ventricular ICD leads are categorized as single-coil or dual-coil leads. Both types include a distal coil, while a dual-coil lead also has a proximal coil along its length, ideally ending in the superior vena cava (SVC) when this configuration is desired. Consequently, a ventricular defibrillator lead has a total of either two or three pins that connect to the ICD: one for pacing and sensing and one or two additional pins for defibrillation, depending on whether it is a single-coil or dual-coil lead. The ventricular ICD lead is essentially a bipolar lead, utilizing either an integrated bipolar configuration with the lead tip and distal coil for pacing and sensing purposes, or a true bipolar configuration with the tip and a ring electrode [28].

Leads: Leads are cables or silicone-coated wires that connect the generator to the myocardium and house the electrodes facilitating conduction between the device and the myocardium. They remain inside the heart, with screws at one end hooked to the device, and the cardiac end secured in place inside the respective cardiac chamber using screws or tines.

Shocking Coil: The lead within the ICD contains a shocking coil responsible for delivering the necessary charge for defibrillation or electrical cardioversion. On plain radiograms, ICD leads can be identified by the presence of well-demarcated spiral thickenings, known as shocking coils. These coils are usually located in the region of the superior vena cava (SVC) on the atrial lead and near the ventricular end of the right ventricular (RV) lead.

Technique

The implantation procedure for an ICD involves draping and prepping the patient in a sterile manner, similar to pacemaker implantation. The patient's pectoral region is exposed, and the procedure is typically performed under local anesthesia with conscious sedation. The device is implanted in a subcutaneous pocket, usually created in the pectoral region, where the device generator is placed. The leads are then connected to the generator on one end, and the other end of the lead is advanced via the subclavian vein and positioned in the respective cardiac chamber. While there is no consensus on defibrillation threshold testing (DFT), some operators perform DFT after device implantation to assess defibrillation thresholds and optimize device settings. Finally, the pocket is closed. The entire procedure usually takes up to 90 min.

Complications
Short-term complications (2–3%) are typically immediate and procedure-related, including:

1. Access complications, such as bleeding, thrombosis of the subclavian/axillary vein, and inadvertent puncture of lung tissue leading to pneumothorax or hemothorax.
2. Pocket complications, such as pain, pocket hematoma, and a condition known as Twiddler syndrome, where the device twists and becomes misplaced due to pressure from an expanding hematoma, leading to device malfunction by disrupting lead and generator connections.
3. Device-related infections (1–2%), including endocarditis, have also been reported.
4. Serious complications (<1%) like pulseless electrical activity (PEA) and death can occur during DFT.

Long-term complications (up to 4%) may include:

1. Device-related pain
2. Anxiety
3. Lead fracture
4. Inappropriate shock delivery
5. Phantom shock, where patients perceive receiving shock therapy, but device interrogation does not reveal any such shock or event
6. Device erosion through the skin
7. Device infection, more common during replacement or generator change procedures rather than new device implantation
8. Immunologic rejection, although rare

Clinical Significance
Patients with structural heart disease and cardiomyopathies, both ischemic and non-ischemic, benefit from ICD therapy, particularly when combined with resynchronization therapy. This combination not only provides a mortality benefit but also improves the quality of life (QOL) and has shown potential for aiding in the recovery of left ventricular ejection fraction (LVEF).

4.1.3 Wearable Cardioverter Defibrillator

The wearable cardioverter-defibrillator (WCD) is an external device designed for the automatic detection and defibrillation of ventricular tachycardia (VT) and ventricular fibrillation (VF). It is used in situations where the implantation of an internal cardioverter-defibrillator (ICD) may be deferred or deemed unnecessary, such as temporary arrhythmias or when the risk of ICD implantation is high. The WCD provides temporary protection against cardiac arrest and is typically worn for several months [6].

Indications for WCD Use Include [6]

1. Recent myocardial infarction or coronary revascularization with severely reduced left ventricular ejection fraction
2. Newly diagnosed nonischemic cardiomyopathy with severely reduced left ventricular ejection fraction
3. Severe cardiomyopathy as a bridge to heart transplantation or in patients with ventricular assist devices
4. Need for interruption of ICD therapy or temporary inability to implant an ICD (e.g., due to infection)
5. Syncope with a high risk of ventricular tachyarrhythmias
6. Ambulatory event monitoring to determine the cause of syncope

Device Description
The WCD consists of four nonadhesive monitoring electrodes and three defibrillation electrodes incorporated into a chest strap assembly (Fig. 4.2). The monitoring electrodes are placed around the chest and held in place with an elastic belt, providing two surface ECG leads. Proper fitting of the vest is crucial to ensure adequate skin contact and minimize false alarms [6].

Mechanism [6]
The mechanism of the wearable cardioverter-defibrillator (WCD) involves programmed arrhythmia detection based on rate criteria. When an arrhythmia is detected, the device emits audible and vibration alarms. Patients are trained to hold response buttons during these alarms to prevent a shock while awake. If a shock is necessary, a voice warns the patient and bystanders and the device charges, extrudes

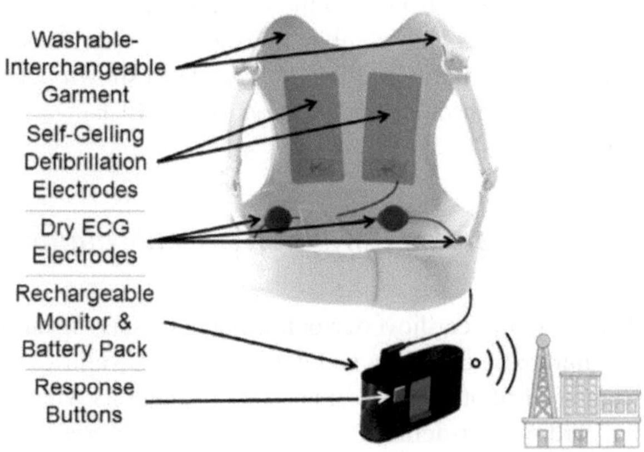

Fig. 4.2 Wearable cardioverter-defibrillator [29]The LifeVest Model 4000 consists of two components: a chest-worn vest assembly and a monitor unit carried at the waist. The monitor contains a rechargeable battery, defibrillation capacitor, response unit/buttons, processor, and display

gel from the defibrillation electrodes, and delivers up to five biphasic shocks at preprogrammed energy levels (maximum output of 150 J).

The WCD allows programmable tachycardia detection rates and shock delays for ventricular fibrillation (VF) and ventricular tachycardia (VT). VF shock delay can be programmed from 25 to 55 s, while VT detection rate is programmable between 120 bpm to the VF setting with a VT shock delay of 60–180 s. Additional shock delays (up to 30 s) may be allowed during sleep. Synchronized shock delivery on the R wave is possible with VT signals, but unsynchronized shocks are delivered if the R wave cannot be identified. The shock energy is biphasic and can be programmed from 75 to 150 J, with up to five shocks delivered per event.

The WCD has the capability to store data on arrhythmias, asystole, patient compliance, noise, and interference. This information is stored and transmitted to the manufacturer network via modem for clinician review.

Efficacy

In terms of efficacy, a postmarket study in the USA involving 3569 patients reported a 100% success rate of the first shock in terminating VT/VF in unconscious patients, with a survival rate of 86%. However, death after a successful first shock occurred in some cases due to recurrent VT/VF, bystander interference, disruption of the electrocardiogram signal from a fall, and inhibition of detection due to pacing stimulus artifact from a unipolar pacemaker. The study also found that long-term survival was similar between WCD and ICD patients [28]. However, it is important to note that the WCD does not offer pacemaker functions, and a small percentage of patients (0.6%) experienced asystole, leading to a mortality rate of 74%.

Error

One potential error is the delivery of inappropriate shocks by the WCD, which may occur due to electronic noise, device malfunction, or the presence of supraventricular tachycardia. The rate of inappropriate shocks by the WCD is comparable to that of ICDs [30, 31]. However, the ability to abort shocks by pressing response buttons while awake and the device's alarm system for ECG noise can potentially reduce inappropriate shocks [6].

Limitations

Limitations of the WCD include the absence of pacemaker functions, the requirement for patient interaction and compliance, and the need to remove the device for bathing, during which a caregiver should be present. A German study reported a mean daily use of the WCD of 21.3 h, and common complaints associated with the device include its weight, sleep disturbances during noise alarms, and skin rash or itching [31].

4.2 Cardiac Ablation Catheters [32]

Catheter ablation, specifically radiofrequency ablation, has significantly transformed the treatment of tachyarrhythmia. It has rapidly advanced over the years and is now considered the primary therapeutic approach for many tachycardias in patients who experience recurring symptoms that impede their productivity and quality of life. Catheter ablation has become the cornerstone of treatment for various arrhythmias, offering an improved option for individuals suffering from recurrent arrhythmias. It is particularly effective in managing symptomatic accessory pathway, atrial flutter, atrial fibrillation, and drug-refractory ventricular tachycardia (VT) or cardiomyopathy induced by premature ventricular contractions (PVCs).

Indications

- Definitive treatment of symptomatic supraventricular tachycardia (SVT) caused by atrioventricular re-entrant tachycardia (AVRT), atrioventricular nodal re-entrant tachycardia (AVNRT), unifocal atrial tachycardia, or atrial flutter
- Treatment of atrial fibrillation (AF) with lifestyle-impairing symptoms and inadequate response to antiarrhythmic medications
- Management of symptomatic idiopathic ventricular tachycardia (VT)

However, when dealing with VT in the context of structural heart disease, catheter ablation is generally reserved for cases where drug therapy has failed or as an adjunctive therapy for patients experiencing frequent discharges from an implantable cardioverter-defibrillator (ICD).

Contraindications
There are no absolute contraindications for catheter ablation, but some relative contraindications may include:

- Vascular access contraindications such as deep vein thrombosis (DVT) for femoral vein access, peripheral arterial disease (PAD), or aortic dissection in the case of retrograde aortic approach
- Presence of intracardiac thrombi to prevent the risk of embolization
- Bleeding diatheses and coagulopathy, which pose significant risks during catheter ablation. However, under certain circumstances, an electrophysiologist may proceed with the procedure if the patient's international normalized ratio (INR) is well controlled up to a level of 3.

Equipment
This nonsurgical procedure employs an electrode catheter, a specialized insulated electrical wire. After localizing the abnormal site in the heart responsible for arrhythmia, radiofrequency energy is applied through the catheter for the purpose of ablation.

Catheters: A variety of catheters are available, typically featuring at least two ring electrodes for bipolar stimulation and recording. These catheters can be made of woven Dacron or newer synthetic materials like polyurethane.

1. Woven Dacron catheters are preferred due to their superior durability and physical properties. They offer a range of options with varying numbers of electrodes, electrode spacing, and curves to accommodate different purposes. These catheters possess excellent torque characteristics and are sufficiently rigid to maintain their shape while softening at body temperature to ensure they are pliable for various vascular configurations.
2. Synthetic catheters, on the other hand, cannot be manipulated or molded within the body, making them less desirable. However, they are cheaper and come in smaller sizes (2–3 French). Currently, most electrode catheters used range from size 3 to size 8 French, with smaller sizes reserved for pediatric patients. In adults, catheters sized 5–7 French are typically employed. Some diagnostic catheters feature a deflectable tip for reaching and recording from specific sites such as the coronary sinus, crista terminalis, or tricuspid valve. Nevertheless, in most cases, the standard woven Dacron catheters are sufficient and significantly more cost-effective. Mapping catheters generally fall into two categories:
   (i) Deflectable catheters to facilitate positioning for mapping and delivering ablative energy
   (ii) Catheters with multiple poles (8–64) that allow simultaneous acquisition of activation points

Catheter ablation has seen advancements in various types of catheters used, including those with cooled tips. These catheters allow for the infusion of saline, enhancing tissue heating without causing surface charring or internal cooling. Typically, ablation catheters have tips ranging from 3.5 to 5 mm in length, although longer tips up to 10 mm are also used. Ongoing research is being conducted on catheters that deliver microwave, laser, cryothermal, or pulsed-ultrasound energy to destroy tissue.

In the category of mapping catheters, there are standard catheters with up to 24 poles that can be deflected to map large or specific areas of the atrium, such as the coronary sinus or tricuspid annulus. Shaped catheters, such as "halo" catheters, record from around the tricuspid ring, while lasso catheters on a deflectable shaft record from 10 to 20 electrodes in the pulmonary vein ostia. Basket catheters, which have up to 64 poles or prongs that spring open, are used to acquire simultaneous data from within a specific cardiac chamber.

One available catheter consists of five flexible splines, with four electrodes on each spline, enabling the acquisition of 20 activation sites. The 2 mm interelectrode distance allows for high-density mapping. However, the flexibility of the splines can sometimes result in variable contact and misinterpretation of data.

Recently, a new catheter has been developed—a 64-pole roving catheter. This mapping catheter, also known as a mini basket catheter, features an 8 French bidirectional deflectable shaft and a basket electrode array with eight splines, each containing eight small (0.4 mm^2), low-impedance electrodes (totaling 64 electrodes). The interelectrode spacing along the spline is 2.5 mm (center-to-center). Mapping can be performed with the basket in various degrees of deployment, ranging from a diameter of 3 to 22 mm. The location of each of the 64 electrodes is identified using a combination of a magnetic sensor in the distal region of the catheter and impedance sensing on each of the 64 basket electrodes. This location identification is possible whether the basket is fully or only partially deployed.

Junction Box

The junction box is a rectangular box that receives intracardiac signals from the catheters and serves as the interface to the physiologic recorder. Multiple switches within the junction box are designated to a recording and stimulation channel, which can be selected through the recording apparatus. To minimize noise on the channels, the junction box is mounted close to the patient's foot.

Recording Apparatus

The physiologic recorder records, displays, and stores intracardiac and surface recordings. It consists of filters, amplifiers, display screens, and recording software. Physiologic signals from the junction box are introduced into the recorder. These signals, typically low in amplitude, are amplified before being displayed and recorded. The recording system individually amplifies and filters each input channel, with most current systems supporting 64 or more channels. The amplifiers can automatically or manually adjust gain control. Ideally, the amplifiers should be positioned close to the patient table to minimize cable length for intracardiac connections and surface ECGs, reducing signal noise. The amplifier is then connected to the main physiologic recorder through a separate channel, preferably running separately from electric power cables. Filters are used to eliminate unnecessary signals that may distort electrograms (EGMs). High-pass filters remove signals below a specified frequency, while low-pass filters remove signals above a given frequency. Most intracardiac electrograms are best identified when the signal is filtered between a high pass of 40 Hz and a low pass of 500 Hz.

Filters are used to eliminate unnecessary signals that may distort electrograms (EGMs). High-pass filters remove signals below a specified frequency, while low-pass filters remove signals above a given frequency. Most intracardiac electrograms are best identified when the signal is filtered between a high pass of 40 Hz and a low pass of 500 Hz.

Stimulator

A programmable stimulator is necessary to obtain electrophysiologic data beyond conduction interval measurements. Stimulators are capable of various pacing modes, including rapid pacing, delivery of single or multiple extra stimuli following a paced drivetrain, and timed delivery of extra stimuli following sensed beats. Stimulators can deliver variable currents ranging from 0.1 to 10 mA. With catheters properly positioned, current thresholds below 2 mA (with a 2 ms pulse width) are typically achieved in both the atrium and ventricle. Higher outputs may be required in cases of diseased myocardium, within the coronary sinus, or when using antiarrhythmic medications. The output is usually set at twice the diastolic threshold. Pacing at higher outputs is discouraged as it may lead to far-field capture and alter the QRS and EGM configuration, potentially misleading the diagnosis. It is recommended to either start with low amplitude until capture is achieved or start with high output and gradually decrease until failure of capture, then increase by a step.

Cardioverter/Defibrillator

During electrophysiology studies, it is essential to have a primary and backup cardioverter/defibrillator readily available. Defibrillators deliver energy in a biphasic waveform, which offers enhanced defibrillation success. Defibrillation pads are attached to the patient and electrically grounded.

Personnel

The following personnel are required for an electrophysiology study/ablation:

1. Cardiac electrophysiologist
2. Cardiac electrophysiology laboratory technician
3. Nursing staff for administration of drugs
4. Radiographer
5. An anesthesiologist may be required in rare and complex cases

Complications

Complications associated with catheter ablation depend on the type of arrhythmia and the site of ablation. Some potential complications include:

1. Death, myocardial infarction, or stroke (ranging from 0.05% to 0.01%). The risk of stroke is higher in curative atrial fibrillation ablation.
2. Heart block requiring a permanent pacemaker (0.5%), primarily based on the proximity of the ablation lesion to the atrioventricular node.
3. Cardiac trauma and perforation leading to tamponade (1–2%).
4. Thromboembolic complications, including systemic and venous embolism (such as pulmonary embolism), which vary depending on the procedure.
5. Vascular access complications, such as arteriovenous fistula, aneurysm, and retroperitoneal bleeding, which are more common (ranging from 2% to 4%).
6. For atrial fibrillation ablation, there are rare complications like pulmonary vein stenosis and phrenic nerve damage. However, identifying the site of the phrenic nerve can help avoid these complications.
7. Incidence of thromboembolic events is common, especially with increased ablation time, leading to the risk of atrial-esophageal fistula.
8. Pulmonary vein stenosis and atrio-esophageal fistulae are rare complications of atrial fibrillation ablation.

Success Rate

Catheter ablation is considered a permanent treatment with a success rate of over 90% for AVRT (atrioventricular re-entrant tachycardia) and AVNRT (atrioventricular nodal re-entrant tachycardia) ablation. The success rates can vary depending on the type of arrhythmia being treated and the specific technique used during the procedure.

It is important to note that the success rate also depends on several factors such as the experience and skill of the electrophysiologist performing the ablation, the characteristics of the arrhythmia, the location of the abnormal tissue causing the arrhythmia, and the presence of any underlying structural heart disease.

For other arrhythmias like atrial fibrillation, the success rates of catheter ablation can vary and may require multiple procedures or additional treatments. Success rates for these arrhythmias are generally lower than for AVRT and AVNRT ablation.

4.3 Stents

Stents are cylindrical implants designed to provide mechanical support to narrow arteries or nonvascular conduits until the risk of complete closure is eliminated [33]. Coronary stents (CS) are used in cases of coronary artery stenosis caused by underlying atherosclerosis. This medical procedure, known as percutaneous coronary intervention (PCI) or coronary angioplasty with stent placement, involves the insertion of coronary stents into the affected arteries. Since their development in the 1980s, coronary stents have undergone continuous advancements in terms of shape, structure, and materials. Ongoing research in this field includes bare-metal stents, drug-eluting stents, bioresorbable stents, and polymer-free stent [34].

Types of Coronary Stents

1. Bare metal stents (BMS)
2. Drug-eluting stents (DES)
3. Bioresorbable scaffold system (BRS)
4. Drug-eluting balloons (DEB)

Bare Metal Stents (BMS)

Bare metal stents (BMS) were the earliest form of stents and were primarily constructed using metallic materials. The first-generation stents are referred to as bare-metal stents because they lacked additional coatings. These permanent stents were typically made of stainless steel and cobalt-chromium (CoCr) alloys for balloon-expandable stents and nickel-titanium alloys (nitinol) for self-expanding stents.

BMS have certain complications, including an increased risk of thrombosis and restenosis. In-stent restenosis (ISR) can occur as a result of intravascular injuries during the stenting procedure. ISR, which leads to the blockage of the artery over time, is the primary cause of stent failure [33]. According to early reports by Fischman et al. [35], approximately 15–20% of patients with implanted BMS required re-intervention within 6–12 months due to ISR. Although the introduction of new generations of stents has reduced the rate of in-stent restenosis, this issue still persists across all types of stents.

Drug-Eluting Stents (DES)

The first drug-eluting stent (DES) introduced in the market was Dexamet, which was coated with Cortizone.

Components

A DES consists of a metallic stent platform, an active pharmacological drug agent, and a carrier vehicle. Stainless steel or cobalt-chromium is commonly used as the metal for its long-term mechanical stability in countering vascular recoil.

Mechanism

The release of the drug from the polymer substrate can be classified into two major mechanisms: physical and chemical. In the physical mechanism, the drug is released through a permanent polymer layer via dissolution or degradation of the polymer, permeation pressure, or ion exchange process. Chemical drug release occurs due to the breakage of covalent bonds, which can happen through chemical or enzymatic degradation. The advantage of the physical mechanism is that it allows control over drug release through the designed stenting system. In contrast, chemical drug release may result in new chemical bindings, which are disadvantages to the system.

Commonly used drugs: Common drugs used in DES act to block signal transduction and cell cycle progression at different phases, thereby inhibiting smooth muscle cell (SMC) proliferation or intimal hyperplasia at the stented arterial site. Rapamycin agents, for example, bind to the intracellular protein FKBP-12, inhibiting the protein kinase mammalian target of rapamycin (mTOR). This complex increases the expression of p27 and blocks the cell cycle progression from the G1 phase to the S phase (DNA synthesis). Another category of drugs, taxanes, interferes with microtubule function, which is necessary for the M phase (mitosis), causing cell arrest in the G2 phase of the cell cycle.

Polymer Coating

To increase the surface area and enable sufficient drug loading and release over time, a carrier vehicle matrix or a polymer coating is used. The polymer coating typically consists of repeating units of biodegradable poly-L-lactide, poly-D and L-lactic-co-glycolic acid in a regular pattern. These polymers degrade into lactic acid and glycolic acid, ultimately converting into water and carbon dioxide [34].

The roles of polymer coating on the stent surface include inhibiting drug wash-off, providing a scaffold for drug loading, allowing engineered control over drug release, and ensuring satisfactory biocompatibility after the drug has been released [33].

Evolution of DES

First-generation DESs had sirolimus or paclitaxel coating on a stainless-steel base. Second-generation DESs feature zotarolimus or everolimus coating on a biocompatible cobalt-chromium or platinum-chromium platform. Drug release occurs through diffusion through pores in the polymeric coating. Despite significant progress, the ideal DES device has yet to be established. The primary goals for new DESs are eliminating potentially lethal consequences and reducing in-stent restenosis [36].

Technical Problems with DES

There are several technical challenges associated with DESs, including delayed endothelialization caused by the locally delivered drugs, inherent thrombogenicity of the stent as a foreign device to the immune system, hypersensitivity and inflammatory reactions due to the metal-based framework and/or polymeric coatings, insufficient drug amounts, lack of sustained drug release, and stent displacement.

Bioresorbable Scaffold System (BRS)
A BRS provides temporary support to the artery and fully biodegrades once its functionality is complete. The complete degradation of the stent allows the artery to revitalize and makes any re-intervention or treatment at the affected site easier.

Mechanism
The functionality of a BRS involves three overlapping phases: revascularization, restoration, and resorption. Revascularization focuses on reopening narrowed vessels. Restoration is the phase where the BRS gradually loses mass, resulting in a reduction of molecular weight. Resorption is the complementary phase where the vascular structure fully recovers its initial normal function.

Advantages of BRS
BRS offers several advantages over bare metal stents (BMS) and drug-eluting stents (DES). These advantages include adaptive shear stress, late luminal gain, late expansive remodeling, reduced restenosis and late stent thrombosis, the possibility of re-intervention at the site of injury, and improved invasive imaging. BRS also has the capacity to restore natural vascular function and provides higher flexibility compared to metal stents. A study demonstrated that BRS outperformed DES stents, with the implanted artery reverting to its previous geometry within 6–12 months of implantation [37].

Disadvantages
BRS has some disadvantages, including a lack of radio-opacity, which is necessary for precise placement and monitoring of the stent within the vessel. It also has reduced radial strength and flexibility compared to metallic stents. Additionally, BRS requires a longer time for predilatation [33].

Other types of stents: Drug-eluting balloons (DEB) only have an antiproliferative drug coating without an underlying metallic structure. BRS, on the other hand, is completely devoid of a metallic structure and fully resorbs within a few months after fulfilling its purpose. There are also specialized stents, such as bifurcation stents and covered stents, designed for specific circumstances like lesions over vascular bifurcations or coronary artery perforations, respectively.

New Advances in Stents

Gene-eluting stents (GES): GES involves using stents as delivery scaffolds for localized genetic exchange in the vascular system. This technique overcomes challenges in cardiovascular gene therapy, such as insufficient gene propagation and systemic immune reactions to vectors. GES allows for prolonged elution and may prevent multisystem immune responses. The stent coatings must be free from inflammation and thrombogenicity while maintaining biocompatibility.

DNA (Plasmid) Eluting Stents: GES and customized coronary stents with self-reporting stent sensors are promising alternatives to existing stent approaches.

These advancements, including gene-eluting stents and 4D printing technologies for customized coronary stents, along with integrated self-reporting stent sensors, should be considered for future developments in optimal coronary stent devices [33].

Vascular Endothelial Growth Factor (VEGF)/VEGF-Paclitaxel Co-eluting Stent: This approach involves releasing VEGF genes followed by paclitaxel. Restenosis is significantly reduced when cell multiplication is inhibited, and complete re-endothelialization is achieved within 4 weeks after implantation [36].

Shape-memory stents: These stents have the ability to self-expand within the body's temperature range.

Polymer-free drug-eluting stents: While drug-eluting stents have shown favorable effects, studies have revealed the inflammatory triggering effects of toxic ions released from polymeric coatings or degradable metals/alloys used as surface coatings. One alternative is to develop polymer-free stents. Polymer-free stents have a faster drug elution rate, but their efficacy and safety are comparable to first-generation DES [33].

4.4 Cardiac Pacemakers

Pacemakers are devices that generate electrical activity and are used to treat patients who have a slow heart rate, symptomatic heart blocks, or heart failure. They consist of two main components: a pulse generator that produces the necessary electrical current to stimulate the heart muscles, and one or two electrodes, also known as leads, which transmit the generated electrical activity from the pulse generator to the heart muscles [38].

Pacemakers can be broadly categorized as external or internal. External pacemakers are typically employed for temporary stabilization of patients or to assist in certain surgical procedures. On the other hand, implantable pacemakers are usually permanent and often more intricate compared to their temporary, external counterparts [39] (Fig. 4.3).

The North American Society of Pacing and Electrophysiology (NASPE) and the British Pacing and Electrophysiology Group (BPEG) collaborated to create a universal pacemaker code that is used worldwide. This code enables healthcare providers and manufacturers to describe the characteristics of the pacemaker device [41].

The code consists of five letters that convey specific information about the pacemaker's functionality. The first letter indicates which chamber of the heart is being paced. The second letter denotes which chamber the device is sensing. The third letter signifies whether the device responds to the sensed signals. The fourth position in the code reveals if the pacemaker modulates or adjusts its programmed rate independently of the patient's cardiac activity, such as during exercise. Finally, the fifth and last letter of the code indicates if the pacemaker supports additional multisite pacing. It is worth noting that the last two letters of the code, in the fourth and fifth position, are rarely utilized in regular terminology [41].

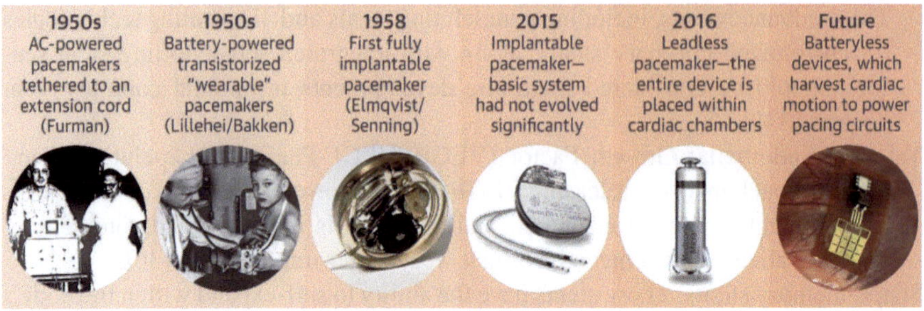

Fig. 4.3 History of cardiac pacemakers [40]. In the evolution of cardiac pacemakers, it initially involved large external devices powered by alternating current (AC). The transition ensued into transistorized battery-powered wearable pacemakers, marking the era of external devices. A pivotal shift occurred with the advent of entirely implantable pacemakers. Currently, leadless pacemaker is gaining rapid momentum and is clinically accessible. Batteryless pacemakers that harvest kinetic energy from heart to power the device are being studied

The most recent generation of pacemakers offers a multitude of capabilities. The simplest settings include AAI and VVI modes. In the AAI mode, the pacemaker paces and senses the atrium, and each sensed event triggers the pacemaker generator to emit a signal within the P wave. On the other hand, the VVI mode paces and senses the ventricle and is inhibited by a sensed ventricular event [39].

Mechanism

They emit pulses at regular intervals, typically ranging from 0.5 to 25 ms, with voltages varying from 0.1 to 15 V. The frequency of these pulses can reach up to 300 min^{-1}. Whether the device is temporary or permanent, the cardiologist or pacemaker technologist possesses the ability to examine and regulate crucial factors such as the pacing rate, pulse width, and voltage [39].

Indications [38]

Common Indications for Pacemaker Placement:

1. Sinus node dysfunction (Class I indication)
2. Acquired AV block
3. Postmyocardial infarction

Less Common Indications:

1. Congenital complete heart block
2. Long QT syndrome
3. Hypertrophic cardiomyopathy
4. Heart failure

Components

Pacemakers are composed of two main components: a pulse generator, which houses the battery and electronics, and leads that extend from the can to connect with the myocardium. These leads serve the purpose of delivering a depolarizing pulse and sensing the heart's natural activity. Currently, pulse generators are typically positioned in the infraclavicular region of the anterior chest wall. Insulating materials are utilized to keep the conductor cables separate from the lead tip electrodes [40].

Cardiac pacing systems predominantly rely on transvenous electrodes to transmit electrical impulses from the pulse generator to the heart muscles. These electrodes are inserted through veins, allowing for effective stimulation. In contrast, epicardial systems involve direct stimulation of the heart's surface through the pulse generator. However, these systems are now less commonly used and have been largely replaced by transvenous pacing. Recent advancements have introduced leadless systems as an alternative to overcome certain limitations associated with transvenous and epicardial pacing methods [38].

Types of Pacemakers [38]

1. *Single chamber*: One lead attach to the upper or lower heart chamber.
2. *Dual chamber*: Uses two leads, one for the upper and one for the lower chamber.
3. *Biventricular pacemakers* [38].

The mortality rate associated with pacemaker insertion can range from 1% to 4%, while complications may occur in approximately 4–15% of patients. Several factors can influence both mortality and complications, such as the presence of renal failure, a high NYHA class indicating advanced heart failure, a low ejection fraction indicating weakened heart function, a low platelet count, a history of stroke, and a higher body mass index. Unfortunately, reprogramming of pacemakers is often neglected over time, which may render subsequent pacemaker visits redundant.

Pacemaker Related Complications [38, 39]

1. *Pacemaker* syndrome: It refers to the negative effects on a patient's health caused by inadequate coordination between the atria and ventricles, or AV dyssynchrony, regardless of the type of pacemaker used. Common symptoms associated with pacemaker syndrome include irregular cannon A-waves, chest discomfort, mental confusion, feelings of light-headedness, persistent fatigue, heart palpitations, difficulty breathing, and fainting episodes.
2. Pneumothorax
3. Cardiac perforation
4. Significant pocket hematoma
5. Lead dislodgement
6. Venous thrombosis and obstruction
7. Mechanical lead complications
8. Pericarditis
9. Skin erosion
10. Failure to sense, capture or output
11. Pacemaker mediated tachycardia
12. Twiddler syndrome

13. Pacemaker pseudomalfunction

Toxicokinetics

Magnet Inhibition: When a patient with a pacemaker requires surgery or a procedure where electrocautery or related devices will be used, this can interfere with the function of the device. Thus, magnet inhibition is commonly done. Placing the magnet over the pacemaker temporarily reprograms the device into an asynchronous pacing mode but does not completely switch it off. After the procedure, the cardiologist has to reprogram the device. In most cases, almost any pacemaker can be inhibited this way. Further, there are some pacers, in which the magnet function can be disabled [39].

Basic Pacemaker Troubleshooting [40]

1. Malfunction may occur with battery depletion.
2. Device-related arrhythmias: pacemaker-mediated tachycardia and upper rate behavior resulting in high or low ventricular rates.

4.5 Implantable Pulse Generator

A pulse generator is composed of several components, including a battery, circuitry, can, antenna, reed switch, and connectors. The primary power source used in modern pacemakers is lithium-iodine. The circuitry contains microprocessors that control various functions such as sensing, output, telemetry, and diagnostics. Connectors are used to attach the pacing or high-voltage leads to the pulse generator. An antenna facilitates communication with a programmer. The can, which acts as a conductor, houses all the components and prevents the entry of bodily fluids. In defibrillators, there are additional components like high-voltage capacitors for energy storage, a high-voltage transformer to convert small battery voltage to high voltage, activity sensors, and audible alarms [42].

The pulse generator generates the electric current necessary to stimulate the myocardium using leads that are inserted into the right atrial and ventricular myocardium through a vein. Typically, the pacemaker is implanted beneath the clavicle, between skin and pectoralis major muscle. The pulse generator's shell is made of titanium, which is well tolerated by the surrounding tissues. The leads are inserted into one of the larger veins, usually the subclavian vein, and guided towards the heart until they make contact with the endocardium. Active and passive lead fixations are also visualized in the image [43].

4.6 Endocardial Leads

Over the past 50 years, there have been significant advancements in pacemakers and pacing electrodes, resulting in improved patient outcomes. Initially, epicardial pacing leads were replaced with transvenous endocardial leads, eliminating the need for thoracotomy and reducing the risks associated with bradycardiac arrhythmia. This led to a substantial decrease in morbidity and mortality associated with pacemaker implantations. In the 1970s, coaxial bipolar pacing leads were introduced, offering a softer, thinner, and easier-to-insert alternative to the initial endocardial leads.

Continued progress in the field has focused on developing smaller lead diameters, new insulation types, and steroid-eluting electrodes to optimize the application of cardiac pacing therapy. Another important aspect of lead development is the advancement in fixation technology, which plays a crucial role in ensuring long-term effective pacing. The introduction of passive alary pacing leads allowed for easy fixation to the right ventricular trabecular muscles in the apex, reducing operation time and receiving positive clinical feedback. While the right ventricular apex remains the most common pacing site, the introduction of active fixation steroid-eluting leads with stable performance and a low rate of dislodgement has made it possible to select pacing sites other than the right ventricular apex [44].

Pacemaker leads consist of conductors wrapped in insulation. Most pacemakers utilize bipolar pacing, where the lead tips are equipped with two electrodes: a positive electrode (anode) and a negative electrode (cathode). By applying a voltage difference between the anode and cathode, the pulse generator causes electrons to flow from the anode to the cathode, depolarizing the myocardium and triggering an action potential that spreads through the myocardium. In addition to pacing, electrodes are also used to record electrical activity, a function known as sensing. Typically, pacemakers employ two leads, one in the right atrium and one in the right ventricle, both of which can be used for pacing and sensing if necessary [43].

Fixation Technology

1. *Active fixation leads*

Active-fixation pacing leads offer the added advantage of potential lead extraction, making them a convenient choice. These leads have gained significant popularity in Europe and the USA due to the increasing number of pacemaker implants and the growing senior population. Among the various pacing sites, the right ventricular outflow tract (RVOT) is widely utilized, apart from the apex. By utilizing screwable fixation at the distal end of the pacing leads, physicians can pace either in the septum or the free wall of the RVOT [44].

2. *Passive fixation leads*

Passive fixation technology, on the other hand, has certain drawbacks, such as the lack of reliable and secure retention of left ventricular (LV) leads in the proximal or middle segments of coronary veins. While passive fixation leads are stable in the atrial appendage, patients with prior cardiac surgery require active fixation leads to prevent dislodgement. Lead dislodgement can lead to increased pacing thresholds, failure to capture, or failure to sense [45].

In Western countries, there is a common preference for active-fixation pacing leads over passive ones. However, in China, although active-fixation pacing leads have been introduced, their use is still limited to several senior clinical cardiac centers.

A long-term prospective study was conducted to investigate the 5-year performance of both active-fixation and passive pacing leads. The study revealed that active-fixation pacing leads did not result in any adverse lead-related events and performed just as stably as passive leads over the 5-year observation period. There was no difference in the electrical performance of the leads [44]. However, reports discussing long-term observations of active-fixation leads and their performance compared to passive leads are scarce.

4.7 Leadless Cardiac Pacemakers (LCP)

The leadless pacemaker represents a remarkable breakthrough in pacemaker technology, as it consists of a self-contained generator and electrode system that can be implanted directly into the right ventricle. What sets it apart is the absence of leads and the need for surgical chest access, making it a highly innovative solution. It is important to note that leadless pacemakers are currently indicated only for single-chamber right-ventricular pacing.

Currently, there are two leadless pacemaker devices available: the Nanostim and the Micra. While both devices are already in clinical use in Europe, only the Micra device has received approval from the FDA. However, it is anticipated that the Nanostim device will also gain FDA approval in the near future. To implant these devices, an endovascular femoral venous approach is employed. The devices function as ventricular-only pacing systems and are inserted into the apex of the right ventricular wall. Furthermore, the distal tips of the devices are designed to release steroids, thereby reducing inflammation [46].

Micra, a leadless pacemaker, is positioned within a steerable catheter delivery system. The leadless pacemaker is introduced via a femoral vein. Progressing into the right ventricle, the device securely attaches to the myocardium using four electrically inactive nitinol tines situated at its distal end. After confirming device fixation and ensuring satisfactory electrical measurements, the tether is severed, and the delivery system is subsequently extracted [47].

Advantages

1. Absence of the need for leads, a large generator, or the creation of a subdermal surgical pocket in the chest wall.
2. Complications commonly associated with transvenous cardiac pacemakers, such as surgical pocket infections, hematomas, and lead fracture or dislodgment, can be eliminated.

3. Leadless pacemakers have an impressive battery life of 5–15 years, depending on the device and settings, while transvenous cardiac pacemakers typically last about 10 years.
4. Leadless pacemakers are compatible with MRI scans and have not demonstrated any damage when exposed to magnetic energy waves.

Indication

1. Leadless pacemakers are suitable for patients with permanent atrial fibrillation (AF) accompanied by bradycardia, as well as those with bradycardia-tachycardia syndrome. They are also appropriate for patients who are anticipated to require infrequent pacing.
2. In post-TAVR patients, leadless pacemakers can help minimize the complications associated with transvenous cardiac pacemakers.

Procedure for Insertion
Leadless pacemakers are inserted directly into the wall of the right ventricle, eliminating the necessity for pacing leads. These devices are approximately 90% smaller compared to transvenous cardiac pacemakers. The insertion process involves a catheter being guided through a single-access site at the femoral vein in the groin. A steerable catheter is advanced through the inferior vena cava, into the right atrium, and then through the tricuspid valves into the right ventricle. The pacemaker is implanted in the septal aspect of the apex of the right ventricle. After confirming the device's fixation to the myocardium through a tug test and verifying adequate electrical measurements, the tether is cut, the catheter delivery system is removed, and the femoral vein is tied off.

Complications

1. Femoral vascular complications.
2. In some cases, intraoperative device repositioning may be necessary.
3. Modest risk of cardiac perforation, potentially leading to pericardial effusion.

Disadvantage
Since leadless pacemakers are relatively new, research on the feasibility of their removal after long-term use is lacking. Consequently, it remains unknown how to replace the batteries in a leadless pacemaker. However, to reduce the need for device removal upon completion of its service life, the device is equipped with an off switch that can be activated during device interrogation [46].

Related Studies
A case report by Fudim et al. suggests that early implantation of a leadless device after high-risk procedures like TAVR (for patients suitable for single-chamber pacing) could reduce the common pacemaker complications associated with transvenous cardiac pacemakers [48].

Omdahl and colleagues investigated the possibility of implanting multiple devices in the heart. They conducted serial implantations of up to three Micra devices in human cadaver hearts and artificially pumped the hearts to demonstrate the feasibility and absence of complications associated with multiple devices. This may offer therapeutic pacing options with leadless devices for three to four decades, extending beyond the longevity of a single device [49].

4.8 Prosthetic (Artificial) Heart Valves

The introduction of valve replacement surgery in the early 1960s brought about a significant improvement in the prognosis of patients suffering from valvular heart disease. Presently, around 280,000 valve substitutes are being implanted worldwide each year, with mechanical valves and bioprosthetic valves accounting for approximately equal shares. Despite considerable advancements in the design of prosthetic valves and surgical techniques in recent decades, valve replacement does not offer a definitive cure for patients. Rather, it replaces native valve disease with what is known as "prosthetic valve disease," and the outcome of individuals undergoing valve replacement is influenced by factors such as prosthetic valve hemodynamics, durability, and thrombogenicity. Nevertheless, by carefully selecting the appropriate prosthesis for each patient and providing meticulous medical management and follow-up after implantation, many complications associated with the prosthesis can be prevented or their impact can be minimized [50].

Types of Prosthetic Heart Valves

1. *Mechanical valves*:
   - (i) Ball and cage
   - (ii) Tilting disk
   - (iii) Bileaflet
2. *Bioprosthetic valves*:
   - (i) Porcine aortic
   - (ii) Bovine pericardial
   - (iii) Stentless porcine
3. *Homograft valves*

Mechanical Valves
Mechanical valves consist of three primary components: an occluder, an occluder restraint, and a sewing ring. Over time, mechanical valve designs have evolved into three main classes: ball and cage, disc valves (tilting and nontilting variants), and bileaflet valves.

Ball and Cage Valve
The original ball and cage valve, known as the Starr-Edwards valve, was used extensively since the 1960s but was retired in 2007. This valve had a ball occluder retained within a silicone-coated stainless steel cage. The sewing cuff included a Teflon ring

to reduce the risk of thrombosis. During diastole (or systole for the mitral valve), the ball occluder would position itself in the sewing ring, creating a seal to prevent backward flow.

Major Drawback However, the ball and cage design had a significant drawback in terms of hemodynamics. The ball occluder disrupted the central blood flow seen in a natural heart valve. Subsequent designs aimed to improve hemodynamics. The nontilting disc valve, consisting of a flat, circular disc within a cage, did not provide significant improvements. However, the later bileaflet design performed better in replicating the natural hemodynamics of a native heart valve [51].

Monoleaflet Valves

Monoleaflet valves feature a single disk secured by lateral or central metal struts. The opening angle of the disk relative to the valve annulus ranges from 60° to 80°, resulting in two distinct orifices of different sizes.

Bileaflet Valves

Bileaflet valves, on the other hand, have two semilunar disks connected to a rigid valve ring by small hinges. The opening angle of the leaflets relative to the annulus plane ranges from 75° to 90°. When the valve is open, it consists of three orifices: a small, slit-like central orifice between the two open leaflets and two larger semicircular orifices on the sides [50]. Fig. 4.4 depicts the bileaflet prosthetic heart valve.

Lifespan of Mechanical Valves

Compared to biological prostheses, mechanical heart valve prostheses have a significantly longer lifespan. Mechanical valves can last for several decades, while biological valves typically last around 10–12 years.

Bioprosthetic Valves

Bioprosthetic valves are composed of three xenograft tissue leaflets made from either porcine valvular leaflets or bovine pericardial tissue. These xenografts are

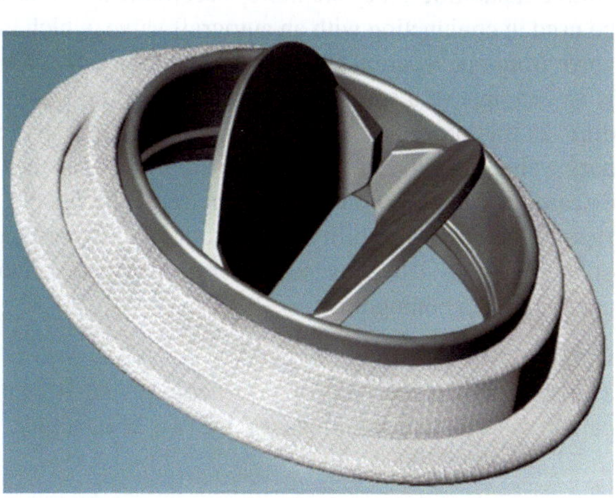

Fig. 4.4 Bileaflet prosthetic heart valve [52]Bileaflet valves consist of two semilunar disks connected to a sturdy valve ring through small hinges

treated with a glutaraldehyde solution to reduce the risk of thromboembolism by inhibiting collagen denaturation. In stented bioprosthetic valves, the leaflets are sewn to a cloth-covered metallic support structure, similar to mechanical valves. Nonstented bioprosthetics require the prosthetic leaflets to attach directly to the aortic annulus.

Advantage: They offer improved hemodynamics compared to mechanical valves. However, they generate greater pressure gradients due to their smaller annulus size [51].

Stented Bioprostheses

Stented bioprostheses are designed to mimic the anatomy of the native aortic valve. They consist of three porcine aortic valve leaflets cross-linked with glutaraldehyde and mounted on a metallic or polymer supporting stent. Pericardial valves are made from bovine pericardium sheets mounted either inside or outside a supporting stent.

Stentless Bioprostheses

To enhance valve hemodynamics and durability, various types of stentless bioprosthetic valves have been developed. These valves are manufactured from whole porcine aortic valves or fabricated using bovine pericardium.

Percutaneous Bioprostheses

Percutaneous bioprostheses are emerging as an alternative to standard aortic valve replacement (AVR) for patients with symptomatic aortic stenosis who are considered at high or prohibitive operative risk. These valves are typically implanted using a percutaneous transfemoral approach. In some cases, a transapical approach through a small thoracotomy may be used to overcome vascular access issues and associated complications. Although this procedure shows promise, it is still considered experimental and undergoing further investigation [50].

Homograft Valves

Homograft valves are obtained from human cadavers. As these valves lack regenerative capabilities, they are more susceptible to cyclic stress. Homograft valves can be used in conjunction with an autograft valve, which involves transplanting a heart valve from one location to another within the same individual. The Ross procedure is an example of a surgical intervention that utilizes a homograft/autograft procedure. In this procedure, a cadaver valve (homograft) replaces a functional pulmonary valve, and the healthy pulmonary valve is then transferred to replace the diseased aortic valve (autograft). The rationale for the autograft component of the Ross procedure is its ability to withstand the higher pressure system of the left side of the heart, as a healthy pulmonary valve has sufficient regenerative capability compared to a homograft.

Choice of the Prosthetic Type

The selection of the type of prosthetic device is centered around the individual patient's needs and is influenced by two key factors: the concern for the risk of bleeding associated with anticoagulation and the potential for valve deterioration. The longevity of the tissue valve is inversely correlated with age, indicating that younger patients experience a higher rate of deterioration compared to the elderly population. Consequently, bioprosthetic valves are often considered a more suitable option for older patients who may have a reduced tolerance for anticoagulation therapy, while mechanical valves may be a preferable choice for younger patients, as bioprosthetic valves may deteriorate more rapidly in their case. It is worth noting, however, that even though bioprosthetic valves may be favored in older patients to avoid complications related to anticoagulation, the occurrence of atrial fibrillation within the same age group can effectively negate any potential benefit of this management approach.

Delivery Method of the Prosthetic Valve

The choice between surgical and transcatheter delivery methods for a prosthetic valve should be considered when discussing management options. Surgical aortic valve replacement (SAVR) is a highly invasive procedure that carries the typical risks associated with an open thoracotomy. On the other hand, transcatheter aortic valve replacement (TAVR) is a less invasive approach but has its own drawbacks, such as valve oversizing, difficulties in intravascular manipulation, and positioning challenges.

Furthermore, any injury that occurs during the deployment of the prosthesis can negatively impact its lifespan. While perioperative imaging can help reduce these risks, it does not completely eliminate them. CT imaging is particularly useful as an imaging tool throughout all stages of TAVR.

Complications

1. *Thromboemboli and thrombotic obstruction*: Mechanical valves carry a significant risk of thromboemboli and thrombotic obstruction. As a result, long-term anticoagulation therapy is required, which in turn raises the likelihood of bleeding complications. However, modern mechanical valves exhibit excellent long-term durability. In contrast, bioprosthetic valves pose a low risk of blood clot formation without the need for anticoagulation, but their durability is limited due to the deterioration of calcific or noncalcific tissue.
2. *Structural Valve Deterioration*: This is the most common reason for reoperative valve replacement in patients with a bioprosthesis. Whenever feasible, the reoperative procedure should be performed early in the disease process, before significant deterioration in left ventricular function and symptomatic status occurs.
3. *Prosthetic valve endocarditis*: Even with appropriate antibiotic prophylaxis, the incidence of prosthetic valve endocarditis is approximately 0.5% per patient-year. This condition is highly serious and associated with high mortality rates of 30–50%.
4. *Paravalvular regurgitation*: Paravalvular regurgitation typically occurs due to infection, loosening of sutures, or the formation of fibrous and calcified tissue around the natural heart valve, resulting in inadequate contact between the sewing ring and the heart's annulus.

5. *Hemolysis*: A significant percentage (50–95%) of patients with mechanical valves experience some degree of red blood cell destruction within blood vessels, known as intravascular hemolysis [50].
6. *Valvular failure, tissue hyperplasia, and overgrowth* [51].

Issues of Concern

Concerns regarding prosthetic heart valve design revolve around four key aspects: optimizing hemodynamic performance, ensuring mechanical or biological durability, minimizing the body's biological response to the prosthetic valve, and developing an optimal delivery system.

While prosthetic heart valves generally offer satisfactory hemodynamics, there is still room for improvement, as the introduction of a foreign object into the body increases the risk of blood clot formation [51].

4.9 Ventricular Assist Devices (VADs)

Heart failure serves as the ultimate convergence point for numerous chronic heart conditions. While the majority of patients experience stability for an extended period through the use of conventional medications and surgical interventions, an increasing subset develops indications of advanced heart failure, warranting evaluation for potential heart transplantation. However, for a specific group of patients, such as those too debilitated to await a heart donor or individuals deemed ineligible for transplantation due to age or concurrent medical issues, ventricular assist devices (VADs) present a life-saving therapeutic option. VADs have emerged as a widely employed treatment modality for end-stage heart failure. Initially conceived as a transient mechanism facilitating heart recovery or transplantation, VADs have received approval from the US Food and Drug Administration over the past decade to provide long-term or even permanent support for patients grappling with end-stage heart failure. The progressive evolution of device design, coupled with advancements in surgical techniques and medical management, has enabled VAD recipients to reclaim their lives by returning home, reintegrating into the workforce, and actively participating in their communities, all while maintaining an exceptional quality of life. Primarily, VADs fulfil their purpose by unburdening the faltering heart and aiding in the sustenance of vital organ perfusion. A recent study involving 34 patients implanted with the HeartWare LVAD demonstrated an overall mortality rate of 24% and an aggregate survival rate of 56% after a 2-year period [53]. As advancements in pump technology continue to refine patient outcomes, the future design of LVADs will likely concentrate on the minimization of long-term morbidities, presenting a new challenge and focal point for improvement.

The Three Major Components of the VAD

1. The inflow cannula
2. The outflow cannula
3. The pump

The *inflow cannula* functions as a sizable tube responsible for draining blood from the heart and transporting it into the pump. On the other hand, the *outflow cannula* redirects the blood either to the aorta, in the case of a left ventricular assist device (LVAD), or to the pulmonary artery, in the case of a right ventricular assist device.

Exiting the skin, typically on the right side of the abdomen, is a driveline that houses the power wires. This driveline connects to a controller worn on a belt, which in turn links to either a power-based unit that can be plugged into the wall or large batteries that can be comfortably worn in a holster or vest, allowing for portable use.

First-Generation LVADs

The initial generation pumps were pulsatile in nature, incorporating artificial heart valves, and had the capability to expel blood at a frequency of approximately 80–100 times per minute. These pumps operated using either forced air or electricity [54]. Referred to as volume displacement devices, the first-generation implantable mechanisms propelled blood flow through a pulse generator, thus earning the name "pulsatile pump." The implantation of these early devices necessitated a median sternotomy, with the placement of cannulas for inflow and outflow at the left ventricular apex and ascending aorta, respectively. These devices relied on battery power, granting a charge that could last between 3 and 5 h.

There were notable drawbacks associated with these devices. Patients often experienced discomfort during their usage, and achieving long-term mechanical durability for the pumps proved to be challenging. Additionally, there was a considerable risk of complications such as infection, formation of blood clots, and damage to blood cells, which needed to be addressed in order to ensure successful long-term support with LVAD therapy [55].

Second-Generation LVADS

The next generation of pumps operate by continuously circulating blood using an internal rotor that rotates at speeds of up to 15,000 revolutions per minute (typically ranging from 8000 to 10,000 rpm). Continuous-flow pumps offer several significant advantages over the older pulsatile pumps. They are smaller in size, emit less noise, are easier to implant, and have a longer lifespan.

The currently approved ventricular assist devices (VADs) are typically implanted just below the diaphragm in the abdominal area. Alternatively, they can be positioned externally on top of the abdomen [54]. These devices incorporate a valveless axial pump, with a rotary motor serving as the sole moving component within the system. The intention behind this design is to minimize the presence of sites that could potentially trigger blood clot formation and reduce the wear and tear associated with multiple moving parts.

Efficiency is further improved by eliminating the need for a reservoir chamber and inflow/outflow valves. To enhance the antithrombotic properties of the blood-contacting surfaces, a specially designed textured titanium lining is employed. It is worth noting that second-generation left ventricular assist devices (LVADs) are projected to have a mechanical lifespan of approximately 5 years, although there have been documented cases of longer-term support [55].

Third-Generation LVADS

The key factor that sets the second- and third-generation LVADs apart is the utilization of different types of bearings. While the former employs contact bearings, the latter utilizes a technology called magnetic levitation (MAGLEV), which enables rotation without any friction or wear. This innovative design aims to reduce the occurrence of blood clot formation and enhance both efficiency and longevity [55]. The HeartMate 3, a third-generation LVAD featuring active magnetic levitation, is surgically placed within the pericardial cavity through either sternotomy or lateral thoracotomy. The inflow of the pump also serves as an inflow cannula and is implanted through the left ventricular wall, while the outflow graft is anastomosed to the ascending aorta [56].

Reasons for VAD Implantation

1. *Bridge to recovery*: Intended for individuals requiring short-term assistance (ranging from several days to a few weeks), this intervention allows the heart to heal from an acute injury before the VAD is subsequently removed. For instance, this approach may be suitable for patients who have suffered a substantial heart attack or experienced severe cardiac inflammation following a viral infection.
2. *Bridge to transplant*: The most extensive cohort to receive VAD consists of individuals who meet the criteria for a heart transplant but suffer from such advanced illness that they cannot afford to await the availability of a suitable donor heart.
3. *Lifetime or destination therapy*: Patients who require destination therapy may be ineligible for heart transplantation due to factors such as advanced age, typically over 70 years, or the presence of chronic medical conditions. Instead, they receive a permanent VAD as a treatment option for end-stage heart failure.
4. *Bridge to candidacy*: For individuals who meet the criteria for potential transplantation but require a duration of VAD assistance to assess the potential enhancement of essential organ function, nourishment, and physical resilience, allowing for a viable and prosperous heart transplant.

Constant and Safe Power Source

Before being discharged following the implantation of a VAD, it is imperative to ensure the availability of a reliable electric supply at home. This entails conducting a comprehensive safety inspection to verify the adequacy of the grounding system for the purpose of charging the VAD battery. Furthermore, it is crucial to inform the electric company about the VAD patient's situation, requesting their inclusion on a priority list for power restoration in the event of an electrical outage. Planned power outages are to be avoided, and it is kindly requested that the electric company refrains from disconnecting the electricity supply due to late or nonpayment issues.

In the event of any complications regarding billing matters, a VAD financial counsellor is readily available to provide assistance and help resolve these concerns.

Complications of VADs

According to recent studies, the survival rates following VAD implantation have shown promising results, with approximately 80–90% of patients still alive after 1 year, and around 60–70% continuing to live at the 2-year mark. Some individuals have undergone successful heart transplants, while others have been successfully weaned off the VAD as part of a bridge to recovery program. Unfortunately, there have also been cases where patients have passed away while relying on the VAD for support.

In the early stages postsurgery, patients may experience complications such as chest bleeding and right heart failure. As time progresses, additional complications can arise, including bleeding from other sources, infections, strokes, or malfunctioning of the device.

Device Failure

The malfunctioning of first-generation pulsatile-flow pumps posed a significant risk to VAD patients, leading to hospitalization and even fatalities. Originally designed for short- to medium-term support, these devices aimed to bridge patients until transplantation. However, as pumps were approved for destination therapy and patients faced longer wait times for transplants, the need for prolonged support exceeding 1–2 years became evident. Regrettably, during this extended period, up to two-thirds or more of the pumps experienced failure. Fortunately, the advent of newer continuous-flow pumps with simpler operations and fewer components has shown remarkable improvement in durability. Although occasional issues with wires, controllers, and batteries still arise, these problems are relatively minor and can be easily addressed as they prompt alarms. In rare cases, it may be necessary to surgically replace the entire VAD [54].

References

1. Goyal A, Chhabra L, Sciammarella JC, Cooper JS. Defibrillation. In: StatPearls [Internet]. Treasure Island: StatPearls Publishing; 2023. [cited 2023 May 29]. Available from: http://www.ncbi.nlm.nih.gov/books/NBK499899/.
2. Wik L, Hansen TB, Fylling F, Steen T, Vaagenes P, Auestad BH, et al. Delaying defibrillation to give basic cardiopulmonary resuscitation to patients with out-of-hospital ventricular fibrillation: a randomized trial. JAMA. 2003;289(11):1389.
3. Cobb LA. Influence of cardiopulmonary resuscitation prior to defibrillation in patients with out-of-hospital ventricular fibrillation. JAMA. 1999;281(13):1182.
4. Holmberg M, Holmberg S, Herlitz J. Incidence, duration and survival of ventricular fibrillation in out-of-hospital cardiac arrest patients in Sweden. Resuscitation. 2000;44(1):7–17.
5. Eisenberg MS, Horwood BT, Cummins RO, Reynolds-Haertle R, Hearne TR. Cardiac arrest and resuscitation: a tale of 29 cities. Ann Emerg Med. 1990;19(2):179–86.
6. Delgado H, Toquero J, Mitroi C, Castro V, Fernandez I. Principles of external defibrillators. In: Erkapic D, editor. Cardiac defibrillation [Internet]. InTech; 2013. [cited 2023

May 29]. Available from: http://www.intechopen.com/books/cardiac-defibrillation/principles-of-external-defibrillators.

7. Fain ES, Sweeney MB, Franz MR. Improved internal defibrillation efficacy with a biphasic waveform. Am Heart J. 1989;117(2):358–64.
8. Lown B, Amarasingham R, Neuman J. New method for terminating cardiac arrhythmias. Use of synchronized capacitor discharge. JAMA. 1962;182:548–55.
9. Dickey W, Dalzell GW, Anderson JM, Adgey AA. The accuracy of decision-making of a semi-automatic defibrillator during cardiac arrest. Eur Heart J. 1992;13(5):608–15.
10. Kerber RE, Becker LB, Bourland JD, Cummins RO, Hallstrom AP, Michos MB, et al. Automatic external defibrillators for public access defibrillation: recommendations for specifying and reporting arrhythmia analysis algorithm performance, incorporating new waveforms, and enhancing safety: a statement for health professionals from the American Heart Association Task Force on Automatic External Defibrillation, Subcommittee on AED Safety and Efficacy. Circulation. 1997;95(6):1677–82.
11. Thomas ED, Ewy GA, Dahl CF, Ewy MD. Effectiveness of direct current defibrillation: role of paddle electrode size. Am Heart J. 1977;93(4):463–7.
12. Atkins DL, Sirna S, Kieso R, Charbonnier F, Kerber RE. Pediatric defibrillation: importance of paddle size in determining transthoracic impedance. Pediatrics. 1988;82(6):914–8.
13. Kirchhof P, Mönnig G, Wasmer K, Heinecke A, Breithardt G, Eckardt L, et al. A trial of self-adhesive patch electrodes and hand-held paddle electrodes for external cardioversion of atrial fibrillation (MOBIPAPA). Eur Heart J. 2005;26(13):1292–7.
14. Kudenchuk PJ, Cobb LA, Copass MK, Olsufka M, Maynard C, Nichol G. Transthoracic incremental monophasic versus biphasic defibrillation by emergency responders (TIMBER): a randomized comparison of monophasic with biphasic waveform ascending energy defibrillation for the resuscitation of out-of-hospital cardiac arrest due to ventricular fibrillation. Circulation. 2006;114(19):2010–8.
15. Kerber RE, Kouba C, Martins J, Kelly K, Low R, Hoyt R, et al. Advance prediction of transthoracic impedance in human defibrillation and cardioversion: importance of impedance in determining the success of low-energy shocks. Circulation. 1984;70(2):303–8.
16. Dalzell GWN, Cunningham SR, Anderson J, Adgey AAJ. Electrode pad size, transthoracic impedance and success of external ventricular defibrillation. Am J Cardiol. 1989;64(12):741–4.
17. Kerber RE, Jensen SR, Grayzel J, Kennedy J, Hoyt R. Elective cardioversion: influence of paddle-electrode location and size on success rates and energy requirements. N Engl J Med. 1981;305(12):658–62.
18. Oechslin E. Treatment strategies in mechanical and electrical cardiovascular failure. Ther Umsch Rev Ther. 1996;53(8):646–57.
19. Egan J, Atkins DL. Defibrillation in children: why a range in energy dosing? Curr Pediatr Rev. 2013;9(2):134–8.
20. Wolfe HA, Morgan RW, Zhang B, Topjian AA, Fink EL, Berg RA, et al. Deviations from AHA guidelines during pediatric cardiopulmonary resuscitation are associated with decreased event survival. Resuscitation. 2020;149:89–99.
21. Pugsley W, Baldwin T, Treasure T, Sturridge M. Low energy level internal defibrillation during cardiopulmonary bypass. Eur J Cardiothorac Surg. 1989;3(3):273–5.
22. McDonough M. Treating ventricular tachycardia. J Contin Educ Nurs. 2009;40(8):342–3.
23. Berger RD, Palazzolo J, Halperin H. Rhythm discrimination during uninterrupted CPR using motion artifact reduction system. Resuscitation. 2007;75(1):145–52.
24. Iqbal AM, Butt N, Jamal SF. Automatic internal cardiac defibrillator. In: StatPearls [Internet]. Treasure Island: StatPearls Publishing; 2023. [cited 2023 May 29]. Available from: http://www.ncbi.nlm.nih.gov/books/NBK538341/.
25. Epstein AE, DiMarco JP, Ellenbogen KA, Estes NAM, Freedman RA, Gettes LS, et al. ACC/AHA/HRS 2008 guidelines for device-based therapy of cardiac rhythm abnormalities. J Am Coll Cardiol. 2008;51(21):e1–e62.

26. Jmarchn. Català: Desfibril·lador automàtic implantable [Internet]. 2022 [cited 2024 May 24]. Available from: https://commons.wikimedia.org/wiki/File:Blausen_0543_ImplantableCardioverterDefibrillator.svg.
27. DiMarco JP. Implantable cardioverter–defibrillators. N Engl J Med. 2003;349(19):1836–47.
28. Ghzally Y, Mahajan K. Implantable defibrillator. In: StatPearls [Internet]. Treasure Island: StatPearls Publishing; 2023. [cited 2023 May 31]. Available from: http://www.ncbi.nlm.nih.gov/books/NBK459196/.
29. Cheung CC, Olgin JE, Lee BK. Wearable cardioverter-defibrillators: a review of evidence and indications. Trends Cardiovasc Med. 2021;31(3):196–201.
30. Klein HU, Meltendorf U, Reek S, Smid J, Kuss S, Cygankiewicz I, et al. Bridging a temporary high risk of sudden arrhythmic death. Experience with the wearable cardioverter defibrillator (WCD). Pacing Clin Electrophysiol. 2010;33(3):353–67.
31. Li Y, Bisera J, Tang W, Weil MH. Automated detection of ventricular fibrillation to guide cardiopulmonary resuscitation. Crit Pathw Cardiol. 2007;6(3):131–4.
32. Ghzally Y, Ahmed I, Gerasimon G. Catheter ablation. In: StatPearls [Internet]. Treasure Island: StatPearls Publishing; 2023. [cited 2023 May 31]. Available from: http://www.ncbi.nlm.nih.gov/books/NBK470203/.
33. Borhani S, Hassanajili S, Ahmadi Tafti SH, Rabbani S. Cardiovascular stents: overview, evolution, and next generation. Prog Biomater. 2018;7(3):175–205.
34. Chhabra L, Zain MA, Siddiqui WJ. Coronary stents. In: StatPearls [Internet]. Treasure Island: StatPearls Publishing; 2023. [cited 2023 May 31]. Available from: http://www.ncbi.nlm.nih.gov/books/NBK507804/.
35. Fischman DL, Leon MB, Baim DS, Schatz RA, Savage MP, Penn I, et al. A randomized comparison of coronary-stent placement and balloon angioplasty in the treatment of coronary artery disease. N Engl J Med. 1994;331(8):496–501.
36. Patel S, Patel KB, Patel Z, Konat A, Patel A, Doshi JS, et al. Evolving coronary stent technologies – a glimpse into the future. Cureus [Internet]. 2023;15(3):e35651. Available from: https://www.cureus.com/articles/121816-evolving-coronary-stent-technologies%2D%2D-a-glimpse-into-the-future.
37. Wayangankar SA, Ellis SG. Bioresorbable stents: is this where we are headed? Prog Cardiovasc Dis. 2015;58(3):342–55.
38. Lak HM, Goyal A. Pacemaker types and selection. In: StatPearls [Internet]. Treasure Island: StatPearls Publishing; 2023. [cited 2023 May 31]. Available from: http://www.ncbi.nlm.nih.gov/books/NBK556011/.
39. Puette JA, Malek R, Ellison MB. Pacemaker. In: StatPearls [Internet]. Treasure Island: StatPearls Publishing; 2023. [cited 2023 May 31]. Available from: http://www.ncbi.nlm.nih.gov/books/NBK526001/.
40. Mulpuru SK, Madhavan M, McLeod CJ, Cha YM, Friedman PA. Cardiac pacemakers: function, troubleshooting, and management. J Am Coll Cardiol. 2017;69(2):189–210.
41. Bernstein AD, Daubert JC, Fletcher RD, Hayes DL, Luderitz B, Reynolds DW, et al. The revised NASPE/BPEG generic code for antibradycardia, adaptive-rate, and multisite pacing. Pacing Clin Electrophysiol. 2002;25(2):260–4.
42. Haghjoo M. Cardiac implantable electronic devices. In: Practical cardiology [Internet]. Elsevier; 2018. p. 251–60. Available from: https://linkinghub.elsevier.com/retrieve/pii/B9780323511490000146.
43. ECG & ECHO. Components and construction of a pacemaker [Internet]. [cited 2023 June 6]. Available from: https://ecgwaves.com/topic/how-pacemakers-work/
44. Liu L, Tang J, Peng H, Wu S, Lin C, Chen D, et al. A long-term, prospective, cohort study on the performance of right ventricular pacing leads: comparison of active-fixation with passive-fixation leads. Sci Rep. 2015;5(1):7662.
45. Elvin E, Kayrak M. Common pacemaker problems: lead and pocket complications. In: Das M, editor. Modern pacemakers – present and future [Internet]. InTech; 2011. [cited 2023 June 1].

Available from: http://www.intechopen.com/books/modern-pacemakers-present-and-future/common-pacemaker-problems-lead-and-pocket-complications.
46. Groner A, Grippe K. The leadless pacemaker: an innovative design to enhance pacemaking capabilities. JAAPA. 2019;32(6):48–50.
47. Reynolds D, Duray GZ, Omar R, Soejima K, Neuzil P, Zhang S, et al. A leadless intracardiac transcatheter pacing system. N Engl J Med. 2016;374(6):533–41.
48. Fudim M, Fredi JL, Ball SK, Ellis CR. Transcatheter leadless pacemaker implantation for complete heart block following CoreValve transcatheter aortic valve replacement. J Cardiovasc Electrophysiol. 2016;27(1):125–6.
49. Omdahl P, Eggen MD, Bonner MD, Iaizzo PA, Wika K. Right ventricular anatomy can accommodate multiple micra transcatheter pacemakers. Pacing Clin Electrophysiol. 2016;39(4):393–7.
50. Pibarot P, Dumesnil JG. Prosthetic heart valves: selection of the optimal prosthesis and long-term management. Circulation. 2009;119(7):1034–48.
51. Mathew P, Kanmanthareddy A. Prosthetic heart valve. In: StatPearls [Internet]. Treasure Island: StatPearls Publishing; 2023. [cited 2023 June 2]. Available from: http://www.ncbi.nlm.nih.gov/books/NBK536987/.
52. Komar S. Русский: Двустворчатый протез клапана сердца/English: Bileafter prosthetic heart valve [Internet]. 2014 [cited 2024 May 24]. Available from: https://commons.wikimedia.org/wiki/File:Bileafter_prosthetic_heart_valve.jpg
53. Popov AF, Hosseini MT, Zych B, Mohite P, Hards R, Krueger H, et al. Clinical experience with HeartWare left ventricular assist device in patients with end-stage heart failure. Ann Thorac Surg. 2012;93(3):810–5.
54. Givertz MM. Ventricular assist devices: important information for patients and families. Circulation [Internet]. 2011;124(12):e305-11. Available from: https://www.ahajournals.org/doi/10.1161/CIRCULATIONAHA.111.018226
55. Rodriguez LE, Suarez EE, Loebe M, Bruckner BA. Ventricular assist devices (VAD) therapy: new technology, new hope? Methodist Debakey Cardiovasc J. 2013;9(1):32.
56. Foster G. Third-generation ventricular assist devices. In: Mechanical circulatory and respiratory support [Internet]. Elsevier; 2018. p. 151–86. Available from: https://linkinghub.elsevier.com/retrieve/pii/B9780128104910000059.

Chapter 5
Significant Risk Medical Devices – Dental

Adity Bansal, Urmila Irom, T. Y. Sree Sudha, and K. S. B. S. Krishna Sasanka

5.1 Introduction

The United States (US) regulatory authority in the globe of health-care products, i.e., The Food and Drug Administration (FDA) has produced a document of guidance, which is dedicated to the studies related to the nonsignificant and significant risk medical devices [1]. It dispenses clarifications in regard to applicable governing requirements, along with recommendations to be contemplated by sponsors of the study, investigators, and other involved parties. It advises them and the IRSs (Institutional Review Boards) to ascertain the differences between nonsignificant and significant risk medical devices.

Along with updating the significant and nonsignificant risk devices list, it helps to clarify the responsibilities of IRB while building the risk determination of these investigational devices. Also, it puts together this guidance congruous with Code of Federal Regulations (21 CFR 10.115), i.e., Agency's practice regulations for good guidance.

A. Bansal (✉)
Department of Dentistry, All India Institute of Medical Sciences (AIIMS), Deoghar, Jharkhand, India

U. Irom
Manipur Health Services, Government of Manipur, Ukhrul, India

T. Y. Sree Sudha
Department of Pharmacology, All India Institute of Medical Sciences (AIIMS), Deoghar, Jharkhand, India

K. S. B. S. Krishna Sasanka
Department of ENT and Head & Neck Surgery, All India Institute of Medical Sciences (AIIMS), Deoghar, Jharkhand, India

P. S. Timiri Shanmugam et al. (eds.), *Significant and Nonsignificant Risk Medical Devices*, https://doi.org/10.1007/978-3-031-52838-5_5

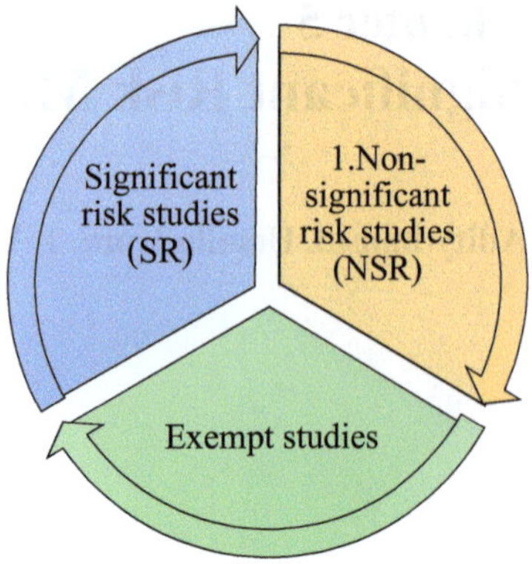

Fig. 5.1 Types of device studies

5.1.1 Investigational Device Exemptions (IDE)

As per 21 CFR 812 (IDE regulation) delineates three types of device studies (Fig. 5.1):

5.2 Significant Risk Medical Devices: What Is It?

It is defined as an investigational device that:

- Represents a potential for serious health risk, welfare, or safety of a patient/subject.
- Is claimed for sustaining or supporting a human life and presents a serious risk to the safety, health, or the welfare of the patient/subject.
- Is considered of significant importance with respect to diagnosis, cure, management, and/or disease treatment.
- Apart from the above, if it presents a potentially serious risk to safety, health, or welfare of a subject.

Any device study, which does not meet the SR device study definition, is termed as ***nonsignificant risk device study***.

5.3 Significant Risk Devices: Dental

The devices which have been termed as significant risk devices in dental field (Fig. 5.2) are as follows:

1. **Absorbable materials:** to aid periodontal defects healing, along with other maxillofacial applications,

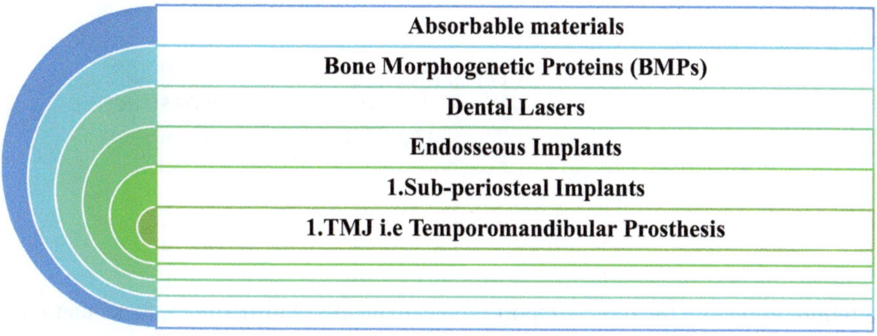

Fig. 5.2 Significant risk dental devices

2. **Bone morphogenetic proteins (BMPs):** either with or without bone. For example, hydroxyapatite (HA).
3. **Dental lasers:** hard tissue applications.
4. **Endosseous implants:** along with accompanying augmentation and bone-filling materials, which are used in concomitance with implants.
5. **Subperiosteal implants.**
6. **TMJ, i.e., temporomandibular prosthesis.**

5.4 Absorbable Materials

5.4.1 Periodontium and Healing

A functional periodontium is constituted of cementum, alveolar bone, and the periodontal ligament (PDL). It is considered essential for the following:

- Biomechanical function of the tooth.
- Structural support of the junctional epithelium and gingival tissues.
- Connective tissue attachment, thus forming a barrier in opposition to the infiltrating bacteria [2].

Destruction via inflammation of the periodontium occurs in response to the biofilms caused by bacterial invasion on the surface of the tooth [3]. While the removal of the biofilm attenuates the progression of the disease, the destruction of the cementum and alveolar bone surrounding the tooth surface is irreversible. Subsequent to the removal of the biofilm, there is formation of the epithelial tissue against the root of the tooth, along with the filling of the residual bone defect with fibrous connective tissue [4]. These tissues, which are reparative in nature, are prone to breakdown in the future, which can further lead to periodontitis [5].

There are broadly two types of therapies in such conditions (Fig. 5.3). One is the ***"resective therapies,"*** with the aim to eradicate the defects of the bone, but there is

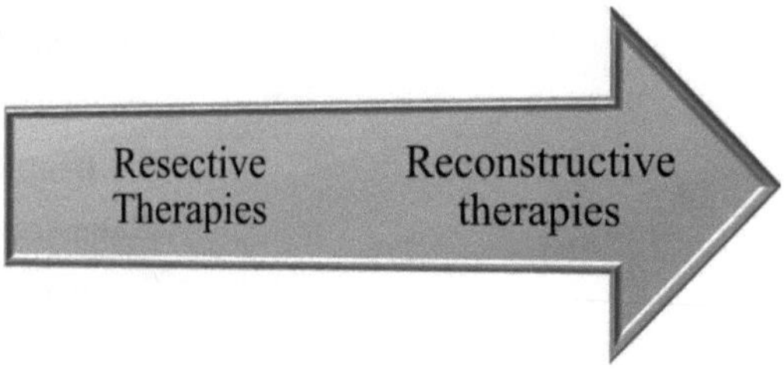

Fig. 5.3 Two types of therapies in periodontitis

a failure to restore the tooth support biomechanically. Furthermore, these therapies lead to the exposure of the roots in the oral cavity [6]. Alternatively, an attractive approach which helps in the restoration of both function and form, thus ensuring the long-term health of the periodontium is called as ***"reconstructive therapies."***

5.4.2 Guided Tissue Regeneration

Guided tissue regeneration, i.e., GTR was described in the1980s and gave an early approach for engineering of the periodontal tissues [7]. With time, regenerative potential with respect to the periodontal tissues has been uncovered. GTR helps in restoration of periodontal tissues and has been in use for decades. Initial human and animal studies use GTR barriers, which promoted new cementum, bone, and periodontal ligament (PDL) formation; however, regeneration of complete PDL was seen to be inconsistent [8, 9]. Factors that can constrain the outcomes of GTR can be:

- Characteristics of the surface in the roots affected by periodontitis
- Capacity of regeneration of remaining PDKL and bone in larger defects of the peridontium [10]
- Surgical factors
- Barrier design
- Configuration of the defect

The ***fundamental principles*** of successful GTR have been explored extensively, especially in model of canine supra-alveolar periodontal defect, where there is surgical creation of defined defects, followed by occurrence of spontaneous regeneration of the tissues [11]. The initial wounds consist of vitronectin and fibronectin-rich granulation tissues, along with the formation of collagen type I bands at the tooth and bone surfaces by 2 weeks [12]. By 4 weeks, tissues show increase in the content of type I and type III collagen. Formation of the new cementum began at residual PDL margin, which took place independent of the bone formation, which was observed to happen first along the surface of the root, followed by withing the residual defect of the bone. Though new tissue bulk could be observed at 4 weeks, however, longer time was required for PDL maturation completely.

5.4.3 *Periodontal Tissue Regeneration: Emerging and Current Approaches*

There are various approaches for regeneration of periodontal tissues which can be as follows:

(i) **Demineralization of the root surface**
(ii) **Bone grafts**

- Autogenous
- Allografts
- Synthetic grafts
- Xenografts

(iii) **Barriers (Guided tissue regeneration)**

- **Nonresorbable**

(a) ePTFE, i.e., expanded polytetrafluoroethylene
(b) dPTFE, i.e., dense polytetrafluoroethylene

- **Resorbable**

(a) Collagen
(b) Synthetic polymers

(iv) **Cell transplantation and recruitment**
(v) **Bioactive factors**

- Differentiation factors (none morphogenetic proteins-2, i.e., BMP-2; BMP-7, growth/differentiation factor-5, i.e., GDF-5)
- Growth factors (fibroblast growth factor-2, i.e., FGF-2; platelet-derived growth factor-BB, i.e., PDGF-BB)
- Peptides
- Enamel matrix derivative
- Platelet concentrate (autologous)
- Targeting factors (Wnt pathway)

(vi) **Combination therapy**
(vii) **Scaffolds (tissue engineered)**

5.4.4 *Engineered Barriers (Enhanced-GTR)*

GTR continues to remain a feasible strategy with documentation of long-term success [13]. Overall, the various advantages of GTR have been observed to be:

- Decreased recurrence of periodontal disease

- Tooth survival improvement
- Lowering of the overall cost of the treatment

Furthermore, current "barrier techniques" have varied disadvantages like:

- Demanding technically
- Postoperative complications frequently

5.4.4.1 Nonresorbable Barriers

Most widely studied GTR barrier over years has been ePTFE [9]. Overall, they are considered biocompatible; however, they can elicit mild reaction to the foreign bodies [14]. The porous surface of the barrier can result in rapid colonization of the bacteria when exposed to the oral cavity. Therefore, dPTFE containing submicron pores prevents biofilm formation, but lacks in the stability of the wound, as minimal adhesion is shown by overlying tissues [15, 16].

An additional procedure is required by such nonresorbing barriers, which is ordinarily performed after 4–6 weeks of barrier placement. Long-term regeneration of the tissues is predicted mainly by the tissue volume formed during initial healing [17, 18]. In this phase, maintenance of the space and stability of the wound is required for the optimal outcomes of the GTR.

5.4.4.2 Resorbable Barriers

They have shown outcomes similar to ePTFE barrier [19]. Also, they do not require a second procedure for removal. The bacterial colonization is also seen to be minimal following exposure to the oral cavity due to dissolution of the barrier and subsequent epithelialization of the wound [20].

Collagen Barriers They have shown multiple properties to its advantage like supports binding with host cell, thus showing integration with the periodontal tissues, and furthermore allowing infiltration of the vessels [21]. Cross-linking between fibers of the collagen created a material which degrades at a slower pace, but has improved properties mechanically [22]. On the contrary, these cross-linked barriers produce an increased reaction to the foreign bodies, along with delayed integration with the tissues and vascularization [23, 24].

The examples of collagen-based resorbable barriers are as follows:

- Porcine skin type-I collagen
- Bovine tendon type-I collagen
- Cadaveric human skin type-I collagen

Synthetic-Polymer Barriers Comprised of aliphatic polyesters so as to produce varied profiles of degeneration [25]. They are as follows:

- Poly DL-lactic/co-glycolic acid
- Poly DL-lactide and its solvent (*N*-methylpyrrolidone)
- Polyglactin 910

Resorbable barriers can occasionally result in severe inflammatory reaction at the local site [26]. However, both the types of resorbable barriers can maintain their function over elongated time periods and are biocompatible generally [27, 28]. However, their function is usually questionable beyond 6 weeks, as if retained for long, they might interfere with remodeling and maturation of the periodontal tissues [29].

Since these resorbable barriers cannot ensure maintenance of space with respect to large periodontal defects due to lack of mechanical properties, particulate grafts are used conjointly with them to enhance their stability [30]. Though grafts can help support formation of bone, cementum, and PDL, but these can cause interference with the formation and migration of the provisional tissue, thus hindering regeneration [29].

Literature has extensively reviewed the benefits of the use of GTR in comparison to the conventionally used debridement via open flap, and its findings supported use of GTR in situations clinically, i.e., intra-bony defects and furcation involvement [31]. A review in Cochrane compared the utilization of GTR for the management of intra-bony defects and it resulted in improvement with respect to the reduction in the pocket depth, gain in attachment, reduced gingival recession, along with attaining gain in probing the hard tissues at the time of re-entry [32].

Conversely, it has been observed that GTR causes improvements that are variable, modest and might not result in long-term retention of teeth as per the published systematic reviews [33, 34]. The achieved gain in the attachment with GTR may be maintained for years; however, no major difference is observed after 12 years between GTR and debridement via open flap [35].

5.4.4.3 Various Engineered Barriers

Numerous investigations have been done for variety of GTR. An ideal barrier should be stiff to allow space maintenance and should resorb in a suitable time causing minimal inflammation [36]. An in vivo testing was done for a barrier designed bi-layered, with the under surface having scaffold of porous calcium, which helps in retention of the clot and thus enhancing stability of the wound [37]. It was a ***combination of PLGA (polylactic-co-glycolic acid) and Cap (i.e., calcium phosphate)***, which creates a degradable and moldable material, where the porous layer faces the surface of the tooth root, and the outer smoother layer acts as a barrier opposed to infiltration of the gingival cells. It sufficiently promoted regeneration of periodontium in canine periodontal defects, along with restoration of large volume of bone.

Another designed barrier incorporated ***HAp (i.e., hydroxyapatite) and polyhydroxybutyrate*** in order to provide porous inner surface in a degradable and stiff barrier [38]. However, it led to the recession of the gingiva along with exposure of the barrier and that too in the early healing process. Also, there was formation of minimal periodontal tissues. Additionally, it was considered to be too stiff; hence, infiltration of the bacteria might have occurred following exposure.

Varied "***Electrospun Barriers***" have also been tested in vitro, with the aim to enhance microstructure and composition control [39, 40]. These fibers can be used deliver and load growth factors, antibiotics, or molecular drugs via degradation of the surface or from hollow fibers. However, there is limited potential as this is an in vitro study. One such barrier was tested in a clinical trial in porcine animal model and was composed of beta-tricalcium phosphate and polylactic acid with a coating of polyethylene oxide, to escalate hydrophilicity [41]. The performance was observed to be similar to PLA barrier, thus recommending that hydrophilic barrier might promote attachment of connective tissue cell and thus stability of the tissues.

5.5 Bone Morphogenic Proteins

Bone morphogenic proteins (BMPs) are a group of multifunctional regulatory glycoproteins belonging to transforming growth factor beta (TGF-beta) superfamily [42]. Studies have revealed that these molecules are associated with the regulation of proliferation of cells, their survival, differentiation, and apoptosis processes. They have also been associated with the process of homeostasis. But the most important ability is their part in the induction of formation of bone, cartilage, tendons, and ligaments at the orthotopic as well as heterotopic sites [40]. BMPs in embryonic stage helps in development of heart, nerves, and cartilage, while in postnatal phase is mainly concerned with bone formation and homeostasis. They stimulate the mesenchymal cells to differentiate into osteoblasts and chondroblasts [42].

BMPs were first discovered in the 1960s by an American orthopedic surgeon, ***Marashal R. Urist*** and his colleagues in the demineralizing bone in rodent model and found it to be the active element that induced new bone formation at an ectopic intramuscular location. It was also found to induce formation of bone in the subcutaneous sites by ***Reddy and Huggins*** in 1972. ***Wozney*** in 1988 for the first time isolated the BMPs and cloned their cDNAs [43]. Using the recombinant DNA technology, a variety of BMPs have been identified and cloned; however, the most important one has been identified as recombinant human bone morphogenic protein-2 (Rh-BMP-2) [44]. It was also the first and only substitute to the bone grafts approved by FDA in orthopedic procedures in 2002, and its use in oral maxillofacial reconstructive surgeries began in 2007 [44].

Hydroxyapatite (HA) is a calcium phosphate compound in which the ratio of calcium to phosphorus is 1:67 and has the molecular formula $Ca_{10}(PO_4)OH_2$. Of all the types of calcium phosphate compounds available in nature, the most stable and the least soluble compound is hydroxyapatite. HA has been found to possess an excellent biocompatibility as well as a good bioactivity [45]. The disadvantage of this material however is its property of being highly porous and the poor mechanical strength [46].

In the dental tissues, 70–80% of the dentin and enamel is made of the HA crystals, constituting a major part to its mechanical strength. Enamel is the hardest structure in the human body, and HA is the major component of it, responsible for covering of the pores on the surface of the enamel reflecting of the light and hence its semi-translucency [47].

It is used in varied areas in dentistry owing to its biocompatibility and its similarity in the structure to the nonorganic portion of bone. It is also bioactive and can induce osteo-induction, osteo-integration, and osteo-conduction. It has been used in implantology, surgical procedures, aesthetics, preventive procedures as well as periodontal surgeries [45].

5.5.1 Mechanism of Action

BMPs have been shown to have the capability of inducing ectopic bone and cartilage formation, like in the embryonic stage where the endochondral bone forms from the mesenchymal cells. Studies have proven the role of BMPs in regulating the embryonic chondrogenesis and osteogenesis [48]. While there are different BMPs, those with the highest capacity of osteogenesis are BMP2, BMP4, BMP5, BMP6, BMP7, and BMP9. BMP2 are seen in areas of cartilage formation, osteogenic zones, and periosteum. BMP4 is found only in the perichondrium, while BMP6 in the hypertrophic chondrocytes. Mutations in BMP5 have been shown to cause skeletal deformities. BMP7 are concentrated around the perichondrium while they are absent in the areas of joint formation. BMP2 and BMP7 act by inducing Runx2 and Osterix transcriptional factors in the mesenchymal cells, thereby regulating them in osteoblast differentiation [49].

5.5.2 Other Functions of BMPs

The studies available have shown that BMPs help in regulation of the stem cells. However, the functions are varied in the various stem cell compartments [50]. Studies have also stated that BMPs can produce pleiotropic effects including oncogenesis and mutagenesis and have been reported widely in musculoskeletal oncology cases, the most common being osteosarcoma [51]. BMP2 and BMP4 have also been identified in immunohistochemistry and in situ hybridization studies of bone tumors [52].

The role of BMPs have also been identified in various steps in the development and differentiation of the vertebrate nervous system [53]. In the period of embryo development, BMPs are concerned with apoptosis, or the programmed cell death which results in the ideal morphogenesis [42]. The role of each BMPs is listed as described below in Table 5.1:

Table 5.1 Depicting function of the BMP family

BMP	Functions
BMP1	Its action is on pro-collagen type I, II, III, and it is a metalloproteinase Involved in the development of the cartilage
BMP2	Responsible for induction of bone and cartilage formation Role in osteoblasts differentiation
BMP3	Bone formation induction
BMP4	Regulation of bone, teeth, and limb formation Involvement in repair of fractures
BMP5	Role in cartilage development
BMP6	Maintenance of joint stability in the adults
BMP7	Function in differentiation of osteoblasts Major role involved in the repair and development of renal tissues
BMP8	Formation of bone and cartilage
BMP9	Promotes chondrocyte differentiation
BMP10	Embryonic stage: helps in heart trabeculation
BMP15	Role in development of follicles and oocyte

HA-coated surfaces promote osteoblastic cell adhesion, differentiation and growth, and a new bone, and it is formed through creeping substitution from the adjacent living bone. HA promotes osseointegration by encouraging a rigid anchorage between the adjacent tissue and an implant, preventing a fibrous tissue growth, increasing the success of the implant and restoring the function [54]. Their scaffolds are also used as vehicle to deliver cytokines. They also have the capacity to bind with BMPs in vivo.

5.6 Uses in Dentistry

5.6.1 Use of Bone Morphogenic Proteins

BMPs, the osteogenic proteins, have been available in the United States as INFUSE bone graft since the Food and Drug Administration approved it in July 2002 for their use in the anterior lumbar interbody fusion. Since its approval for usage in the oral and maxillofacial regions, it has also been used in periodontal regeneration and dental regeneration in addition to its main use in bone regeneration.

5.6.2 Bone Regeneration

BMPs have been extensively used in bone regeneration in the recent years owing to its inductive osteogenic potential. BMP7 have been used to induce new bone generation from the adipose tissues which are widely used undifferentiated stem cells source for tissue engineering [55, 56]. BMP2 and BMP7 have proven their effective bone regeneration potential and have been used for filling the defects while placing bio-implants and restoring mandibular defects [57].

5.6.3 Sinus Lift Augmentation

The effectiveness of BMP has been studied extensively in the area of sinus floor augmentation [58].

The methods for sinus augmentation using BMPs included the following:

(a) 10 mg of BMP7 noncollagenous protein in a poloxamer carrier
(b) 25 mg of BMP7 noncollagenous protein in a poloxamer carrier
(c) 10 mg of BMP7 noncollagenous protein combined with demineralized bone matrix in poloxamer carrier

In 2-week follow–up, it was seen that a good amount of new bone was formed as compared to autogenous bone grafts used in the study.

5.6.4 Implants

The rehabilitation of the dentition using the implants is also an area where the use of BMPs have been studied enormously and found to be highly effective. Bone formation around the implants depend on the material of the implant; however, the chemically treated surfaces with BMP7 produce greater cell quantity and osteo-inductive proteins [59]. The implants where the surface has been treated with BMP7 and BMP4 were seen to show clinical proof of better osseointegration and vertical augmentation of the alveolar ridge [60].

5.6.5 Periodontal Regeneration

The regenerative therapy of the periodontal tissues is done in conditions such as trauma to periodontium and other periodontal diseases causing loss of tooth supporting structure to produce a functional substitution of the destroyed tissue. The latest studies in this area use BMP2, BMP7, and BMP14. BMP7 has also been seen to have its effects on cementoblasts, leading to the root-cementum formation [61]. The application of BMP7 along with insulin-like growth factor-1 has demonstrated to be have the potential for periodontal reconstruction [62].

5.6.6 Dental Regeneration

It refers to the formation of new enamel or dentin through induction of the odontoblasts and the ameloblasts cells. The BMPs target the pulp tissues which are rich in mesenchymal cells, where they bind to the BMP-specific surface receptors following which there is initiation of the cellular cascade that results in cell differentiation and new reparative dentin formation [63]. These in vitro studies have shown that enamel matrix derivative contains both insulin-like growth factor and BMP6 like molecules while Rh-BMP7 promotes the formation of reparative dentin. So, BMP7 gene therapy may be used to promote mineralization when used in pulp capping procedures [64].

5.6.7 Use of Hydroxyapatite

HA, due to its composition mimicking the inorganic substance of the bone, has been increasingly getting popularity in orthopedics and its use in dentistry has been rising. The various areas where HA are used include the following:

5.6.8 Implantology

Nano-HA is the most common coating material used in titanium and stainless steel implants because of the better bonding with the adjacent bone and ability for formation of new bone resulting in superior bone to implant contact [65]. It has also been seen that these HA coating inhibits microbial growth and reduces the early inflammatory reaction [66]. In areas of bony atrophy, HA containing alloplast materials may be used for augmentation prior to implant placement [67].

5.6.9 Maxillofacial Surgeries

HA can be used in treating bony defects as macroscale blocks or as injectable nanoparticles [68]. When in combination with stem cells or growth factors, it can help in bone and cementum regeneration and can be used in cleft lip and palate repair or other periodontal procedures. A composite structure of the HA with a scaffold can also be used in post-extraction sockets to delimit the alveolar bone height loss [69].

5.6.10 Dentin Hypersensitivity

HA in the nanoparticle form can be used as they will penetrate into the dentinal tubules and stop the circulation of the dentinal fluid by clogging the tubules. It is much better than the available desensitizing agents as it adds a layer of apatite on the teeth surface owing to its remineralizing ability providing a protective layer [70].

5.6.11 Bleaching

Seventy percent of the patients complain of sensitivity after bleaching. So, by adding HA or other remineralizing agents such as fluoride calcium, they can fill the minute defects of the enamel post bleaching hence reducing post-bleaching sensitivity [71].

5.6.12 Caries Prevention

HA are added to the toothpastes to provide ions which will either directly replace the lost mineral or can act as a carrier of the lost material for the early-stage caries [36]. It also helps in caries prevention by penetration of the porosities of the tooth surface while also providing a protective layer [72].

5.6.13 Adverse Effects

5.6.13.1 Side Effects of Bone Morphogenic Proteins

A phase 1 study on BMP7 was done in knee osteoarthritis where there was no report of dose limiting toxicity. No reporting of ectopic bone formation in the serial radiographic evaluation was also seen [73].

BMP2 however has species-specific dosage for osteogenesis. The FDA approved dosage for human use is 1.5 mg/ml. A clinical study on the dose efficacy of RhBMP 2 in post-extraction socket augmentation showed significant difference in the 0.75 mg/ml and 1.5 mg/ml groups. Another study on sinus floor augmentation using the same dosages showed similar results [74]. The most recognized side effect of BMP2 is the ectopic bone formation from its leakage from the site of the implant. In preclinical in vivo studies, it has been reported to cause bone cyst formation, while no report of the same is available in clinical studies [75]. Another side effect of BMP2 use is the local inflammation. Life-threatening cervical inflammation has also been reported post its implantation in cervical spine [76]. ***Smucker*** **et al.** had reported six cases of such fatal swelling of the cervical region [77]. Complications related to postoperative healing including conditions such as wound dehiscence, fever, hemorrhage have been observed in orthopedic studies [78]. Many in vivo and in vitro studies have been done on BMPs to see the effect on cancer cells. While BMPs promote bone and cartilage formation and help in neural and cardiac morphogenesis, their overexpression has been observed in pancreatic, gastric, prostrate, and breast cancer [79]. When 1 ng/ml RhBMP2 cocultured with fibroblast were used, it was seen that the viability of the cancer cells were not affected, but the invasion ability by the cancer cells was increased significantly in oral squamous cell cancer models [80].

5.6.13.2 Side Effects of Hydroxyapatite

Not much side effects have been reported regarding the use of HA in terms of acute or chronic toxic effects [81]. Since HA has inherent porosity, during the application of an HA-based implant, it can produce a crack which may propagate leading to failure. Also, application of HA in bulk amount can produce impropriate load sharing due to modulus discrepancy between the bone and the implant [81]. Upon degradation of the HA implant, it can cause chronic inflammatory reaction and implant failure [82]. Also, rapid degradation will increase the local concentration of calcium. Although high calcium content promotes bone regeneration, rapid degradation will give away the structural integrity and lead to implant failure [83]. A possibility for carcinogenic effects has been investigated in some in vivo studies; however, the results were not positive. No reports on formation of tumors have been reported in literature [84].

5.7 Dental Lasers

The concept use of laser energy along with the components of the oral tissues has been inspected in the literature since a long time period. ***Nd:YAG laser*** (1064 nm) and carbon dioxide laser (10,600 nm) were the first wavelengths of laser to be used and practiced in general dentistry [85]. Though their configuration in the emission mode was used for soft tissue ablation certain times, Nd:YAG laser was found suitable in preparation of the tooth cavity.

Multiple studies claimed the use of 1064 nm wavelength for initial carious lesions, which are pigmented; however, the exposure of enamel and dentine to the energy of this laser caused melting of HA (hydroxyapatite) along with thermal cracking, due to the long width of the pulse and its accompanying heat transfer and absence of the water spray [86, 87]. Literature has also shed light on the resistance of the amorphous HA to the dissolution of the acid post-laser application [88, 89]. The wavelength of the ***carbon dioxide*** also has been termed as impractical for the procedures of restoration, even though, there is high peak of absorption of this wavelength by CHA, i.e., carbonated hydroxyapatite, because the continuous emission of the wave of the laser energy, along with the paucity of the water coolant caused carbonization, followed by cracking and tooth tissue melting [90, 91].

Er,Cr:YSGG and ***Er:YAG*** lasers were then investigated on the dental tissues, giving significance to true ablation, and reducing mechanical or thermal damage to the pulp or tooth [92, 93]. The micro-pulse mode of emission in the ***Erbium YAG laser*** is due to the primary chromophore being water, which results in expansive and rapid vaporization. It further causes vaporization of minimal amounts of water present in dentine and enamel, leading to gross structure dislocation. The popping sound which is audible in the surrounding air is due to the change in pressure, and the tissue with greater content of water, i.e., carried followed by dentine and enamel respectively, the louder the sound [94].

If the water coolant is insufficient, the continuous dispersion of ablation products causes eschar to build-up, which if gets super-heated can cause thermal energy to be conducted to the surrounding tissues of the tooth, thus causing cracking, melting leading to pain and damage to the tooth pulp [95]. Another laser considered useful for tooth ablation is ***Er,Cr:YSGG***, which is 2780 nm wavelength. It also causes ablation via interstitial water vaporization [85]. Furthermore, there has been literature regarding the use of atomized water spray along with ***Erbium YSGG***, which is called as hydrokinetic effect [85]. It has been observed that the rate of ablation of erbium YSGG is marginally lower when compared to Erbium YAG with the enamel [85]. This was further confirmed from study by Apel et al. although both types of lasers permit preparation of cavity within the acceptable parameters clinically [96, 97].

"Laser" cavity essentially looks like a crater form when seen in gross or micro-appearance. This cavity is significantly unlike the **"classical"** form of the cavity which is obtained via rotary instruments [98]. Furthermore, the edges of the cavity which is visually observed as "etched" appearance is due to the mineral micro-dislocation, and can aid in composite bonding.

5.7.1 Marginal Integrity

Literature observed weak stability of restoration done via laser, which was explained by weakness in the enamel with respect to the margins due to ablation [99, 100]. However, one study did considered laser as a better option for marginal integrity [101]. Also, the treatment of the cut surface with acid etching can improve the restoration longevity, along with enhancing its bond strength [102]. This method can be employed when composite resin is used for restoration of incisal or facial surfaces or for orthodontic bracket placement.

5.7.2 *Pain Perception*

The patients perceive pain during the dental procedures, and hence it remains a major determining factor for patient acceptance. Many studies have been carried out for the evaluation of the above [103, 104]. It has also been developed that use of Nd:YAG laser causes pulpal analgesia perhaps due to the "gate theory" interference for propagation of the neural stimulus [105].

It is claimed that preparation of the tooth via laser leads to avoidance of the pain due to the lack of thermal and tactile stimulation as opposed to rotary instruments. Also, there is conditioning and emotional state of the patient associated with the procedure too. Literature has varied studies demonstrating the minimal requirement of local anesthesia when laser is used [106, 107]. In a study by Keller and Hibst of 194 teeth in 103 patients, who required 206 preparations, the request for local anesthesia was demanded by only 6%, and the patients in the study rated treatment with laser as much more comfortable than conventional use of bur technique [108]. Matsumoto et al. gave a similar report [109]. Overall, it continues to be a topic for debate.

5.7.3 *Erbium Lasers: Tooth Cavity Preparation*

5.7.3.1 Enamel and Erbium Lasers

Predominantly, the composition of the enamel is 85% carbonated HA, i.e., mineral, with 3% organic proteins and 12% water (Fig. 5.4).

Of all the hard tooth tissues, the greatest resistance toward ablation by laser is exhibited by enamel. Furthermore, the combination of fluorapatite mineral along with hydroxyl group replacement with fluoride in the fluoridated enamel provides even wider resistance. In the cavities where there is less prismatic density like in

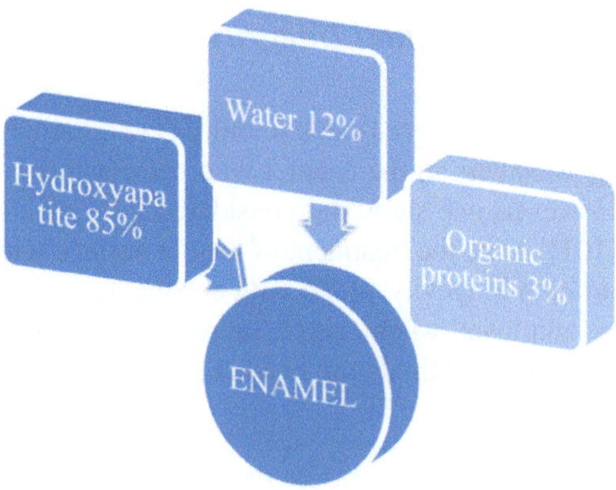

Fig. 5.4 Composition of enamel

Class III or IV or V, it is observed that the rate of ablation of the laser is comparable to the rotary instruments [110]. Efficient and fast preparation of the cavity can be attained via 400–700 mJ/10–20 pps power levels, along with ample cooling [111].

5.7.3.2 Dentine and Erbium Lasers

It is composed of greater content of water in contrast to enamel, i.e., 20%, 47% carbonated HA, and protein being 33%, which is chiefly collagen (Fig. 5.5).

Hence, the rate of ablation is faster in comparison to enamel, with parameters for power being lower [85]. When dentine is carious, the laser beam has the potential to pass swiftly via layer on the surface, thus causing dehydration in layers which are deep. However, it is always advisable to make use of an excavator for removal of bulk volume for prevention of damage due to heat and for expedition of cavity preparation.

5.7.4 Erbium Lasers: Bone Ablation

Studies have shown that these lasers have a thermally produced explosive process, similar to the ablation of dentine and enamel [112]. It is crucial to prevent damage due to heat and hence requires “co-axial” water spray. It has also been perceived that if temperature rises above 74 °C, i.e., the critical temperature, there will be rapid denaturation of the collagen causing tissue coagulation [113]. If the temperature goes above 100–300 °C, ascending dehydration will be observed, which will cause carbonization of lipids and proteins. Currently, the mid-infrared lasers show poor hemostatic effect, thus these can be used for bone ablation, as it will ensure perfusion of the blood to the surgical site.

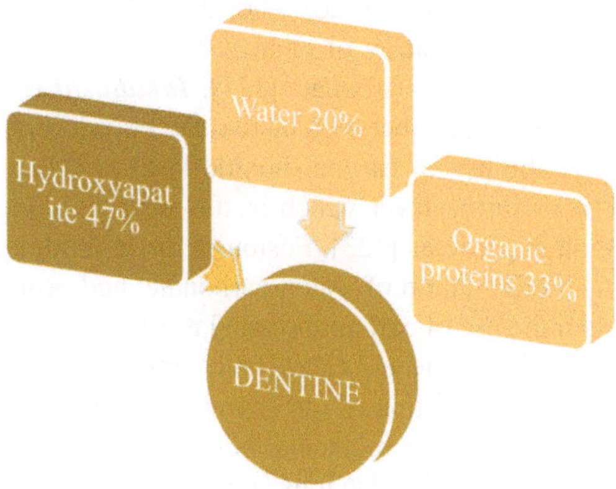

Fig. 5.5 Composition of dentine

5.7.4.1 Risks Involved

- It results in notable blood spatter, hence requires precautions like mask and protection of the eye.
- Use of water spray which is air-induced runs the risk of creating air embolism.
- Bone ablation has higher acoustic levels (100–120 dB) unlike ablation of the tissues.

5.7.4.2 Advantages

- Less traumatic to the patient.
- No vibration unlike intense vibration observed with surgical burs of slow speed.
- Required speed of laser cutting is slightly lower in mandible due to the considerable cortical composition of the bone.

Excessive parameters of the power should be kept away for minimization of the spattering of the blood, along with reduction of debris "stall-out." Minimal evidence exists for the thermal damage when the cut surface was analyzed microscopically, which was restricted to just 20–30 micrometers in depth [114]. When healing was studied for the lased bone, it was supported that reduced trauma physically, along with reduced heating and decreased contamination of the bacteria led to healing process that is not complicated and with osteogenic potential, which is dissimilar to the surgical bur conventional use [115, 116].

The study by ***Ramalho KM*** **et al.** also describes the successful use of lasers for hard tissue application, along with citing the importance of specific knowledge concerning the interaction of such laser with the biological tissues [117]. ***Lesniewski*** **et al.** did a literature review comparing the use of lasers in hard and soft tissue procedures [118]. ***Braun*** **et al.** analyzed the microcracks of the hard tissues of the dentition after irradiation with 970 nm Diode laser, and hence proposed a protocol for laser which will prevent these side effects [119].

The study by ***Xue*** **et al.** reviewed the effects of carbon dioxide laser on the hard tissues of the teeth [120]. They quoted multiple studies like ***Featherstone*** **et al.** who observed alteration of enamel crystal composition, along with decomposition of hydroxyapatite carbonate [121]. ***Takahashi*** **et al.** noticed production of crater like structures, together with increase in phosphorous and calcium content of the enamel with the use of carbon dioxide laser [122]. Its effect on dentine produces small cracks in the layer which is sub-surface, along with production of solidified and molten particles [123]. Fusion of the individual crystallites of the dentine, together with elimination of water, carbonate, and protein amide is observed in the dentine with the use of carbon dioxide laser [124].

Both dentine and enamel have mineral HA as common, whereas carbonated HA is the composition of the enamel. These are crystalline structures with a resistant, strong, and ionic molecular bonding. Water, which acts as an interstitial medium or as alkaline hydroxide ions and acidic hydrogen ions, is susceptible to the process of vaporization when erbium wavelength is applied. This leads to a creation of temperature and pressure change causing dislocation of the crystalline structure and explosive ablation at the microscopic level at the application point, which is called as ***"spallation."*** In order to prevent undesirable collateral damage, the management of both osseous and dental tissues needs thermal containment and accuracy. This requires the use of appropriate wavelength of the laser and photonic power, adjunct with water cooling. This allows selective and predictable removal of carious lesion.

With the help of multiple randomized clinical trials and use of visual analogue scoring, has defined the speculation of cavity preparation which is anesthesia free, i.e., painless [125]. Additionally, both carbon dioxide and erbium laser require co-axial spray of water while using for dispersion of ablation products, along with providing cooling of hard tissues [126].

5.7.4.3 Factors Affecting Absorption: Laser

It is important for the clinician using laser photonic energy to know the factors which can affect laser light absorption [127]. They are as follows:

- Laser wavelength
- Composition of the tissues
- Mode of laser emission
- Thickness of tissues
- Laser beam: incident angle
- Surface wetness: due to saliva/water
- Time of exposure
- Thermal relaxation factors: which can be endogenous (blood supply, tissue density and type) and exogenous (pre-cooling of the tissues, pulsing laser emission, water spray, and high-speed suction)
- Modes between the tissue and laser delivery tip, i.e., noncontact versus contact

Consequence of the above factors is to choose appropriate laser for use, and to avoid possible thermal deleterious effects. There are three fundamental elements for consideration when it comes to the use of lasers:

1. Appropriate/correct wavelength of laser
2. Appropriate power density of light delivery
3. Appropriate process of thermal relaxation

It is important to observe the protocols as mentioned above when using lasers, as otherwise it might result in unwanted collateral rise in temperature, along with causing change with respect to the optical properties of the tissues, thus altering the optimal interaction between laser and the tissues.

The extensive application of laser wavelengths has endorsed their incorporation into surgical and restorative dentistry. Early caries intervention, besides embracing of micro-retentive cavities of tooth unalike direct metal restorations, the lasers have and will find growing acceptance.

5.8 Endosseous Implants

The loss of teeth has been a major problem since time immemorial. In ancient times, the main problem was just the inability to masticate properly threatening their survival. With modern advances in food processing, the importance has shifted from survival to enjoying the textures of the various food items. In the modern world, apart from functionality, the maintenance of dentition is highly important for the aesthetics. Hence, replacement of the missing teeth became a very important part of the dental procedures.

The methods of replacing the missing teeth have evolved over time from transplanting teeth from another individual to the present-day implants, along with the fixed and removable prosthesis.

A dental implant is a material which is used to replace a missing tooth or teeth in cases of partial or complete edentulism. Its use and popularity have been increasing due to its superiority over the conventionally used fixed partial dentures or removable complete dentures. It is constructed of an alloplastic biocompatible substance and is implanted into the oral tissue under the mucosa and/or periosteum or placed inside the bone. A fixed or removable prosthesis is then placed over the implant, and it provides a good retention and support to the former.

Endosseous implants are those in which the implant is placed into the maxillary or the mandible bone. They are the most used type of dental implants and have been fabricated in different shapes—screws, helical wires, tripods, blades, and vents [128].

5.8.1 Types of Implants and Implant Materials

It can be divided based on position of placement into:

- Endosteal
- Transosteal
- Subperiosteal

Endosteal Implant These are the implants where pierce only one of the cortices in the maxillary or mandibular bone. Root form implant is the commonest endosteal implant used.

Transosteal Implant These are the endosseous implants where it engages both the cortices of the maxillary and mandibular bone.

Subperiosteal Implant In this type, a custom-made cast framework consisting of a substructure and superstructure is placed inside the periosteum.

Implants can also be classified based on the material of its fabrication into:

- Metal
- Ceramic

- Polymers

Based on its potential to generate a biologic response in the adjacent host tissue, it is divided into:

1. ***Biotolerant***
   - **Metals**: Stainless steel, tantalum, niobium, gold Co-Cr alloys
   - **Polymers**: Polyamide, polytetrafluroethylene, polyethylene, polymethylmethacrylate, polyurethane
2. ***Bioinert***
   - **Metals**: Titanium alloy: commercially pure: Ti-6AL-4U
   - **Ceramics**: Zirconium oxide, aluminum oxide
3. ***Bioactive***
   - **Ceramics**: Bioglass, hydroxyapatite, carbon-silicon, tricalcium phosphate

Biotolerant implants are held in the surrounding tissue by fibrous tissue formation, e.g., stainless steel.

Bioinert implants do not react with the surrounding tissue and are held by a mechanical rigid bond, e.g., titanium and its alloys.

Bioactive implants where a layer of regenerative material is coated onto the implant surface allowing the formation of new bone around it, e.g., hydroxyapatite [129].

The goal of placing an implant today is achieving osseointegration between the bone and the implant. Osseointegration is defined as the direct contact between the vital bone and the implant placed [130]. Branemark introduced the concept of osseointegration after his multiple studies on animals and human. Prior to this, the materials used were inert and a layer of fibrous tissue encapsulation was preferred [131]. Today the presence of this fibrous layer indicates failure of the implant [132].

The prognosis of the placed dental implants depends on the composition and surface roughness which will define the amount of implant tissue interaction and osseointegration [133]. However, when there is an insufficiency in the bone available or bony defects of maxilla and mandible, placement of the implant becomes very difficult or impossible as the three-dimensional support it requires cannot be attained. In these situations, bone grafts and bone regeneration come into play. Scaffold materials, growth factors, and stem cells can be used for the regenerative treatment for the defects [134].

5.8.1.1 Bone Grafts and Augmentation Materials

These are the biomaterials that are used to replace the bony defects and build up the atrophic bone regions. A study by ***Chiapasco* et al.** has shown that bone grafting is required in one out of every four implants placed [135]. In regard to implants, bone augmentation procedures are done in block bone grafting and maxillary sinus floor elevation.

Implants should have a suitable amount of bone coverage, i.e., 1.5–2 mm in buccal and lingual sides of the implant shoulder; otherwise, it results in loss of the crestal bone with inflammation and infection of the peri-implant tissue due to exposure and invasion of the implant surface by the biofilms [136]. Thus, proper bone augmentation should be done when placing the implant.

Bone graft materials act via three mechanisms: osteo-induction, osteogenesis, and osteo-conduction. Osteo-induction is the process through which the undifferentiated mesenchymal cells are induced to transform into osteoblasts and chondroblasts to form new bone around the implant. The healing of the implant by forming a direct contact with the adjacent bone is osteo-conduction [135].

Bone grafts can be identified into autogenous, allografts, xenografts, and alloplasts.

The various bone augmentation materials used in implantology are as follows:

- **Autogenous bone:** Particulate bone, block graft
- **Allogenic graft:** Freeze dried bone, deproteinized bone, fresh frozen bone, demineralized freeze-dried bone
- **Xenogenic bone:** Coralline material, porcine/bovine bone, calcifying coralline algal material
- **Alloplastic bone:** Polymers, glass ceramics, metals, calcium phosphates

Autogenous bone grafts can be obtained from either a neighboring region or an intraoral site or an extraoral region. Allogenic bone grafts are usually obtained from another human from the iliac or tibial bone. Xenogenic bone substitutes are usually obtained from algae, corals, equine, porcine, while the most used is from a bovine source. Alloplastic graft materials include hydroxyapatite, beta-tricalcium phosphate, calcium silico-phosphate, polymers, titanium particles, bioglasses, or their combinations.

Barrier membranes are used when the procedure of guided bone regeneration (GBR) is used. They can be traditionally made of polytetrafluoroethylene (PTFE), but newer membranes are made of collagen I, III, or a combination of the two [137]. The collagen membranes are obtained from porcine or bovine sources [138]. In large defects, the membrane needs to be supported by bendable titanium framework into a PTFE material, or by tenting with the help of screws [137]. Guided bone regeneration is a method to regenerate adequate bone prior to placement of an implant. It uses a barrier membrane and may or may not use bone grafts or substitutes.

5.8.1.2 Mechanism of Action

Implant

The method by which the implant amalgamates with the bone creating an intact contact is called osseointegration. It is the direct constructional and functional connection between the vital bone and the surface of the load bearing implant. It is a time-dependent procedure. The histologic appearance should be an ankylosis between the two with no fibrous or connective tissue in between. It is highly critical for the stability and success of the implant. The process involves an initial engagement of the implant body into the bone, followed by continuous bone apposition and remodeling. The process is activated when the implant is placed and a lesion is formed in the bone matrix. Upon exposure of the matrix to the extracellular fluid, noncollagenous proteins and growth factors are released and the process of bone repair commences. Osseointegration occurs in the following three stages [139]:

- Woven bone formation
- Adaptation of the bony mass to the load—bone deposition
- Adaptation of the bone structure to the load—bone remodeling

5.9 Augmentation Materials Used with Implants

The three mechanisms by which the bone graft material act are as follows:

Osteogenic grafts contain inherent osteoblasts which form the new bone. Autogenous bone grafts work through this process. Stem cell harvests are newer development through which osteogenic bone grafts can be prepared without harvesting from the patient's own bone.

Osteo-induction is seen in bone morphogenic proteins whereby they stimulate the undifferentiated mesenchymal cells to convert into osteoblast cells and which in turn forms the new bone.

Osteo-conduction is seen when a scaffold is used as the bone substitute and it helps in the migration of the osteogenic cells and their establishment over the scaffold and then formation of the new bone.

Autogenous grafts while being osteogenic also acts as a scaffold for the host osteoblasts to migrate into, thus enabling osteo-conduction. Allogenic grafts such as fresh frozen bone or freeze-dried bone usually work through osteo-conduction but possess some osteo-inductive potential. Demineralized freeze-dried bone allograft however works primarily through osteo-induction.

In GBR, new bone regeneration is due to migration of the pluripotent and osteoblast cells from the periosteum and/or bone marrow and/or adjacent bone to the intended defect site while excluding the cells which impedes bone formation [140].

5.9.1 Uses in Dentistry

5.9.1.1 Implants

1. Partial edentulism, to replace one or more teeth without affecting the adjacent teeth.

2. In patients who do not want the conventional complete dental prosthesis.
3. To preserve the already in use removable partial prosthesis.
4. As anchorage during orthodontic treatment.
5. In severe compromised ridges where conventional prosthesis will not be supported.
6. Patients with poor oral muscular coordination.
7. In reconstruction after resection of the maxilla or mandible for future rehabilitation.

Absolute Contraindication
Acute illness, large bony defects, uncontrolled metabolic disease, bone or soft tissue pathology and infection.

Relative Contraindications
Diabetes, osteoporosis, parafunctional habits, HIV, AIDS, bisphosphonate usage, post-chemotherapy, post irradiation of head and neck, psychiatric disorders.

5.9.1.2 Bone Augmentation Materials

1. Ridge augmentation
2. Sinus grafting
3. Grafting the extraction sockets for maintenance of bone volume

5.9.2 Adverse Effects/Complications

5.9.2.1 Implants

1. Fracture of the implant component, loosening of implant—Due to biomechanical overloading because of poor implant position and angulation or lack of posterior support, bruxism or inadequate bone, it can lead to loosening or fracture of the implant component [141, 142]. The treatment then requires a second procedure for the complete removal of the implant. It can be replaced with another wider or longer implant; however, if the bone loss is significant, bone augmentation needs to be done first.
2. Infection and inflammation—Poor oral hygiene maintenance is the main reason for the microbial build up around the implant and hence the resulting inflammation. Signs and symptoms include erythema, induration, and painful inflammation of the peri-implant gingiva. Following the infection, there will be erosion and resorption of the bone around the implant. The treatment will depend on the severity of the infection. If initial infection, antibiotic coverage is given. In advanced cases, the implant needs removal and proper curettage followed by bone grafting for replacement of implant after healing.

3. Local toxicity and foreign body reaction—The cements used for placement of the prosthesis can be forced subgingivally leading to an acute or delayed reaction. The patient will complain of focal gingival edema, erythema, and discomfort [143]. The reaction to the excess cement can also cause resorption of the bone around the implant.
4. Injury to the surrounding structures—When placing an implant in the maxilla or mandible, there are many vital structures in close proximity to the implant—the mandibular canal, maxillary sinus, nasal cavity, and incisive foramen which can be harmed during the procedure. A proper evaluation of the structures is very important prior to the surgical placement. If the canal is violated, paraesthesia and pain may be the result [144]. If injury occurs to the maxillary sinus or nasal cavity, it can cause infection and oro-antral fistula or even inadvertent loss of the implant into the sinus, requiring a trans-antral endoscopic surgery for retrieval.
5. Metal hypersensitivity—Titanium, even if considered an inert material can induce clinically significant hypersensitivity when exposed chronically to potential metal allergens [145]. Symptoms of chronic hypersensitivity may be chronic fatigue, fibromyalgia, and depression. The most common presentation will be an oral lichenoid reaction.
6. Other complications: The following events can also occur while doing the implant surgery [144]:

   - Mandible fracture
   - Ingestion/aspiration of implant
   - Hemorrhage
   - Injury and devitalization of the adjacent teeth

5.9.2.2 Bone Augmentation Materials

1. Infection—Localized infection of the recipient sites can be seen leading to failure of osseointegration and implant [146].
2. While using bio membranes, PTFE membranes get exposed as compared to resorbable ones and lead to infection [147].
3. Wound dehiscence—This can occur due to lack of osteogenic potential of the recipient site and reaction of the surrounding soft tissue to the bone graft material [25].

Late resorption—Some amount of resorption and remodeling is inevitable; however, if the amount is large, it can jeopardize the bone supporting the implant, resulting in dehiscence, peri-implantitis, and failure of the implant [146].

5.10 Subperiosteal Implants

Subperiosteal implants are those implants which are placed beneath the periosteum, on top of the maxilla or mandible but not directly inserted into the bone. These are custom-made implants usually given in cases of advanced alveolar ridge resorption. Rehabilitation of a severely atrophic jaw is a challenge as placing of endosteal implant in an atrophic bone is bound to fail unless other measures such as bone augmentation is done. Subperiosteal implants acts as a perfect alternative in these conditions for full mouth rehabilitation and an immediate prosthesis can be placed.

The concept of subperiosteal implants was first conceptualized by ***Gustavo Dahl*** [148]. These implants were however associated with problems of implant exposure, implant mobility, and failure. It was further refined by ***Goldberg and Gershkoff*** in 1948 with the introduction of the attached framework. ***Cerea and Dolcini*** have used direct metal laser sintering titanium subperiosteal implants and reported 95.8%success rate in 2-year follow-up and very less complications [148]. David F Angelo and Jose R V Ferreira used a custom-made bimaxillary subperiosteal implant with both subperiosteal support and endosseous support [149].

Nowadays with the advancement in dental technologies, CAD-CAM can be used to for customizing the implants, improving the placement, and favoring osseointegration and implant success. Also, BMP- and HA-coated subperiosteal implants have been used with the purpose to regenerate bone in the surrounding areas [150]. Complications after placement includes pain, implant framework exposure, bone resorption, and mobility of implant [151].

5.10.1 Structure of Subperiosteal Implants

A subperiosteal implant has a framework with per mucosal extensions and may or may not consist of bars and struts. Struts can be primary, secondary, and peripheral. The implant can be a complete arch, unilateral, and is loaded immediately. The procedure of subperiosteal implant placement is of two types—a dual stage surgical technique and a single stage surgical technique.

Dual stage surgical technique—an exposure of the bone followed by impression is made. Cast of the impression is made and an implant is fabricated. The implant is a cobalt chrome or a titanium alloy. In the second surgery, it can be placed onto the bone and loaded immediately with a removable or a fixed prosthetic denture.

Single stage surgical technique—In this one, a CT (computed tomography) scan or a CBCT (cone beam volumetric tomography) is taken, and through computer modeling software, a model of the jaw bone is fabricated and the implant is then made in accordance with the model. Then, the surgical procedure will be the same as the second stage of the dual technique.

5.10.2 Uses in Dentistry

1. Extensive bone resorption of the maxilla or mandible where bone augmentation procedures are contraindicated or patient is unwilling for the same
2. Aged patient with metabolic diseases that reduce regeneration capacity
3. Pathology of the maxillary sinus

5.10.3 Contraindications

1. Patients with calcium metabolism disorders where there is progressive bone resorption
2. Poor bone quality of the recipient site
3. Post-radiotherapy patient

5.10.4 Benefits

1. Reduces the number of surgical steps and period of treatment
2. Excludes the need for bone grafting

5.10.5 Disadvantages

1. Slow, predictable implant rejection
2. Bone loss and implant failure

5.10.6 Adverse Effects

1. Wound dehiscence—It can occur after the first or second surgery, when the incision is placed at too much distance from the vestibule. Improper vascularization and impropriate suturing or infection can increase the chances. This will lead to bone resorption and implant failure [152].
2. Calculus formation—Due to prolonged biofilm formation on the abutments, natural teeth, and insufficient cleaning of the prosthesis, calculi deposits can occur hampering the gingival tissues and leading to peri-implantitis [152].
3. Pocket formation—A primary pocket formation can occur at the surgical site, requiring a second gingivectomy procedure to eliminate the pocket. A secondary pocket can also form when there is a failure for the soft tissue attachment with the abutment [152].

4. Abscess and fistula formation—Periodontal abscess can form in repetitive subacute form or chronically. Pus discharge can occur around the abutment or a fistula can form around the implant [152].
5. Bone resorption—Due to the previous abovementioned conditions, one resorption around the implant is an obvious physiological result [152].
6. Carcinogenicity—Due to chronic mechanical irritation of the ulcer as a result of the improper subperiosteal implant and chronic peri-implantitis, it can promote carcinogenesis [91].
7. Chronic irritation can act as a cofactor for carcinogenesis while peri-implant infection will promote it [153].

5.11 Temporomandibular Joint (TMJ) Prosthesis

Temporomandibular joint is a ginglimo-diarthroidal synovial joint, which is considered as a compound joint with four articulating surfaces. The two main bony surfaces involved are glenoid fossa and mandibular condyle.

5.11.1 TMJ Ankylosis

Complete or significant limitation of the temporomandibular joint movement due to union of the head of the mandibular condyle and glenoid fossa, which can be either bone or fibrous, thus causing obliteration of the normal TMJ articulation [154].

5.11.2 Etiology: TMJ Ankylosis

There is a multitude of etiological factors for TMJ ankylosis. They are as follows:

- **Trauma:** This is the most common cause for ankylosis of TMJ [155]. Trauma can be due to intra-capsular fracture especially in the children. Furthermore, it can be due to trauma during birth.
- **Infection:** Due to suppurative arthritis or otitis media.
- **Surgical:** Which can be due to orthognathic or TMJ surgery.
- **Inflammation or systemic diseases:** It can be due to psoriatic or rheumatoid arthritis, ankylosing spondylitis, or Still's disease.
- **Idiopathic**.

According to a study by ***De Roo N*** **et al.** trauma constitutes 84.3% of the cases of TMJ ankylosis, followed by infection in 7.4% cases [156]. Similar results were given by ***Bhatt*** **et al.** [157], where trauma caused 79% cases, followed by infection (11.8%), idiopathic (6.5%), and birth trauma (2.7%).

5.11.3 Goals of Reconstruction

The fundamental goals in cases of reconstruction of TMJ are as follows:

1. To restore ramal length of the mandible along with its morphology
2. To attain normal range of motion
3. To attain normal occlusion and jaw relations
4. Also, the growth of the ramal condylar unit, i.e., RCU should be similar to that of the opposite TMJ, and also in conjunction with entire maxillofacial skeleton especially in pediatric patients.

5.11.4 Methods of Reconstruction

The TMJ can be reconstructed via autogenous or alloplastic methods (Fig. 5.6).

5.11.5 Autogenous Reconstruction

Reconstruction via autogenous methods first began in 1860 where temporalis flap was first employed, followed by use of autogenous fat, or use of free muscle [158]. Reconstruction is done of either RCU or glenoid fossa lining. The various methods for ***RCU reconstruction*** can be:

- Costochondral graft (CCG)
- Sternoclavicular graft (SCG)
- Metatarsal graft
- Calvarial graft

Glenoid fossa lining can be reconstructed via:

- Auricular graft
- Dermis graft
- Temporalis myofascial graft

Other methods of autogenous reconstruction can be:

- Transport disc distraction osteogenesis (TDDO)
- Ramus sliding osteotomy

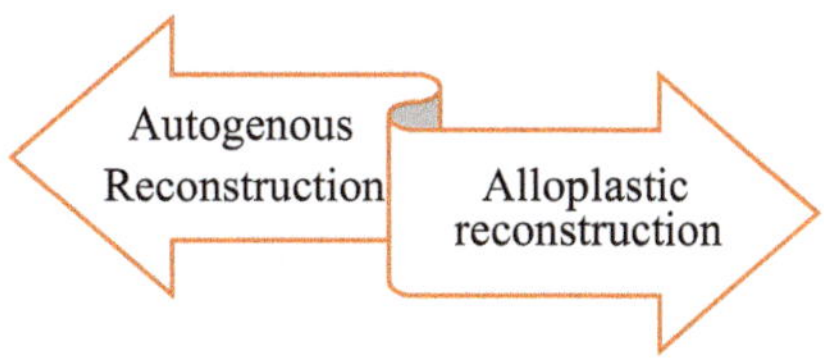

Fig. 5.6 Method of temporomandibular joint reconstruction

5.11.6 Alloplastic Reconstruction

First idea about inter-positioning in case of TMJ ankylosis was given by John Murray, who placed a wood block between skull base and the condyle which is osteotomized [157]. Furthermore, people used ivory TMJs which are prosthetic, or a gold foil which is interposed in the gap or tantalum foil [158].

5.11.7 Indications of Alloplastic TMJ Reconstruction

- Degenerated, ankylosed, or resorbed joints having anatomic abnormalities which are severe
- Autogenous grafts failure in patients who underwent multiple operations causing poor vascularization, scarring of the tissue bed
- Severe inflammatory diseases of the joint resulting in joint components mutilation and thus disability in function, e.g., rheumatoid arthritis
- Pre-existing reactions of the foreign bodies causing autogenous graft destruction
- Formation of heterotopic bone following recurrent TMJ ankylosis
- Condylar fractures which are irreparable
- Avascular necrosis
- Extensive resection of TMJ secondary to neoplasia
- Congenital disorders like hemi-facial microsomia

5.11.8 Contraindications of Alloplastic TMJ Reconstruction

- Material allergy
- Infection with respect to the site of implantation (active/chronic)
- Insufficient support of bone with respect to either quantity or quality
- Systematic diseases which cause increased infection susceptibility
- Mandibular fossa perforations which are extensive
- Deficiency of bone: zygomatic arch or articular eminence
- Partial reconstruction of TMJ
- Patients with neurological or mental conditions who are unable to follow instructions postoperatively
- Teeth grinding or clenching habits which are hyper-functional

5.11.9 Advantages

- Patient can start physical therapy immediately
- Donor site not required
- Decreased surgery time
- Possibly decreased surgery time
- Can maintain occlusion stably post-surgery as there will be no changes in dimensions of the implant, unlike resorption observed in autogenous grafts
- Better adaptation to bony surface and can be customized

5.11.10 Disadvantages

- Growth of the patient cannot be predicted or followed
- Might require second surgery, which cannot be predicted
- Prosthesis cost
- Wear and failure of the material of prosthesis
- reactions to foreign body

5.11.11 Total Joint Prosthesis

The various types of total joint prosthesis are as follows (Fig. 5.7):

1. Kent-Vitek prosthesis
2. Christensen prosthesis
3. TMJ concepts prosthesis
4. Lorenz-Biomet prosthesis

1. ***Kent-Vitek Prosthesis***

It is the first ever TMJ prosthesis. It consists of **"VK-1"** implant for glenoid fossa. It has a bilaminate layer of PTFE, and proplast, i.e., hydroxyapatite [159].

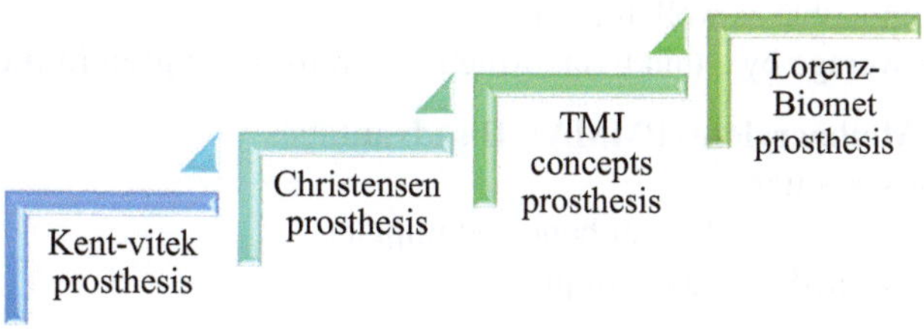

Fig. 5.7 Types of total TMJ prosthesis

The prosthesis for the mandibular condyle is made up of cobalt-chromium. Proplast encourages hard and soft tissue growth rapidly.

With advancements came **"VK-2 Fossa,"** which is three-layered. A study reported 170 patient series with around 182 joints, showing higher rate of success with VK-2, when compared to VK-1 [160]. Another study by Kiehn and associates also reported TMJ reconstruction as successful with fossa with VK-1 [161].

Complications/Risks Involved

1. PTFE fragmentation and wear
2. Reaction to foreign body
3. Fossa and condyle erosion

It was concluded by ***Sir John Charnley*** that use of PTFE should be discontinued in the field of orthopedics as it causes tissue reaction secondary to the wear debris, which further led to acetabular component loosening [162]. A retrospective study by ***Kent*** **et al.** on 49 patients reported a success rate of 62.2% with VK-1 fossa [163]. However, there was a need for removal of ten prosthesis due to erosion, infection, or displacement of the implant.

2. ***<u>Christensen's Prosthesis</u>***

It was the most widely used prosthesis. Thus, he devised a prosthesis for the mandibular condyle which helps compliment the eminence of the fossa. ***Chase DC*** **et al.** published a report of use of this prosthesis in 40 joints, with a mean follow-up time of 2.2 years [164]. It showed an increase in the ability to function and decreased report of pain.

The original head of the condyle consisted of Cobalt-Chromium component with respect to the ramus along with **"Polymethyl methacrylate (PMMA)"** head. Study by ***Peter D Quin*** **et al.** reported the requirement of revision surgeries in 55% of the patients where prosthesis of the fossa was reconstructed against the natural mandibular condyle post meniscectomy [165].

Contraindications

- It is contraindicated in the degenerative diseases of the joint.

Christensen's II

The modification was done to retain the head, i.e., PMMA, but the strut of the ramus was made thicker, along with computer sketched "offset" screw holes.

Polymethyl Methacrylate (PMMA): Advantages

- It is biocompatible as well as strong.
- There is coverage by a thin tissue which is fibrous, i.e., **"pseudo-membrane."**

Polymethyl Methacrylate (PMMA): Disadvantages

- Wear debris reaction
- Osteolysis: at the interface of bone and implant
- Probable elevated frictional torque

3. ***Tech Medica System/TMJ Concepts Prosthesis***

It is custom designed via CAD-CAM technology, especially delineated for patients who have been operated multiple times and for patients with severe anatomical distortion [166]. The articulating surface of this fossa component is made of ultra-high molecular weight polyethylene, i.e., UHMWPE; with accompanying backing of the mesh with titanium. The condylar head of the mandibular component is composed of cobalt-chromium-molybdenum, i.e., Co-Cr-Mb, and titanium being the ramal component [167].

Advantages of UHMWPE

- It has considerably good resistance to wear

Disadvantages of UHMWPE

- The stock prosthesis is fitted appropriately with respect to the base of the skull with PMMA cement
- Presence of osteolysis: at the interface of prosthesis and bone

4. ***Lorenz Prosthesis***

It was approved in Food & Drug Administration (FDA) in the year 2006. It also has fossa prosthesis as UHMWPE, with thickness being minimum at the midpoint, i.e., 4 mm. There is presence of exaggerated 'lipping' in order to protect mandibular condyle from encroachment by the formation of heterotropic formation of bone, and also to avoid dislocation.

Methods of fossa fixation: Articular eminence flattening, followed by tripod stability and fixation with cement.

Condylar prosthesis is made of cobalt-chromium-molybdenum and is spherical in shape. The neck design has swan shape to avoid the obstruction with respect to implant bone interface, unlike the "right angled" design of metallic prosthesis of mandibular condyle.

Leandro **et al.** reported an experience of 10 years, with 300 patient follow-up, reconstructed with the Lorenz/Biomet prosthesis [168]. It represents that prosthesis is effective and safe option for proper joint re-establishment, and stomatognathic function. Also, long-term significant improvements were observed in range of motion of mandible.

Study by ***O'Connor*** **et al.** reported that in patients with severe damage to TMJ, use of TMJ concepts results in good outcomes in these patients, which further validated this procedure [169].

5.11.12 Recent Advances: Tissue Engineering

It is promising that TMJ prosthesis can be reconstructed via bioengineering method which will be biocompatible and can withstand physiologic joint load.

Three critical elements that make tissue engineering possible are as follows:

- Scaffold
- Cells
- Bioactive molecules

Growth of the cells is observed when biomechanical stimulus is applied, which eventually causes secretion of substances that leads to the production of desired tissue, which is then used for the replacement of missing or diseased tissues. Engineering of TMJ is a complicated process as it involves multiple cellular components, which are required at the right place.

However, TMJ reconstruction has undergone major re-thinking over one last decade, and literature shows significant divergence as to the ideal reconstruction method. Total alloplastic reconstruction is now appraised as the "gold standard."

5.12 Conclusion

It is of prime importance that dental devices should be in accordance with Food & Drug Administration (FDA), FDA informed consent, Investigational Drug Exemption (IDE), and Institutional Review Board (IRB). The determination of the risk is not based solely on the device, but it is also based upon the proposed device use in an investigation. Over the past years, it has been an outstanding expansion of technologies and dental devices in the health sector; however, there is paucity of scientific research which are well-conducted, which is required to determine and study the associated or probable risks with the devices. Thus, it is important to know about the various dental devices which pose a significant risk to human life.

References

1. Health C for D and R. IDE approval process. FDA [Internet]. 2020 Nov 25 [cited 2023 Mar 13]. Available from https://www.fda.gov/medical-devices/investigational-device-exemption-ide/ide-approval-process
2. Nanci A, Bosshardt DD. Structure of periodontal tissues in health and disease. Periodontol 2000. 2006;40:11–28.
3. Kinane DF, Stathopoulou PG, Papapanou PN. Periodontal diseases. Nat Rev Dis Primers. 2017;3:17038.
4. Sculean A, Gruber R, Bosshardt DD. Soft tissue wound healing around teeth and dental implants. J Clin Periodontol. 2014;41(Suppl 15):S6–22.
5. Cortellini P, Buti J, Pini Prato G, Tonetti MS. Periodontal regeneration compared with access flap surgery in human intra-bony defects 20-year follow-up of a randomized clinical trial: tooth retention, periodontitis recurrence and costs. J Clin Periodontol. 2017;44(1):58–66.
6. Carnevale G, Kaldahl WB. Osseous resective surgery. Periodontol 2000. 2000;22:59–87.
7. Karring T, Nyman S, Gottlow J, Laurell L. Development of the biological concept of guided tissue regeneration—animal and human studies. Periodontol 2000. 1993;1(1):26–35.

8. Caton J, Wagener C, Polson A, Nyman S, Frantz B, Bouwsma O, et al. Guided tissue regeneration in interproximal defects in the monkey. Int J Periodontics Restorative Dent. 1992;12(4):266–77.
9. Gottlow J, Nyman S, Karring T, Lindhe J. New attachment formation as the result of controlled tissue regeneration. J Clin Periodontol. 1984;11(8):494–503.
10. Caton JG, Greenstein G. Factors related to periodontal regeneration. Periodontol 2000. 1993;1(1):9–15.
11. Wikesjö UM, Kean CJ, Zimmerman GJ. Periodontal repair in dogs: supraalveolar defect models for evaluation of safety and efficacy of periodontal reconstructive therapy. J Periodontol. 1994;65(12):1151–7.
12. Christgau M, Caffesse RG, Schmalz G, D'Souza RN. Extracellular matrix expression and periodontal wound-healing dynamics following guided tissue regeneration therapy in canine furcation defects. J Clin Periodontol. 2007;34(8):691–708.
13. Stavropoulos A, Bertl K, Spineli LM, Sculean A, Cortellini P, Tonetti M. Medium- and long-term clinical benefits of periodontal regenerative/reconstructive procedures in intrabony defects: systematic review and network meta-analysis of randomized controlled clinical studies. J Clin Periodontol. 2021;48(3):410–30.
14. Zhao S, Pinholt EM, Madsen JE, Donath K. Histological evaluation of different biodegradable and non-biodegradable membranes implanted subcutaneously in rats. J Craniomaxillofac Surg. 2000;28(2):116–22.
15. Carbonell JM, Martín IS, Santos A, Pujol A, Sanz-Moliner JD, Nart J. High-density polytetrafluoroethylene membranes in guided bone and tissue regeneration procedures: a literature review. Int J Oral Maxillofac Surg. 2014;43(1):75–84.
16. Marouf HA, El-Guindi HM. Efficacy of high-density versus semipermeable PTFE membranes in an elderly experimental model. Oral Surg Oral Med Oral Pathol Oral Radiol Endod. 2000;89(2):164–70.
17. Tonetti MS, Prato GP, Cortellini P. Factors affecting the healing response of intrabony defects following guided tissue regeneration and access flap surgery. J Clin Periodontol. 1996;23(6):548–56.
18. Tonetti MS, Pini-Prato G, Cortellini P. Periodontal regeneration of human intrabony defects. IV. Determinants of healing response. J Periodontol. 1993;64(10):934–40.
19. Murphy KG, Gunsolley JC. Guided tissue regeneration for the treatment of periodontal intrabony and furcation defects. A systematic review. Ann Periodontol. 2003;8(1):266–302.
20. Benic GI, Hämmerle CHF. Horizontal bone augmentation by means of guided bone regeneration. Periodontol 2000. 2014;66(1):13–40.
21. Behring J, Junker R, Walboomers XF, Chessnut B, Jansen JA. Toward guided tissue and bone regeneration: morphology, attachment, proliferation, and migration of cells cultured on collagen barrier membranes. A systematic review. Odontology. 2008;96(1):1–11.
22. Delgado LM, Bayon Y, Pandit A, Zeugolis DI. To cross-link or not to cross-link? Cross-linking associated foreign body response of collagen-based devices. Tissue Eng Part B Rev. 2015;21(3):298–313.
23. Rothamel D, Schwarz F, Sager M, Herten M, Sculean A, Becker J. Biodegradation of differently cross-linked collagen membranes: an experimental study in the rat. Clin Oral Implants Res. 2005;16(3):369–78.
24. Schwarz F, Rothamel D, Herten M, Wüstefeld M, Sager M, Ferrari D, et al. Immunohistochemical characterization of guided bone regeneration at a dehiscence-type defect using different barrier membranes: an experimental study in dogs. Clin Oral Implants Res. 2008;19(4):402–15.
25. Gentile P, Chiono V, Tonda-Turo C, Ferreira AM, Ciardelli G. Polymeric membranes for guided bone regeneration. Biotechnol J. 2011;6(10):1187–97.
26. Hürzeler MB, Quiñones CR, Schüpbach P. Guided bone regeneration around dental implants in the atrophic alveolar ridge using a bioresorbable barrier. An experimental study in the monkey. Clin Oral Implants Res. 1997;8(4):323–31.

27. Sheikh Z, Qureshi J, Alshahrani AM, Nassar H, Ikeda Y, Glogauer M, et al. Collagen based barrier membranes for periodontal guided bone regeneration applications. Odontology. 2017;105(1):1–12.
28. Wang J, Wang L, Zhou Z, Lai H, Xu P, Liao L, et al. Biodegradable polymer membranes applied in guided bone/tissue regeneration: a review. Polymers (Basel). 2016;8(4):115.
29. Susin C, Fiorini T, Lee J, De Stefano JA, Dickinson DP, Wikesjö UME. Wound healing following surgical and regenerative periodontal therapy. Periodontol 2000. 2015;68(1):83–98.
30. Sculean A, Nikolidakis D, Schwarz F. Regeneration of periodontal tissues: combinations of barrier membranes and grafting materials—biological foundation and preclinical evidence: a systematic review. J Clin Periodontol. 2008;35(8 Suppl):106–16.
31. Bartold PM, Gronthos S, Ivanovski S, Fisher A, Hutmacher DW. Tissue engineered periodontal products. J Periodontal Res. 2016;51(1):1–15.
32. Stoecklin-Wasmer C, Rutjes AWS, da Costa BR, Salvi GE, Jüni P, Sculean A. Absorbable collagen membranes for periodontal regeneration: a systematic review. J Dent Res. 2013;92(9):773–81.
33. Needleman IG, Worthington HV, Giedrys-Leeper E, Tucker RJ. Guided tissue regeneration for periodontal infra-bony defects. Cochrane Database Syst Rev. 2006;2:CD001724.
34. Jepsen S, Eberhard J, Herrera D, Needleman I. A systematic review of guided tissue regeneration for periodontal furcation defects. What is the effect of guided tissue regeneration compared with surgical debridement in the treatment of furcation defects? J Clin Periodontol. 2002;29(Suppl 3):103–16; discussion 160–162.
35. Nickles K, Ratka-Krüger P, Neukranz E, Raetzke P, Eickholz P. Open flap debridement and guided tissue regeneration after 10 years in infrabony defects. J Clin Periodontol. 2009;36(11):976–83.
36. Bottino MC, Thomas V, Schmidt G, Vohra YK, Chu TMG, Kowolik MJ, et al. Recent advances in the development of GTR/GBR membranes for periodontal regeneration—a materials perspective. Dent Mater. 2012;28(7):703–21.
37. Carlo Reis EC, Borges APB, Araújo MVF, Mendes VC, Guan L, Davies JE. Periodontal regeneration using a bilayered PLGA/calcium phosphate construct. Biomaterials. 2011;32(35):9244–53.
38. Reis ECC, Borges APB, del Carlo RJ, Oliveira PM, Sepúlveda RV, Fernandes NA, et al. Guided tissue regeneration using rigid absorbable membranes in the dog model of chronic furcation defect. Acta Odontol Scand. 2013;71(3–4):372–80.
39. Kishan AP, Cosgriff-Hernandez EM. Recent advancements in electrospinning design for tissue engineering applications: a review. J Biomed Mater Res A. 2017;105(10):2892–905.
40. Cheng H, Yang X, Che X, Yang M, Zhai G. Biomedical application and controlled drug release of electrospun fibrous materials. Mater Sci Eng C Mater Biol Appl. 2018;90:750–63.
41. Chen CC, Lee SY, Teng NC, Hu HT, Huang PC, Yang JC. In vitro and in vivo studies of hydrophilic electrospun PLA95/β-TCP membranes for guided tissue regeneration (GTR) applications. Nanomaterials (Basel). 2019;9(4):599.
42. Chen D, Zhao M, Mundy GR. Bone morphogenetic proteins. Growth Factors. 2004;22(4):233–41.
43. King GN, Cochran DL. Factors that modulate the effects of bone morphogenetic protein-induced periodontal regeneration: a critical review. J Periodontol. 2002;73(8):925–36.
44. Bowler D, Dym H. Bone morphogenic protein: application in implant dentistry. Dent Clin N Am. 2015;59(2):493–503.
45. Bordea IR, Candrea S, Alexescu GT, Bran S, Băciuţ M, Băciuţ G, et al. Nano-hydroxyapatite use in dentistry: a systematic review. Drug Metab Rev. 2020;52(2):319–32.
46. Crisan L, Crisan B, Soritau O, Baciut M, Biris AR, Baciut G, et al. In vitro study of biocompatibility of a graphene composite with gold nanoparticles and hydroxyapatite on human osteoblasts. J Appl Toxicol. 2015;35(10):1200–10.

47. Zakaria SM, Sharif Zein SH, Othman MR, Yang F, Jansen JA. Nanophase hydroxyapatite as a biomaterial in advanced hard tissue engineering: a review. Tissue Eng Part B Rev. 2013;19(5):431–41.
48. Hogan BL. Bone morphogenetic proteins in development. Curr Opin Genet Dev. 1996;6(4):432–8.
49. Zhang D, Ferguson CM, O'Keefe RJ, Puzas JE, Rosier RN, Reynolds PR. A role for the BMP antagonist chordin in endochondral ossification. J Bone Miner Res. 2002;17(2):293–300.
50. Zhang J, Li L. BMP signaling and stem cell regulation. Dev Biol. 2005;284(1):1–11.
51. Yoshikawa H, Nakase T, Myoui A, Ueda T. Bone morphogenetic proteins in bone tumors. J Orthop Sci. 2004;9(3):334–40.
52. Montesano R. Bone morphogenetic protein-4 abrogates lumen formation by mammary epithelial cells and promotes invasive growth. Biochem Biophys Res Commun. 2007;353(3):817–22.
53. Liu A, Niswander LA. Bone morphogenetic protein signalling and vertebrate nervous system development. Nat Rev Neurosci. 2005;6(12):945–54.
54. Park J, Kim BJ, Hwang JY, Yoon YW, Cho HS, Kim DH, et al. In-vitro mechanical performance study of biodegradable polylactic acid/hydroxyapatite nanocomposites for fixation medical devices. J Nanosci Nanotechnol. 2018;18(2):837–41.
55. Al-Salleeh F, Beatty MW, Reinhardt RA, Petro TM, Crouch L. Human osteogenic protein-1 induces osteogenic differentiation of adipose-derived stem cells harvested from mice. Arch Oral Biol. 2008;53(10):928–36.
56. Schwarz F, Ferrari D, Sager M, Herten M, Hartig B, Becker J. Guided bone regeneration using rhGDF-5- and rhBMP-2-coated natural bone mineral in rat calvarial defects. Clin Oral Implants Res. 2009;20(11):1219–30.
57. Ayoub A, Challa SRR, Abu-Serriah M, McMahon J, Moos K, Creanor S, et al. Use of a composite pedicled muscle flap and rhBMP-7 for mandibular reconstruction. Int J Oral Maxillofac Surg. 2007;36(12):1183–92.
58. Iglesias-Linares A, Yañez-Vico RM, Moreno-Fernandez AM, Mendoza-Mendoza A, Solano-Reina E. Corticotomy-assisted orthodontic enhancement by bone morphogenetic protein-2 administration. J Oral Maxillofac Surg. 2012;70(2):e124–32.
59. Mamalis AA, Markopoulou C, Vrotsos I, Koutsilirieris M. Chemical modification of an implant surface increases osteogenesis and simultaneously reduces osteoclastogenesis: an in vitro study. Clin Oral Implants Res. 2011;22(6):619–26.
60. Togashi AY, Cirano FR, Marques MM, Pustiglioni FE, Lang NP, Lima LAPA. Effect of recombinant human bone morphogenetic protein-7 (rhBMP-7) on the viability, proliferation and differentiation of osteoblast-like cells cultured on a chemically modified titanium surface. Clin Oral Implants Res. 2009;20(5):452–7.
61. Hakki SS, Foster BL, Nagatomo KJ, Bozkurt SB, Hakki EE, Somerman MJ, et al. Bone morphogenetic protein-7 enhances cementoblast function in vitro. J Periodontol. 2010;81(11):1663–74.
62. Yang L, Zhang Y, Dong R, Peng L, Liu X, Wang Y, et al. Effects of adenoviral-mediated coexpression of bone morphogenetic protein-7 and insulin-like growth factor-1 on human periodontal ligament cells. J Periodontal Res. 2010;45(4):532–40.
63. Narukawa M, Suzuki N, Takayama T, Shoji T, Otsuka K, Ito K. Enamel matrix derivative stimulates chondrogenic differentiation of ATDC5 cells. J Periodontal Res. 2007;42(2):131–7.
64. Lin ZM, Qin W, Zhang NH, Xiao L, Ling JQ. Adenovirus-mediated recombinant human bone morphogenetic protein-7 expression promotes differentiation of human dental pulp cells. J Endod. 2007;33(8):930–5.
65. Besinis A, De Peralta T, Tredwin CJ, Handy RD. Review of nanomaterials in dentistry: interactions with the oral microenvironment, clinical applications, hazards, and benefits. ACS Nano. 2015;9(3):2255–89.

66. Abdulkareem EH, Memarzadeh K, Allaker RP, Huang J, Pratten J, Spratt D. Anti-biofilm activity of zinc oxide and hydroxyapatite nanoparticles as dental implant coating materials. J Dent. 2015;43(12):1462–9.
67. Mertens C, Wiens D, Steveling HG, Sander A, Freier K. Maxillary sinus-floor elevation with nanoporous biphasic bone graft material for early implant placement. Clin Implant Dent Relat Res. 2014;16(3):365–73.
68. Ryabenkova Y, Pinnock A, Quadros PA, Goodchild RL, Möbus G, Crawford A, et al. The relationship between particle morphology and rheological properties in injectable nano-hydroxyapatite bone graft substitutes. Mater Sci Eng C Mater Biol Appl. 2017;75:1083–90.
69. Canullo L, Wiel Marin G, Tallarico M, Canciani E, Musto F, Dellavia C. Histological and histomorphometrical evaluation of postextractive sites grafted with Mg-enriched nano-hydroxyapatite: a randomized controlled trial comparing 4 versus 12 months of healing. Clin Implant Dent Relat Res. 2016;18(5):973–83.
70. Baglar S, Erdem U, Dogan M, Turkoz M. Dentinal tubule occluding capability of nano-hydroxyapatite; The in-vitro evaluation. Microsc Res Tech. 2018;81(8):843–54.
71. Vano M, Derchi G, Barone A, Genovesi A, Covani U. Tooth bleaching with hydrogen peroxide and nano-hydroxyapatite: a 9-month follow-up randomized clinical trial. Int J Dent Hyg. 2015;13(4):301–7.
72. Souza BM, Comar LP, Vertuan M, Fernandes Neto C, Buzalaf MAR, Magalhães AC. Effect of an experimental paste with hydroxyapatite nanoparticles and fluoride on dental demineralisation and remineralisation in situ. Caries Res. 2015;49(5):499–507.
73. Hunter DJ, Pike MC, Jonas BL, Kissin E, Krop J, McAlindon T. Phase 1 safety and tolerability study of BMP-7 in symptomatic knee osteoarthritis. BMC Musculoskelet Disord. 2010;11:232.
74. Boyne PJ, Lilly LC, Marx RE, Moy PK, Nevins M, Spagnoli DB, et al. De novo bone induction by recombinant human bone morphogenetic protein-2 (rhBMP-2) in maxillary sinus floor augmentation. J Oral Maxillofac Surg. 2005;63(12):1693–707.
75. James AW, Zara JN, Zhang X, Askarinam A, Goyal R, Chiang M, et al. Perivascular stem cells: a prospectively purified mesenchymal stem cell population for bone tissue engineering. Stem Cells Transl Med. 2012;1(6):510–9.
76. Shahlaie K, Kim KD. Occipitocervical fusion using recombinant human bone morphogenetic protein-2: adverse effects due to tissue swelling and seroma. Spine (Phila Pa 1976). 2008;33(21):2361–6.
77. Smucker JD, Rhee JM, Singh K, Yoon ST, Heller JG. Increased swelling complications associated with off-label usage of rhBMP-2 in the anterior cervical spine. Spine (Phila Pa 1976). 2006;31(24):2813–9.
78. James AW, LaChaud G, Shen J, Asatrian G, Nguyen V, Zhang X, et al. A review of the clinical side effects of bone morphogenetic protein-2. Tissue Eng Part B Rev. 2016;22(4):284–97.
79. Clement JH, Raida M, Sänger J, Bicknell R, Liu J, Naumann A, et al. Bone morphogenetic protein 2 (BMP-2) induces in vitro invasion and in vivo hormone independent growth of breast carcinoma cells. Int J Oncol. 2005;27(2):401–7.
80. Kim M-joo, Kim K-mahn, Kim J, Kim K-nam. BMP-2 promotes oral squamous carcinoma cell invasion by inducing CCL5 release. PLoS One. 2014;9(10):e108170.
81. Lode A, Bernhardt A, Kroonen K, Springer M, Briest A, Gelinsky M. Development of a mechanically stable support for the osteoinductive biomaterial COLLOSS E. J Tissue Eng Regen Med. 2009;3(2):149–52.
82. Sun TW, Yu WL, Zhu YJ, Chen F, Zhang YG, Jiang YY, et al. Porous nanocomposite comprising ultralong hydroxyapatite nanowires decorated with zinc-containing nanoparticles and chitosan: synthesis and application in bone defect repair. Chemistry. 2018;24(35):8809–21.
83. Heimbach B, Tonyali B, Zhang D, Wei M. High performance resorbable composites for load-bearing bone fixation devices. J Mech Behav Biomed Mater. 2018;81:1–9.

84. Landi E, Logroscino G, Proietti L, Tampieri A, Sandri M, Sprio S. Biomimetic Mg-substituted hydroxyapatite: from synthesis to in vivo behaviour. J Mater Sci Mater Med. 2008;19(1):239–47.
85. Parker S. Surgical lasers and hard dental tissue. Br Dent J. 2007;202(8):445–54.
86. Birardi V, Bossi L, Dinoi C. Use of the Nd:YAG laser in the treatment of early childhood caries. Eur J Paediatr Dent. 2004;5(2):98–101.
87. Harris DM, White JM, Goodis H, Arcoria CJ, Simon J, Carpenter WM, et al. Selective ablation of surface enamel caries with a pulsed Nd:YAG dental laser. Lasers Surg Med. 2002;30(5):342–50.
88. Kwon YH, Kwon OW, Kim HI, Kim KH. Nd:YAG laser ablation and acid resistance of enamel. Dent Mater J. 2003;22(3):404–11.
89. Tsai CL, Lin YT, Huang ST, Chang HW. In vitro acid resistance of CO_2 and Nd-YAG laser-treated human tooth enamel. Caries Res. 2002;36(6):423–9.
90. Malmström HS, McCormack SM, Fried D, Featherstone JD. Effect of CO_2 laser on pulpal temperature and surface morphology: an in vitro study. J Dent. 2001;29(8):521–9.
91. Watanabe I, Lopes RA, Brugnera A, Katayama AY, Gardini AE. Effect of CO_2 laser on class V cavities of human molar teeth under a scanning electron microscope. Braz Dent J. 1996;7(1):27–31.
92. Goodis HE, Fried D, Gansky S, Rechmann P, Featherstone JDB. Pulpal safety of 9.6 microm TEA CO_2 laser used for caries prevention. Lasers Surg Med. 2004;35(2):104–10.
93. Tepper SA, Zehnder M, Pajarola GF, Schmidlin PR. Increased fluoride uptake and acid resistance by CO_2 laser-irradiation through topically applied fluoride on human enamel in vitro. J Dent. 2004;32(8):635–41.
94. Glockner K, Rumpler J, Ebeleseder K, Stadtler P. Intrapulpal temperature during preparation with the Er:YAG laser compared to the conventional burr: an in vitro study. J Clin Laser Med Surg. 1998;16(3):153–7.
95. Lee BS, Lin CP, Hung YL, Lan WH. Structural changes of Er:YAG laser-irradiated human dentin. Photomed Laser Surg. 2004;22(4):330–4.
96. Apel C, Meister J, Ioana RS, Franzen R, Hering P, Gutknecht N. The ablation threshold of Er:YAG and Er:YSGG laser radiation in dental enamel. Lasers Med Sci. 2002;17(4):246–52.
97. Harashima T, Kinoshita JI, Kimura Y, Brugnera A, Zanin F, Pecora JD, et al. Morphological comparative study on ablation of dental hard tissues at cavity preparation by Er:YAG and Er,Cr:YSGG lasers. Photomed Laser Surg. 2005;23(1):52–5.
98. Boyde A. Enamel structure and cavity margins. Oper Dent. 1976;1(1):13–28.
99. Chinelatti MA, Ramos RP, Chimello DT, Borsatto MC, Pécora JD, Palma-Dibb RG. Influence of the use of Er:YAG laser for cavity preparation and surface treatment in microleakage of resin-modified glass ionomer restorations. Oper Dent. 2004;29(4):430–6.
100. Corona SAM, Borsatto MC, Pecora JD, De SA Rocha R a. S, Ramos TS, Palma-Dibb RG. Assessing microleakage of different class V restorations after Er:YAG laser and bur preparation. J Oral Rehabil. 2003;30(10):1008–14.
101. Kohara EK, Hossain M, Kimura Y, Matsumoto K, Inoue M, Sasa R. Morphological and microleakage studies of the cavities prepared by Er:YAG laser irradiation in primary teeth. J Clin Laser Med Surg. 2002;20(3):141–7.
102. Niu W, Eto JN, Kimura Y, Takeda FH, Matsumoto K. A study on microleakage after resin filling of Class V cavities prepared by Er:YAG laser. J Clin Laser Med Surg. 1998;16(4):227–31.
103. Arora R. Influence of pain-free dentistry and convenience of dental office on the choice of a dental practitioner: an experimental investigation. Health Mark Q. 1999;16(3):43–54.
104. Blechman AM. Pain-free and mobility-free orthodontics? Am J Orthod Dentofacial Orthop. 1998;113(4):379–83.
105. Orchardson R, Whitters CJ. Effect of HeNe and pulsed Nd:YAG laser irradiation on intradental nerve responses to mechanical stimulation of dentine. Lasers Surg Med. 2000;26(3):241–9.
106. Hubbard LG. Smile improvement: the laser way. Dent Today. 2000;19(2):94–5.

107. Kato J, Moriya K, Jayawardena JA, Wijeyeweera RL. Clinical application of Er:YAG laser for cavity preparation in children. J Clin Laser Med Surg. 2003;21(3):151–5.
108. Keller U, Hibst R, Geurtsen W, Schilke R, Heidemann D, Klaiber B, et al. Erbium:YAG laser application in caries therapy. Evaluation of patient perception and acceptance. J Dent. 1998;26(8):649–56.
109. Matsumoto K, Hossain M, Hossain MMI, Kawano H, Kimura Y. Clinical assessment of Er,Cr:YSGG laser application for cavity preparation. J Clin Laser Med Surg. 2002;20(1):17–21.
110. Mercer CE, Anderson P, Davis GR. Sequential 3D X-ray microtomographic measurement of enamel and dentine ablation by an Er:YAG laser. Br Dent J. 2003;194(2):99–104; discussion 89.
111. Hibst R. Mechanical effects of erbium:YAG laser bone ablation. Lasers Surg Med. 1992;12(2):125–30.
112. Peavy GM, Reinisch L, Payne JT, Venugopalan V. Comparison of cortical bone ablations by using infrared laser wavelengths 2.9 to 9.2 microm. Lasers Surg Med. 1999;25(5):421–34.
113. Thomsen S. Pathologic analysis of photothermal and photomechanical effects of laser-tissue interactions. Photochem Photobiol. 1991;53(6):825–35.
114. Wang X, Zhang C, Matsumoto K. In vivo study of the healing processes that occur in the jaws of rabbits following perforation by an Er,Cr:YSGG laser. Lasers Med Sci. 2005;20(1):21–7.
115. Pourzarandian A, Watanabe H, Aoki A, Ichinose S, Sasaki KM, Nitta H, et al. Histological and TEM examination of early stages of bone healing after Er:YAG laser irradiation. Photomed Laser Surg. 2004;22(4):342–50.
116. Wang X, Ishizaki NT, Suzuki N, Kimura Y, Matsumoto K. Morphological changes of bovine mandibular bone irradiated by Er,Cr:YSGG laser: an in vitro study. J Clin Laser Med Surg. 2002;20(5):245–50.
117. Ramalho KM, de Freitas PM, Correa-Aranha AC, Bello-Silva MS, Lopes RM da G, Eduardo C de P. Lasers in esthetic dentistry: soft tissue photobiomodulation, hard tissue decontamination, and ceramics conditioning. Case Rep Dent. 2014;2014:927429.
118. Lesniewski A, Estrin N, Romanos GE. Comparing the use of diode lasers to light-emitting diode phototherapy in oral soft and hard tissue procedures: a literature review. Photobiomodul Photomed Laser Surg. 2022;40(8):522–31.
119. Braun A, Hagelauer FJP, Wenzler J, Heimer M, Frankenberger R, Stein S. Microcrack analysis of dental hard tissue after root canal irradiation with a 970-nm diode laser. Photomed Laser Surg. 2018;36(11):621–8.
120. Xue VW, Zhao IS, Yin IX, Niu JY, Lo ECM, Chu CH. Effects of 9,300 nm carbon dioxide laser on dental hard tissue: a concise review. Clin Cosmet Investig Dent. 2021;13:155–61.
121. Featherstone JDB, Fried D, Bitten ER. Lasers in dentistry III. In: Mechanism of laser-induced solubility reduction of dental enamel, vol. 2973. San Jose: SPIE; 1997. p. 112–6.
122. Takahashi K, Kimura Y, Matsumoto K. Morphological and atomic analytical changes after CO_2 laser irradiation emitted at 9.3 microns on human dental hard tissues. J Clin Laser Med Surg. 1998;16(3):167–73.
123. Kimura Y, Takahashi-Sakai K, Wilder-Smith P, Krasieva TB, Liaw LH, Matsumoto K. Morphological study of the effects of CO_2 laser emitted at 9.3 microm on human dentin. J Clin Laser Med Surg. 2000;18(4):197–202.
124. Fried D, Zuerlein MJ, Le CQ, Featherstone JDB. Thermal and chemical modification of dentin by 9-11-microm CO_2 laser pulses of 5-100-micros duration. Lasers Surg Med. 2002;31(4):275–82.
125. Poli R, Parker S. Achieving dental analgesia with the erbium chromium yttrium scandium gallium garnet laser (2780 nm): a protocol for painless conservative treatment. Photomed Laser Surg. 2015;33(7):364–71.
126. Parker S, Cronshaw M, Anagnostaki E, Mylona V, Lynch E, Grootveld M. Current concepts of laser-oral tissue interaction. Dent J (Basel). 2020;8(3):61.

127. Dederich DN. Laser/tissue interaction: what happens to laser light when it strikes tissue? J Am Dent Assoc. 1993;124(2):57–61.
128. Mah C. The evolution of implants over the last fifty years. Aust Prosthodont J. 1990;4:47–52.
129. Triplett RG, Frohberg U, Sykaras N, Woody RD. Implant materials, design, and surface topographies: their influence on osseointegration of dental implants. J Long-Term Eff Med Implants. 2003;13(6):485–501.
130. Brånemark PI, Adell R, Albrektsson T, Lekholm U, Lundkvist S, Rockler B. Osseointegrated titanium fixtures in the treatment of edentulousness. Biomaterials. 1983;4(1):25–8.
131. Lemons JE. Dental implant biomaterials. J Am Dent Assoc. 1990;121(6):716–9.
132. Albrektsson T. Direct bone anchorage of dental implants. J Prosthet Dent. 1983;50(2):255–61.
133. Wu Y, Zitelli JP, TenHuisen KS, Yu X, Libera MR. Differential response of staphylococci and osteoblasts to varying titanium surface roughness. Biomaterials. 2011;32(4):951–60.
134. Jazayeri HE, Tahriri M, Razavi M, Khoshroo K, Fahimipour F, Dashtimoghadam E, et al. A current overview of materials and strategies for potential use in maxillofacial tissue regeneration. Mater Sci Eng C Mater Biol Appl. 2017;70(Pt 1):913–29.
135. Chiapasco M, Casentini P, Zaniboni M. Bone augmentation procedures in implant dentistry. Int J Oral Maxillofac Implants. 2009;24(Suppl):237–59.
136. Chappuis V, Rahman L, Buser R, Janner SFM, Belser UC, Buser D. Effectiveness of contour augmentation with guided bone regeneration: 10-year results. J Dent Res. 2018;97(3):266–74.
137. Caballé-Serrano J, Munar-Frau A, Delgado L, Pérez R, Hernández-Alfaro F. Physicochemical characterization of barrier membranes for bone regeneration. J Mech Behav Biomed Mater. 2019;97:13–20.
138. Gruber R, Stadlinger B, Terheyden H. Cell-to-cell communication in guided bone regeneration: molecular and cellular mechanisms. Clin Oral Implants Res. 2017;28(9):1139–46.
139. Schenk RK, Buser D. Osseointegration: a reality. Periodontol 2000. 1998;17:22–35.
140. Dahlin C, Linde A, Gottlow J, Nyman S. Healing of bone defects by guided tissue regeneration. Plast Reconstr Surg. 1988;81(5):672–6.
141. Marcelo CG, Filié Haddad M, Gennari Filho H, Marcelo Ribeiro Villa L, Dos Santos DM, Aldiéris AP. Dental implant fractures—aetiology, treatment and case report. J Clin Diagn Res. 2014;8(3):300–4.
142. Sakka S, Baroudi K, Nassani MZ. Factors associated with early and late failure of dental implants. J Investig Clin Dent. 2012;3(4):258–61.
143. Pauletto N, Lahiffe BJ, Walton JN. Complications associated with excess cement around crowns on osseointegrated implants: a clinical report. Int J Oral Maxillofac Implants. 1999;14(6):865–8.
144. Misch K, Wang HL. Implant surgery complications: etiology and treatment. Implant Dent. 2008;17(2):159–68.
145. Goutam M, Giriyapura C, Mishra SK, Gupta S. Titanium allergy: a literature review. Indian J Dermatol. 2014;59(6):630.
146. Lee HW, Lin WS, Morton D. A retrospective study of complications associated with 100 consecutive maxillary sinus augmentations via the lateral window approach. Int J Oral Maxillofac Implants. 2013;28(3):860–8.
147. Donos N, Mardas N, Chadha V. Clinical outcomes of implants following lateral bone augmentation: systematic assessment of available options (barrier membranes, bone grafts, split osteotomy). J Clin Periodontol. 2008;35(8 Suppl):173–202.
148. Cerea M, Dolcini GA. Custom-made direct metal laser sintering titanium subperiosteal implants: a retrospective clinical study on 70 patients. Biomed Res Int. 2018;2018:5420391.
149. Ângelo DF, Vieira Ferreira JR. The role of custom-made subperiosteal implants for rehabilitation of atrophic jaws—a case report. Ann Maxillofac Surg. 2020;10(2):507–11.
150. Loperfido C, Mesquida J, Lozada JL. Severe mandibular atrophy treated with a subperiosteal implant and simultaneous graft with rhBMP-2 and mineralized allograft: a case report. J Oral Implantol. 2014;40(6):707–13.

151. Schou S, Pallesen L, Hjørting-Hansen E, Pedersen CS, Fibaek B. A 41-year history of a mandibular subperiosteal implant. Clin Oral Implants Res. 2000;11(2):171–8.
152. Obwegeser HL. Experiences with subperiosteal implants. Oral Surg Oral Med Oral Pathol. 1959;12(7):777–86.
153. Nariai Y, Kanno T, Sekine J. Histopathological features of secondary squamous cell carcinoma around a dental implant in the mandible after chemoradiotherapy: a case report with a clinicopathological review. J Oral Maxillofac Surg. 2016;74(5):982–90.
154. Gupta VK, Mehrotra D, Malhotra S, Kumar S, Agarwal GG, Pal US. An epidemiological study of temporomandibular joint ankylosis. Natl J Maxillofac Surg. 2012;3(1):25–30.
155. Bhatt K, Roychoudhury A, Balakrishnan P. Temporomandibular joint ankylosis: is hypercoagulable state of blood a predisposing factor? Med Hypotheses. 2013;81(4):561–3.
156. De Roo N, Van Doorne L, Troch A, Vermeersch H, Brusselaers N. Quantifying the outcome of surgical treatment of temporomandibular joint ankylosis: a systematic review and meta-analysis. J Craniomaxillofac Surg. 2016;44(1):6–15.
157. Bhatt K, Roychoudhury A, Bhutia O, Pandey RM. Functional outcomes of gap and interposition arthroplasty in the treatment of temporomandibular joint ankylosis. J Oral Maxillofac Surg. 2014;72(12):2434–9.
158. Kumar D, Rajan G, Raman U, Varghese J. Autogenous reconstructive modalities of TMJ ankylosis-a retrospective analysis of 45 cases. J Maxillofac Oral Surg. 2014;13(4):359–65.
159. McBride KL. Total reconstruction of the temporomandibular joint with the Vitek-Kent prostheses. TMJ Update. 1989;7(1):15–8.
160. Kent JN, Homsy CA, Hinds EC. Proplast in dental facial reconstruction. Oral Surg Oral Med Oral Pathol. 1975;39(3):347–55.
161. Kiehn CL, DesPrez JD, Converse CF. Total prosthetic replacement of the temporomandibular joint. Ann Plast Surg. 1979;2(1):5–15.
162. Gardiner JC, Thomas BJ. Total hip replacement: the current perspective after 37 years. Surg Technol Int. 1996;5:365–9.
163. Kent JN, Block MS, Homsy CA, Prewitt JM, Reid R. Experience with a polymer glenoid fossa prosthesis for partial or total temporomandibular joint reconstruction. J Oral Maxillofac Surg. 1986;44(7):520–33.
164. Chase DC, Hudson JW, Gerard DA, Russell R, Chambers K, Curry JR, et al. The Christensen prosthesis. A retrospective clinical study. Oral Surg Oral Med Oral Pathol Oral Radiol Endod. 1995;80(3):273–8.
165. Riegel R, Sweeney K, Inverso G, Quinn PD, Granquist EJ. Microbiology alloplastic total joint infections: a 20-year retrospective study. J Oral Maxillofac Surg. 2018;76(2):288–93.
166. Wolford LM, Pitta MC, Reiche-Fischel O, Franco PF. TMJ concepts/techmedica custom-made TMJ total joint prosthesis: 5-year follow-up study. Int J Oral Maxillofac Surg. 2003;32(3):268–74.
167. Wolford L, Movahed R, Teschke M, Fimmers R, Havard D, Schneiderman E. Temporomandibular joint ankylosis can be successfully treated with TMJ concepts patient-fitted total joint prosthesis and autogenous fat grafts. J Oral Maxillofac Surg. 2016;74(6):1215–27.
168. Leandro LFL, Ono HY, Loureiro CC de S, Marinho K, Guevara HAG. A ten-year experience and follow-up of three hundred patients fitted with the Biomet/Lorenz microfixation TMJ replacement system. Int J Oral Maxillofac Surg. 2013;42(8):1007–13.
169. O'Connor RC, Saleem S, Sidebottom AJ. Prospective outcome analysis of total replacement of the temporomandibular joint with the TMJ concepts system in patients with inflammatory arthritic diseases. Br J Oral Maxillofac Surg. 2016;54(6):604–9.

Chapter 6
Significant Risk Medical Devices – Ear, Nose and Throat

K. S. B. S. Krishna Sasanka, T. Y. Sree Sudha, Megha Chandran, Ruuzeno Kuotsu, and Saurabh Varshney

6.1 General Introduction

Biomaterials have been a part and parcel of modern medicine for decades now: being used as implants (external/internal; temporary/permanent), drug-delivery systems and medical device parts. Their increasing usage has also been met with an inevitable need to improve the quality and design. The rapid rise in biomaterials usage has unfortunately come at a cost of considerable increase in risk of adverse effects to implanted biomaterials including immune reactions, that is, allergies, chronic inflammation, susceptibility to infection, tissue damage and functional loss. Furthermore, the immune reactions to each biomaterial has been found to be of highly personalised nature and need to be taken into account while assessing and promoting new biomaterials.

Over time, with improved biomedical technologies, rapid progress has been made in the manufacture of newer, safer materials increasing their biocompatibility, thus reducing the adverse effects and improving quality of life of user. The new 3D printing technology has also been found beneficial in customizing shapes and improving technical accuracy of implants. Another key aspect is the selection of the

K. S. B. S. Krishna Sasanka (✉)
Department of ENT, AIIMS Deoghar, Deoghar, Jharkhand, India

T. Y. Sree Sudha
Department of Pharmacology, AIIMS Deoghar, Deoghar, Jharkhand, India

M. Chandran
Department of ENT, AIIMS Bhopal, Bhopal, Madhya Pradesh, India

R. Kuotsu
Department of ENT, AIIMS Kalyani, Kalyani, West Bengal, India

S. Varshney
AIIMS Deoghar, Deoghar, Jharkhand, India

P. S. Timiri Shanmugam et al. (eds.), *Significant and Nonsignificant Risk Medical Devices*, https://doi.org/10.1007/978-3-031-52838-5_6

matrix material for the implant; for better results, a huge variety of matrix systems are needed for surface functionalisation in order to release active ingredients. Therefore, it is beneficial to utilise biodegradable and bioabsorbable matrix materials which degenerate completely after releasing active ingredients, thus avoiding any additional surgical procedure for implant removal. In various clinical aspects, innovative biomaterials have been developed not only to replace the demolished tissue and regain the original state of biological functions but also to counterbalance damaged sensory or neuronal cells by electrical stimulation.

Biomaterials used in otorhinolaryngology—head and neck surgery—range from a variety of applications including prostheses (ear implants/voice prosthesis/facial implants), air way devices, nasal packing materials, drug-delivery systems, bone substitution materials to treatment of post-tonsillectomy haemorrhage [1]. The different materials include a variety of ceramics, metals, alloys, polymers and polymer-based composites [2]. The biocompatibility of any implant is based on its technical or surgical requirements, material properties, biological interactions with body tissues and host immune mechanisms. The goal is to maintain optimal biofunctionality over extended time periods. So far, no 100% ideal implant have been discovered while attempts have been continuously made to get closer to it.

Background

Biomaterial is any single or combination of substances either natural or synthetic that is used in whole or as a part in human body which cures, expands or replaces any tissue, organ or function of a part of the body [3]. Biomaterials were founded a few centuries after the Common Era when foreign materials were used as implants in the body. Novel implants based on innovative technologies and improved biomaterials have strongly influenced otolaryngology in recent years. With regard to suitability of biomaterials, major demands include technical functionality, mechanical properties, good biocompatibility over extended periods and high physiological stability [4]. An ideal biomaterial (yet to be discovered) fulfils all criteria and has the minimum adverse effects.

Adverse reactions to biomaterials may be varied and dependent mainly on material characteristics, antimicrobial properties and host tissue response. Implants made of any biomaterial tend to react in an analogous way with the living tissue. Chemical interaction between body niche and implant surface results in generation of entities that influence surrounding tissue morphology and function causing small to major complications or even death. This process is controlled by chemical and thermodynamics. So, selection of material implant should be after consideration of these aspects in relation to material characteristics so as to have minimum biomaterial effect on the body milieu. Polymeric substances like silicon rubber, polytetrafluoroethylene (PTFE), some high molecular weight polyethylene, some carbon forms and ceramics tend to be considerable in the above aspect.

Host tissue immune response following biomaterial implantation is caused by foreign body giant cells and macrophages by their interaction on the synthetic surfaces and mainly consists of inflammatory and wound-healing responses. This is based mainly on the characteristics of the biomaterial surface: physical, chemical and morphological. These influence the biocompatibility of the medical device and may be short term or long term [2]. Kerr [5] and Brown et al. [6] noted the presence of multinucleated giant cells in and around various implant surfaces. Persistence of giant cells in the absence of chemical incompatibility was found to correlate relative movement of an implant well linked by ingrowth of fibrous tissue.

Another risk following incorporation of a biomaterial in the body is the colonisation of the material by opportunistic/pathogenic microorganisms and biofilm formation on the implant surface. Bacterial biofilms have notoriously higher resistance and are extremely difficult to eradicate, being one of the major causes of implant failure and post-operative infections. Fungal biofilm formation has been found to be a major cause of post-operative infection following tympanostomy tube insertion. Hence, increasing the anti-microbial properties via biomaterial modifications is an important aspect of improved implant survival. Various methods for increasing anti-microbial properties include antibiotic coating, anti-microbial coating via nanoparticles, anti-fouling, surface charge modification and so on. Sometimes polymers are used as controlled antibiotic release systems, which results in effective microbicidal activity [2].

6.2 Otology

Biomaterial usage in otology maybe mainly categorised in four fronts: cochlear implants, middle ear implants, tympanostomy tubes and bone anchored hearing aids. In the early times, repeated failures following initial short-term success had halted research process of biomaterials in otology resulting in a slower material development. But now with the promotion of intact canal wall techniques and increased onus on middle ear reconstruction biomaterial study has further gained importance as it is in other surgical disciplines.

6.2.1 Tympanostomy Tubes

Introduction

Eustachian tube block due to any infection, mass, allergy or adenoid enlargement lead to fluid collection in middle ear causing impaired vibration of tympanic membrane and reduced transmission of sound. If not treated in correct time, it leads to complications such as eardrum perforation, cholesteatoma, tympanic membrane atrophy, granulation tissue, myringosclerosis, behavioural changes and sensorineural hearing loss [7]. Surgical treatment is recommended for such a condition to open the eardrum, re-establish ventilation of the middle ear and maintain normal hearing capabilities. Myringotomy or tympanotomy is the surgical procedure to remove fluid in the middle ear. A small surgical slit is made in TM for fluid drainage. In severe cases where entire fluid cannot be drained just by the incision, a small tube is

Fig. 6.1 Grommet

inserted through the slit for drainage called the Tympanostomy tube or Grommet or Ventilation tube (Fig. 6.1).

Mechanism and Indications

It equalises middle ear pressure, provides ventilation to the middle ear and prevents recurrent infections. Indicated in persistent serous otitis media, recurrent acute otitis media, complications of AOM or Eustachian tube dysfunction.

Parts

Body, medial flange, lateral flange.

Materials Used [8]

The most prevalent materials used for tympanostomy tube are plastics and metals (stainless steel, gold and titanium). Others include polymers (PTFE, Teflon, Polylactides, Polydiaxanone) and silicone. Silicone is popular as it is stretchable and soft. Fluoroplastics have strong heat resistance and chemical inertness.

(a) *Fluoroplastic*: Non-sticky surfaces: reduced clogging/adhesions to tube. Rigid material; inexpensive; facilitates easy tube insertions. Good biocompatibility
(b) *Stainless steel*: Rigid; easy insertion; biocompatible
(c) *Silicone*: Soft material; easy manipulation; proven biocompatibility. Compressed easily to aid insertion while still retains its shape
(d) *Titanium*: Less occlusion with body fluids due to micro-polished lumens and flanges. Half the weight of stainless steel; easy insertion; expensive
(e) *Silver coated tubes*: Silver oxide coated on silicone or fluoroplastic material. Silver has antimicrobial properties, reduce the incidence of postoperative otorrhea compared to other materials

Various types of tympanostomy tubes are available. All ventilate the ear canal and help drain the fluid behind the ear drum.

Duration

Four to six months in the eardrum before they eject out. The ideal tube would not come out prematurely and can be easily inserted and removed, causing fewer complications [6]. The duration of retention of ventilation tube is dependent on grommet structure and tympanic epithelial migration rate. Collared tubes ("Shoehorn") or "T" tubes are generally retained longer compared to others. The external tympanic epithelium peripheral migration displaces the tube towards periphery leading to final extrusion [8].

Complications

Tympanostomy tube (TT) occlusion is relatively common and the tube becomes non-functional in 7–37% of patients [6]. Every patient has otorrhea once within the first 4 weeks of operation or later [9]. It is important to treat and prevent both otorrhea and contamination and biofilm formation.

Displacement, encapsulation by host tissue, extrusion and infection with bacterial biofilm formation. Using material with least host-implant inflammatory response is the most important aspect.

Another risk factor for complications is the bacterial colonisation and biofilm formation. Antibiotic incorporation and silver oxide impregnation improve biocompatibility. Joe et al. designed a novel tympanostomy stent (TS) which had minimised and smooth surface area to prevent biofilm adherence while preserving function. It was coated with titanium oxide (TiO2).

6.2.2 Cochlear Implants

Introduction

Cochlear implant (CI) has been considered a gold standard for the treatment of congenitally deaf children and post-lingually deafened children and adults. They are one of the greatest successes among neurobionic prostheses. Bilateral cochlear implantation has become the standard therapy for congenitally deaf children, especially in prelingual candidates.

Mechanism

CI resembles the normal physiological hearing. The external sound is collected by the microphone and sent to the processor which converts it into electrical signals and transmit it via transmitter coil and implanted internal component to the cochlear electrodes. The electrodes stimulate the ganglion cells of scala tympani and the impulses through the auditory nerve reach the auditory cortex.

The delicacy and the importance of the procedure and its possible outcome places extensive demands on these implants especially regarding the biocompatibility and durability of the prostheses' surface components. Other parts face many mechanical challenges like electrode array flexibility, breakage resistance and mechanical strength of implant casket. Its positional vicinity to the middle ear mucosa and inner ear fluids makes microbial colonisation and infection from

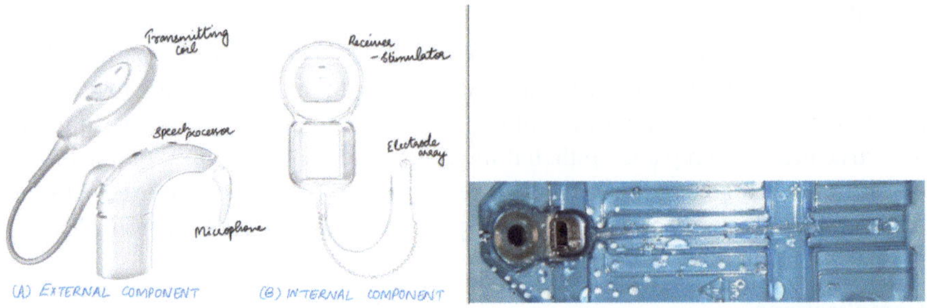

Fig. 6.2 Cochlear implant

contamination a major risk. The current CI systems are based mainly on cardiac pacemaker studies.

Parts: [8] (Fig. 6.2)

(a) *External component: Microphone, speech processor and transmitter.*
Present outside body
(b) *Internal component: Receiver stimulator and electrode array.*
Surgically anchored on the skull.

Models: Clarion; MED—EL combi 40+; Nucleus 24.

Material Demands

The following are requirements of a CI (first described in 1992 by E. Lehnhardt (modified) [10]):

- Biocompatible material with long term stability
- No additional damage during insertion of the electrode array
- Minimally invasive surgical technique
- Efficient, harmless and sustained stimulation of auditory nerve
- Minimal risk of infection by the implant or route of access to cochlear spaces

Other requirements include mechanical stability and transfer of charge to auditory nerve.

Surgical Technique

Surgery involves cortical mastoidectomy followed by posterior tympanotomy and drilling of a cochleostomy. The electrode array is then inserted into the cochlea. A groove is fashioned on the skull bone which it acts as implant bed and avoid excessive mobility of device. A tunnel is made connecting implant bed and mastoid allowing protection of electrodes at point of exit and serving as an additional support. Use of auxiliary materials like osteosynthetic materials, bone cement, dacron sutures and clips, which may have decreased biocompatibility and increased local effects, is thus avoided (e.g. cholesteatoma formation following use of dacron sutures).

The electrode array part requires flexibility and long-term mechanical stability. Silicone carrier material forms the modern CI electrodes with platinum contact embedding and input wires made of platinum-iridium 90/10 with insulating Teflon coating [11].

Materials
The most suited CI materials include titanium and ceramics. Titanium is favoured over ceramics for lodging the implant's electronic components despite ceramics having a better resistance to breakage. This is because ceramics have been found to be associated with greater chances of implant failure due to a lack of tightness and poorer mechanical resistance. A layer of silicone is coated over implant casing.

Silicone

It is used as encapsulation of implant parts and is known for its good biostability, flexibility and compatibility. It has been used in various prosthetic preparations across different branches. An important cause of implant extrusion is silicone allergy.

In a study for Nucleus 22 devices, foreign body reaction and contact dermatitis were found to occur due to one of the silicon components [12, 13]. Positive allergy test results were found for silicone LSR-30 of Nucleus 24 contour device, while silicone LSR-70 of Advanced Bionics device showed negative results. The cochlear device use was expanded and reintegrated with the Advanced Bionics device.

A rare complication is foreign body reaction (FBR) which includes macrophages and giant cells. In a few studies, FBR, contact dermatitis, silicone allergy and *Enterobacter cloacae* growth with CIs have been reported [13]. FBR was observed with the silicone material of a Nucleus 22 device [12]. Allergic reaction to silicone is a common occurrence, and hence, allergy tests (individual compatibility tests) are usually incorporated in evaluation.

Platinum

Precious metals like platinum have low chemical reactivity making them highly corrosion resistant and are preferred contact materials in electrical stimulation. However, the undesired effects like partial dissolution or formation of unstable surface films on electrical stimulation are a major drawback. These depend upon the density of the charge that is transmitted during electrical stimulation and on the polarity of the stimulus [14, 15]. Platinum compared to iridium is softer, is well tolerated by human body and is easier to work with. Platinum is the best electrode material presently available.

Titanium

Titanium is inert and solid. It is particularly suitable in cases where rigidity, low weight and high corrosion resistance are essential. The use of titanium as receiver/stimulator casing started long back similar to cardiac pacemaker technology. While using titanium as casing material, receiver coil must be placed outside of the casing.

Ceramics

They are non-metallic composites consisting of a matrix of a hybrid base material involving different substances. It is more prone to breakage than titanium as casing material.

New Biomaterials

Teflon

PTFE (polytetrafluoroethylene) or Teflon is a highly stable (thermally and chemically) hydrophobic polymer used in cochlear implants. The placement of electrode array close to modiolus within the cochlea is enabled by a movable Teflon strip. It has been experimentally used for stoma sealing also.

Electrically conducting polymers

They are synthetic materials that possess properties of electrical conduction, semiconduction or isolation. The term "polymer electronics" was coined for the use of these materials for electronic applications.

Nanoparticles and cochlear implants

Development of nanoscale level drug carriers has led to studies on cochlear implant-based drug release for local therapy of inner ear using nanoparticles. Similar to non-viral vectors of biogenic agents (e.g., neurotrophic factors, genes, and steroid sequences), nanoparticles are protected from body's metabolic effects and transported specifically to the target location and time-released.

Device failure could be due to various reasons like an internal component short circuit or fluid leakage, requiring a reimplantation. The major risk associated with the implant material would be a localised tissue reaction or formation of reparative granuloma which can even result in implant rejection. Rarely, a skin flap necrosis with infection may commence following the surgery leading to temporary or permanent removal of the implant.

6.2.3 Auditory Brainstem Implant (ABI)

Introduction

They are similar to CI but transmit impulses directly to the brainstem bypassing the non-functioning or absent the inner ear or auditory nerve.

Indications

- Neurofibromatosis type 2 (NF-2)
- Bilateral vestibular schwannomas
- Cochlear nerve congenital aplasia
- Patients with damaged VIII nerves
- Patients ineligible for conventional cochlear implants (post-meningitis patients with ossified cochleas)

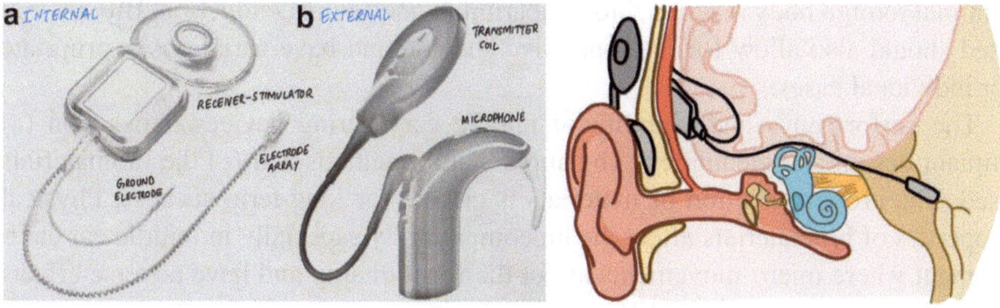

Fig. 6.3 Auditory brainstem implant

Parts: (Fig. 6.3)

The external system: Microphone, battery source, speech processor, transmitter coil, and magnet

The internal system: Receiver-stimulator with magnet, a ground electrode and an electrode array

The receiver stimulator is surgically placed. (For patients needing magnetic resonance imaging (MRI), magnet is removed and replaced with non-magnetic spacer (metallic) during implant placement).

Implant activation occurs approximately 6 weeks after implantation to allow for healing.

Single use.

Materials

Biomaterials used in ABI are similar to that of cochlear implants. Here, the electrodes are embedded in silastic paddle which is flat and rigid lying along the curved surface of the CN in the brainstem. This is in contrast to the thin and flexible CI electrodes following the tonotopic axis of cochlea [16]. The feature of flexible polymers conforming to brainstem surface is the current area of research.

Complications

- CSF leak
- Implant migration
- Non-auditory stimuli in both adults and children

Less common complications are cerebellar contusion, permanent facial palsy, meningitis, damage to the lower cranial nerves, hydrocephalus, pseudomeningocele, headache and tinnitus.

6.2.4 Middle Ear Implants

Introduction

Middle ear implantation site is different due to its aeration and potential microbial colonisation. Hence, any implant in the middle ear should have biostability, minimal foreign body reaction, biocompatibility and sound conduction. Biomaterial used should also allow for intraoperative shaping and have variations appropriate for individual cases.

The performance and viability of middle ear hearing devices depend on (a) implant design and cleanliness, (b) implantation methods and (c) the biomaterials used. The proper selection of materials is critical for long-term success. Physical properties of biomaterials affect the biocompatibility especially in middle ear environment where micro-movements affect the functionality and have adverse effects on implant-tissue interfaces [17].

In middle ear, micro-motion between tissue and implant adversely affects attachment of tissue to the surface. Porosity is another physical property important in biocompatibility, because (soft) porous materials match the properties of host tissues better. Stiffer porous or nonporous implants are generally encapsulated by relatively avascular and acellular tissue, and this tissue encapsulation is a prelude to rejection and possible extrusion of an implant [17]. Titanium, which is relatively stiff, is commonly used in implantable hearing aids.

While performing ossiculoplasty, ossicular chain reconstruction (OCR) protheses must be of biocompatible material like hydroxyapatite (HA).

6.2.4.1 Stapes Prosthesis

Introduction

The first stapes prosthesis was made by Shea and Treace as an alternative for fenestration after stapedectomy for otosclerosis using piston (Teflon) and vein graft over the round window [18]. A secure continuity between the mobile incus and the perilymph in oval window is the core concept of a stapes prosthesis.

Traditionally stapes prosthesis included a Teflon piston (Fig. 6.4) with a wire loop made of stainless steel, titanium or a platinum. The wire loop needs to be manually crimped onto the incus and this process is technically difficult.

Technique

The surgical steps for stapedectomy include exposure of incudo-stapedial joint (IS joint), stapes foot plate, stapedius tendon and facial nerve. After confirming fixation of stapes and ossicular mobility, the IS joint is dislocated and the stapedius muscle is cut. The stapes superstructure is then down-fractured. As per surgeon preference and prosthesis being used, total/partial stapedectomy, footplate drill out, or a stapedotomy may be performed. Instrument usage or suctioning around and further beyond footplate should be minimised to decrease the risk of sensorineural hearing loss and perilymph leak. A vein graft or fat obtained from lobule may be placed upon the oval window before prosthesis placement so as to seal the vestibule immediately. In stapedotomy, oval window is open till prosthesis placement. In total stapedectomy with *bucket handle–type prosthesis*, sometimes the prosthesis is pushed on tissue graft medially until the relative depth to incus is correct; at other times, incus is lifted, prosthesis is placed, and incus is released so that the lenticular process rests in the bucket handle (Fig. 6.4).

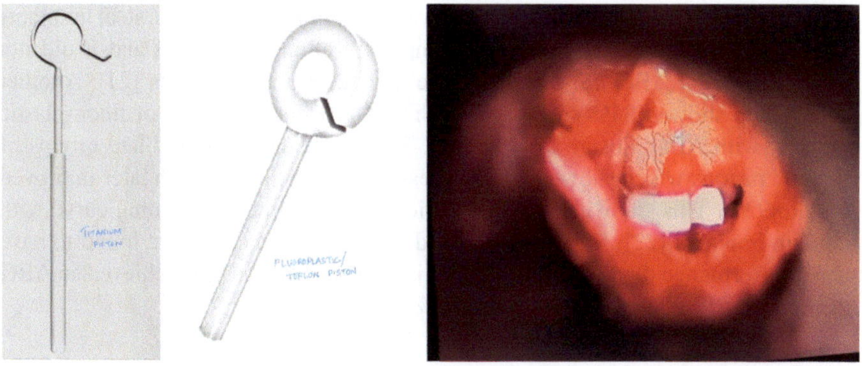

Fig. 6.4 Stapes piston and its placement

Materials

In 1960, a stainless-steel loop prosthesis that may be crimped onto incus long process was popularised by House. The crimp had excellent long-term results. Adhering gel form was tried but was found to have increased complications.

There are mainly two stainless steel variants commonly used: the 300 and the 400 series. 300 series is used typically for implants as it has reduced magnetism due to more random arrangement of micro dipoles. Hence, they are safer for MRI. In ossicular chain reconstruction, titanium has become more popular due to its additional non-ferromagnetic properties [19]. Nitinol, a metal alloy made of nickel and titanium, is another material commonly used in stapes pistons with a special property of returning to the original shape when heated. So using nitinol an internal prosthesis wire may be manufactured to the increased diameter size and during placement, may be stretched to required size. It may then be laser heated so that it returns to its original size and shape allowing a secure circumferential crimping around incus. This avoids need for manual manipulation.

Another way to facilitate appropriate crimping is changing material of the hook to platinum. Hydroxyapatite and inomeric cements were initially used for aiding prosthesis attachment.

Challenges

Most important challenges of a stapedial prosthesis are incus necrosis (polyethylene strut), fistula, wire loop loosening, and granuloma formation. The wire loop must be manually crimped on to incus, which is a technically difficult process and result in delayed failure. Tight crimping may result in incus necrosis, while in too loose crimping, vibration of wire loop can result in incus erosion. To curb defects of manual crimping, heat-activated memory, shape prostheses were developed [19]. Self-crimping prosthesis is mostly nitinol-based shepherd crook and Teflon-based piston.

Relative efficacy of specific prostheses has been studied comparing different studies with varied stapes prostheses placed by surgeons. There was no significant difference between the Gyrus titanium versus Robinson stainless steel implants found by Lippy [20]. Another study compared Teflon and nitinol and it did not detect significant ABG closure difference at 1, 2, 3 and 6 months [21]. Another study group showed 84% success with NiTiBOND but only 36% of fluoroplastic implants with ABG closure less than 10. They included early and late groups of titanium pistons which showed 44% success at the beginning which later improved to up to 92%. This study also revealed evidence that there is a learning curve with titanium implants [22]. In yet another study, nitinol prostheses were found to have better performance than platinum [23]. Rajan and colleagues [24] achieved an ABG less than 10 dB: 89% with nitinol and 74% with titanium.

Failure Rates
At 10 years the failure rates were found to be 11.2% for stainless steel McGee with platinum ribbon and an average failure time of 2.5 years (the crimped platinum ribbon loosened over time and also was displaced laterally), whereas failure rates were 9.5% for stainless steel Robinson bucket handle and had a longer average failure time of 8.6 years. Another study found 11% revision rates with their non-crimp nitinol pistons and 4% with crimped platinum wire prostheses. This may be due to learning curve issues and laser energy selection [19].

Powered Implants Acoustic stimulator for cochlea with implantable electromagnetic transducer has been made. It is for placement on oval window following stapedectomy and later activated by a behind-the-ear processor [25]. Another one is a floating mass transducer which is to be placed on round window with vibratory stimulation attached to an ossicular prosthesis. The results were found to be better with round window placement [26].

6.2.4.2 Ossicular Replacement Prosthesis

Introduction
As earlier mentioned, integration of biomaterial is more difficult in middle ear due to its aeration, reactive mucosal covering and potential for microbial colonisation. This is described by the term “middle ear biocompatibility” [27]. The biomaterial used for ossicular reconstruction should promote mucosal covering of the implant and so as to minimise extrusion [27].

The implants should also have sound conduction properties depending on implant stiffness, mass, design and incorporation [27].

Types
Depending on the presence or absence of the stapes suprastructure, prostheses may be of two types: [8] (Fig. 6.5)

- Partial Ossicular Replacement Prostheses (PORP)
- Total Ossicular Replacement Prostheses (TORP)

They differ in their length and design. TORPs are placed onto stapes footplate like a stamp (columella), while PORPs are fitted on to the capitulum like a bell.

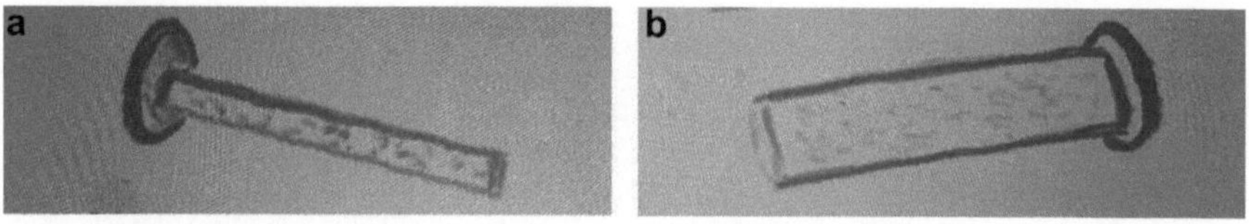

Fig. 6.5 Ossicular prostheses. (**a**) TORP. (**b**) PORP

Materials

There are two types of materials used:

1. Natural: Autografts and allografts, cartilage, bone
2. Biomaterials

    (a) *Hydroxyapatite*: It is one of the popular materials. Dense type (Ceramic calcium phosphate) resembles natural bone. It resists degradation and has good sound conduction. It can also be sculpted and shaped.
    (b) *Titanium*: Light, biocompatible.
    (c) *Platinum*: Biocompatible, non-corrosive, non-magnetic, malleable.
    (d) *Stainless steel*: Good conduction of sound, adhesion resistant.
    (e) *Fluoroplastic* [Teflon]: Excellent sound conduction, proven biocompatibility, smooth, minimal adhesions due to non-sticky surface.
    (f) *Gold*: Malleable, biocompatible.

Mechanism

Ossicular prostheses create new connection between tympanic membrane and inner ear and reestablish optimal sound conduction.

Single use.

Causes of Implant Failure

- Prosthesis displacement
- Extrusion
- Absorption of biomaterial

6.2.4.3 Vibrant Sound Bridge

It is a semi-implantable hearing device that resides beneath the skin without the visibility in the ear canal.

Parts: (Fig. 6.6)

- *Vibrating ossicular prosthesis (VORP)*: Surgically implanted internal part

  Parts

  1. Receiving coil
  2. Conductor link
  3. Transducer.
- *External audio processor*

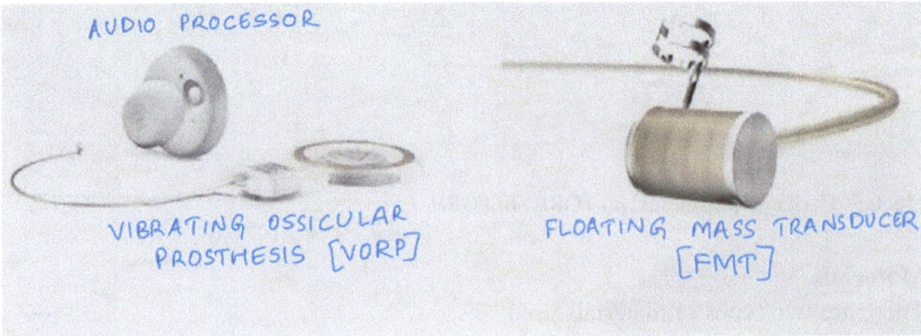

Fig. 6.6 Vibrant sound bridge

Fig. 6.7 Bone anchored hearing aid (BAHA)

Mechanism

The vibrations are produced by small electromagnetic coil and magnet. It is then coupled to the incudal long process and transmit these vibrations.

Indication

Individuals with serviceable hearing looking after sound quality improvements, comfort and cosmesis.

6.2.4.4 Bone Anchored Hearing Aids (BAHA) (Fig. 6.7)

An osseointegrated implant with the titanium abutment is fixed to the skull [8].

Indications

- Bilateral canal atresia (absolute indication)
- Bilateral otitis media
- Congenital conductive hearing loss

Parts

- Screw
- Titanium abutment
- Sound processor

6.3 Nasal Stents and Implants Including the Synthetic Materials Used in Stents for Drug Delivery System

Introduction

Stents are primarily designed to promote wound healing and to relieve obstruction and are placed temporarily in host tissues [28]. Implants are defined as devices that can be placed in the human body and remain for a duration of 1 month or more and promote the quality of life and function.

Mechanism of Instruments

Nasal implants are inserted during nasal or paranasal sinus surgeries. Drug-delivery devices or stents sustainably release drugs into the body for longer durations.

Uses

- These stents are used in conditions such as paranasal infections, inflammations, neoplasms, autoimmune diseases.
- Nasal stents are used for nasal reconstruction of deformities.
- Drug-eluting nasal implants also keep the middle meatus open after FESS.
- Nasal implants can be used to reduce bleeding and prevent synechiae.
- Nasal stents helps in drainage of the sinus mucosa which would in turn help in wound healing.
- In order to prevent blood and mucous filling the ethmoid sinus, nasal stents are used to fill the sinus cavity [29].
- Various middle meatal implants known today are propel implants, Sinu-Foam spacers, Relieva Stratus and micro-flow spacers [30].
- Nasal implants are used for nasal augmentation.

6.3.1 Synthetic Polymer Materials

Controlled Drug Delivery System of Nose and Paranasal Sinuses Using Synthetic Materials: Introduction

Controlled drug delivery allows increased local therapeutic concentrations in a sustained manner, especially in locations that are difficult to access. One such location which is difficult to assess is the sinonasal mucosa. These areas are often found inflamed in patients with rhinitis or rhinosinusitis, which leads to diminished quality of life, significant healthcare expenses, and multiple co-morbidities.

Though various medical therapies with daily administration are available, various factors like the anatomical, physiological, and patient adherence barriers can limit their therapeutic efficacy. Henceforth, biomaterial-based systems were developed which can locally deliver anti-inflammatory, antibiotic, decongestant, and antihistamine medications over an extended duration.

Nasal packs, dressings, sinus stents, polymeric meshes, nanoparticles, microparticles and in situ hydrogels are reviewed below.

Individual Synthetic Materials and Their Benefits

Polyesters of poly lactic acid (PLA) and poly glycolic acid (PGA) are two polyesters that were commonly used in controlled drug delivery.

Poly (ε-caprolactone) (PCL): Because of its surface hydrophobicity and crystallinity, it degrades over a period of months to years.

Carboxymethylcellulose (CMC): It is a material used mostly for nasal packs; it promotes healing and control bleeding which are not massive. In addition, it has the advantage of high biocompatibility and biodegradability.

Polyvinyl acetate (Merocel, Medtronic Inc., Minneapolis): It is a synthetic hydroxylated sponge which is a biomaterial mainly used as a hemostatic nasal tampon.

Adverse Effects of Nasal Packs

- Merocel can cause further inflammation if left in contact with the healing mucosa for too long; it can be incorporated into the tissue and cause further inflammation [31].
- Merocel can cause pain as well as damage to the mucosa.
- Merocel can cause bleeding during removal [32].
- Nasal packs can also cause infection if kept in situ for long; it can also cause headache.
- Toxic shock syndrome can occur if merocel are placed for more than 48 hrs.
- Nasal stuffiness and the need to breath from open mouth can cause dryness in the mouth as well.

Nasal Sinus Stents They are composed of biodegradable polymers which do not require removal and are loaded with therapeutics for local delivery to the sinonasal mucosa as they degrade (Fig. 6.8).

Adverse Effects

Adverse effects most commonly associated with stents are:

- It can cause infection
- Sometimes stents can cause obstruction in the oropharyngeal region.
- Because of the stent being in situ, it can cause headache.
- There is incidences of migration and extrusion of the stent [33].

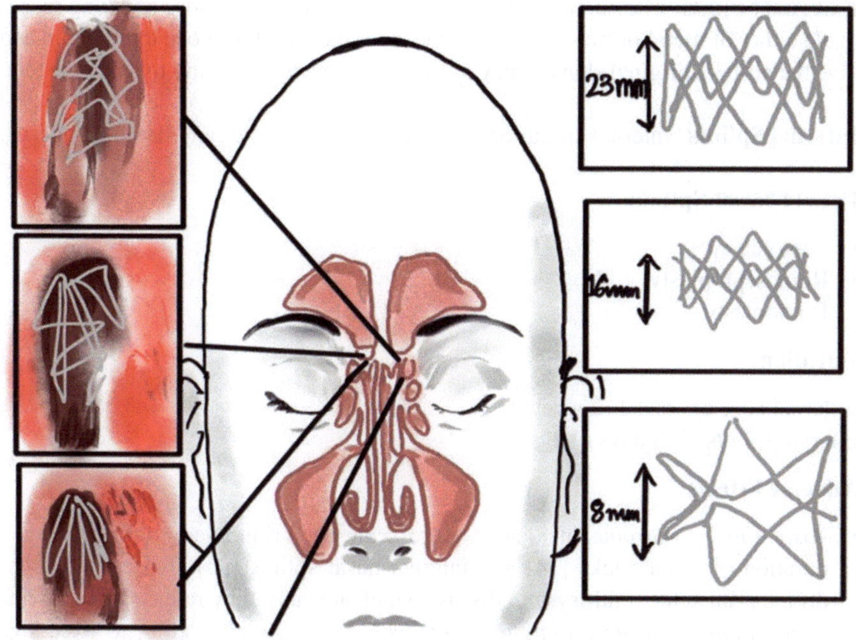

Fig. 6.8 Images showing sinus stents which are degradable and are used for corticosteroid delivery locally to specified sinuses

6.3.2 *Nasal Splints*

Introduction
Nasal splints are mainly used post-surgery like septoplasty, rhinoplasty or post-fracture to the nasal bones; it helps in augmentation and gives a better aesthetic appearance to the nose.

Uses

- It helps in maintaining the contour of the nose.
- It helps to improve the breathing problems.
- It helps to correct deformities.
- It helps to repair the injury and broken tissue.

Parts of Instruments
Nasal splints are composed of two pieces which are made of either plastic, aluminium, silicone or polytetrafluoroethylene (Teflon), and they are positioned to align with the nose.

Two types of nose splints: external and internal.

External Splints They consist of a bandage which wraps the nose and long pieces made either of metal or plastic which runs down the length of nose to maintain and give the nose its original shape. They prevent oedema and swelling to a large extent.

Internal Splints Internal splints are made either of plastic or any flexible material.

Uses of Nasal Splints

- Septoplasty
- Rhinoseptoplasty
- Rhinoplasty

Duration

External nose splints: 1–2 weeks
Internal splints: 3–5 days

Adverse Effects

- *Toxic shock syndrome* incidences were seen after rhinoplasty and prolonged duration of nasal packs [34] and internal nasal splints [35]. Toxic shock syndrome is an acute, multisystem disease which are caused by release of exotoxins from *Staphylococcus aureus* or *Streptococcus pyogenes* leading to excessive activation of inflammatory cells and release of inflammatory cytokines, resulting in tissue damage and organ dysfunction [36].
- *Soft-tissue oedema* in the immediate post-operative period can be prevented by postoperative care like corticosteroid use, head elevation, taping, and application of cold compresses. Most oedema will resolve within 4 weeks.
- Deformities and deviation can occur following use of nasal implants or nasal packing.
- Nasal airway obstruction occurs during nasal packing and post pack removal if nasal crusting develops.
- Biofilm formation depends on the site of application, shape and surface characteristics of the material used.

6.4 Laryngeal Implants

Introduction

An implant is said so when it is kept in the body permanently. Glottis insufficiency is one of the most important indications for using an implant. Glottic insufficiency is a condition wherein during phonating there is a partial closure or a gap of the vocal folds. It can lead to leakage of air through the glottis during phonation causing a higher chances of aspiration [37]. Glottic insufficiency can be due to variety of causes, but the main cause where laryngeal implants can be used is unilateral paralysis of recurrent laryngeal nerve which is lower motor neuron paralysis. Here implants are used to medialise the vocal cords to prevent aspiration and dysphonia [38].

Mechanism of Instrument
The main mechanism of implants is to medialise the position of the vocal folds. Implants can be placed as solid implants by an open surgical approach into the paraglottic space by making a window in the framework of larynx or using injectable implants with high viscosity and are injected into the vocal cord which is paralysed to increase its volume and hence helping in reduction of glottal gap [38].

Parts of Instrument There are two types of implants:

- Solid implants: Titanium, Gore-tex, Silastic keel implants (Fig. 6.1). Silastic keel implants have an external part which is sutured with the thyroid cartilage and an inner part is kept in between the vocal cords.
- Injectable implants: Teflon, fat, glycerine, collagen.

Uses

1. Solid implants are used during thyroplasty type 1 wherein the main aim is for inward displacement with an implant through a window in the thyroid cartilage.
2. Used for medialisation thyroplasty.

Single of Multiple Use Single use.

Benefits

1. Solid implants:
   (a) Titanium: It has better biocompatibility, better fixation, very versatile, a shorter healing time and improved cell adhesion [39]. In addition to the above said benefits, it also has an individually adjustable application during the operation [40].
   (b) Silastic keel implants: Less expensive, better biocompatibility, better fixation and very versatile [39] (Fig. 6.9).
   (c) Gore-tex (expanded polytetrafluoroethylene, ePTFE) is made of micropores and is non-immunogenic, non-toxic, soft and non-degradable [41].
   (d) The pore size of e-PTFE ranges between 10 and 30 μm and there is minimal tissue ingrowth [42].

2. Injectable implants:
   Main advantage over solid implants is that it can be done without the use of general anaesthesia; therefore, it can be done as an OPD procedure.
   (a) Teflon: During the 1960 and 1980, it was used widely the most for injection laryngoplasties [43]. But in view of the reports stating that Teflon particles have relocated to distant organs and also it was found that Teflon particles were causing foreign body granuloma in huge severity, therefore, its use is obsolete now [44, 45].

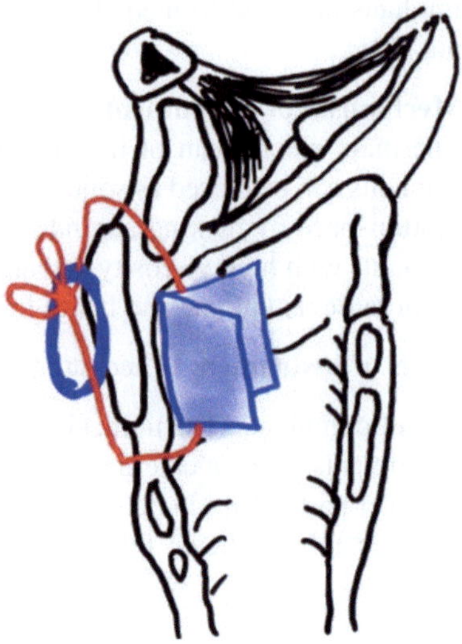

Fig. 6.9 Anterior glottic stenosis treated with silastic sheet

(b) Hydroxylapatite: It is similar to bone tissue; therefore, it has an impressive biocompatibility.
(c) Collagen: It is a permanent implant used in injection laryngoplasty for medialisation of vocal cord, the other benefit is that it is also used widely in aesthetic facial surgery. According to different authors, duration of the augmentation process is ranging from 6 to 18 months. Collagen has low viscosity; therefore, we can use fine needles with low-pressure injection.
(d) Hyaluronic acid: It is being very commonly used in aesthetic surgery mainly used as filler and have a durability of 3–6 months. Nowadays, hyaluronic acid has been used for augmentation of vocal folds in cases of vocal fold palsy with electromyographically showing good prognosis. It is useful in those patients who aren't sure to go ahead with definitive thyroplasty or to go ahead with injection laryngoplasty, as using hyaluronic acid is often helpful to experience the possible result as it is a rapid resorbable, easy-to-use substance without any significant side effects. Another additional benefit of using hyaluronic acid is that allergy testing is not necessary.

- Autologous fat: It is readily procurable, hassle free to harvest and easily injected; since it is autologous in nature, there arises no foreign body reaction and no hypersensitivity reactions [46], it has high biocompatibility, and it is safer [47]. Significant improvement has been seen in phonatory function like in jitter or shimmer, noise-to-harmonic ratio. Use of autologous fat has also showed improvement in maximal phonation time, grade, asthenia and breathiness [13]. Autologous fat is a material that is similar to the viscoelasticity of the vocal fold membranous layer [48].

Duration of Implant in the Body

Temporary injectable materials are hyaluronic acid, collagen, and gelfoam last a few weeks to a few months.

Long-lasting and permanent materials are calcium hydroxyapatite paste, autologous fat, and polytetrafluoroethylene paste (Teflon) which can last for years.

Adverse Effects

I. Goretex implants.

- Goretex implants has high chance of extrusion into the lumen.
- It can cause persistent inflammation leading to granulation formation [49].
- It can cause polyp over the anterior commissure with an overlying granulation tissue.
- Sometimes the edge of the implant can be seen extruding into an otherwise stable airway [50].

II. Hydroxylapetite implants:

- This implants can cause adynamic mucosa and also decrease wave [51].
- Granulomas.
- Tissue inflammation marked by oedema, erythema.
- Hypervascularity.
- Migration of implant.

III. Autologous fat:

- Over-injection can cause dysphonia [52].
- Post-lipoinjection immediate postoperative dysphonia secondary to transient VFs inflammation [53].
- Globus sensation or dyspnea [54].
- Laryngocele formation due to over injection causing airway obstruction and dysphonia.
- Laryngeal oedema requiring a temporary tracheotomy.
- It can lead to the formation of intracordal cyst at the injection site which are secondary to a superficial injection [54].
- Fat extrusion [54].
- It has a high resorption rate which can lead to instability [55, 56].

IV. Collagen:

- Collagen injection can lead to submucosal deposits which can disturb the mucosal waves making it abnormal, leading to dysphonia which are quite significant.
- It can lead to hypersensitivity reactions.
- It can cause abscess formation locally.
- It has possibility of developing collagen vascular disease [7].
- Allergy testing is mandatory because of high risk of hypersensitivity reactions.

V. Teflon:

- It can cause hypersensitivity reactions.
- It can cause inflammation characterised by foreign body reaction leading to granuloma formation.
- Granuloma formation can cause airway obstruction and dysphonia.

Contraindication of Injection Laryngoplasty

Injection laryngoplasty is not indicated in the following cases:

- It should not be used during acute laryngeal inflammation/infection.
- It should not be used in cases of inadequately controlled malignancy.
- It should not be used in cases of disease of the upper aerodigestive tract if it is rapidly progressing.
- It should not be used if both the vocal cords are paralysed.
- Not useful in posterior glottic insufficiency [55].

6.4.1 Laryngeal Stents

Introduction

Laryngeal stents most commonly used is the Montgomery tube and it is mainly used to stabilise the larynx after laryngofissure. It prevents synechia formation and also prevents the formation of stenosis.

Mechanism of Instrument

The Montgomery T-Tube has been introduced since 1965 [32] and it is the most commonly used laryngeal stent. Its main role is that it can stay within the larynx for a year or even more than that and maintains the laryngeal airway patency and also prevents stenosis and synechiae.

Parts of Instrument

It is a T-shaped silicone tube which is uncuffed. It has two long legs, which is kept in the trachea and the subglottic, and also has a short leg which is positioned in the tracheostomy site. There is also another part called tracheostomal leg which has to be unplugged only during cleaning and in case of emergency (Fig. 6.10).

Uses

Management of tracheal stenosis using M-tube to prevent tracheal stenosis and synechiae and eventually helping in maintaining a patent airway.

Single or Multiple Use

Used only in tracheal stenosis post laryngofissure, kept in situ for a complete 1 year.

Benefits

Silicon made laryngeal implants are soft in nature and light weight and can be kept in situ for a year.

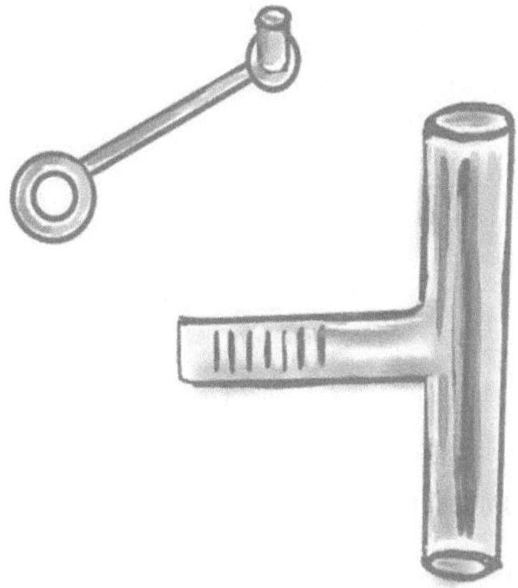

Fig. 6.10 Montgomery tube (M-tube)

Duration of Implant in the Body One year before the removal of it.

Adverse Effects

- The most common complication was minimal surgical emphysema
- Crusting
- Granulation around the stoma
- Subglottic granulations
- Hoarseness of voice

6.4.2 Tracheoesophageal Prosthesis

Introduction

Tracheoesophageal (TE) speech is regarded superior method to oesophageal speech and serves as the mainstream for rehabilitation in patients of total laryngectomy [57].

Mechanism of Instrument

For a tracheoesophageal speech to occur, it utilises a voice prosthesis (VP) (Fig. 6.11) that is inserted through a puncture in the common wall which separates the trachea from the oesophagus. When the patient closes the stoma, air from the lungs is passed through the prosthesis via a one-way valve, which causes the pharyngoesophageal segment to vibrate. This vibration passes into the oral cavity, and the articulators move to create speech. The VP permits air exchange into the pharyngoesophagus for speech production at the same time preventing aspiration of food and liquid into the lungs.

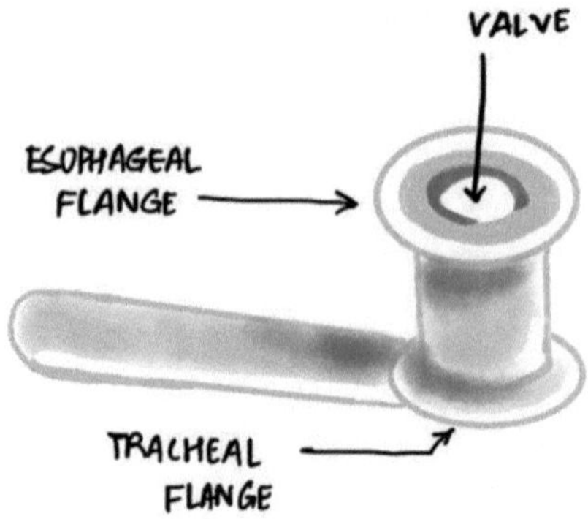

Fig. 6.11 Tracheoesoophageal speech (TEP) with its various parts

Table 6.1 Non-indwelling Vs Indwelling TEPs

Non-indwelling	Indwelling
Can be removed by the patient	Needs to be replaced by the clinician
Daily maintenance includes cleaning and flushing	Robust construction, i.e. longer life
Need manual dexterity; voice prosthesis has a strap and requires a piece of adhesive tape	Success independent of patient age and general health
Less expensive	More expensive
For example, Blom-Singer Duckbill, Panje	Blom-Singer Indwelling, Provox, Groningen Voice Button

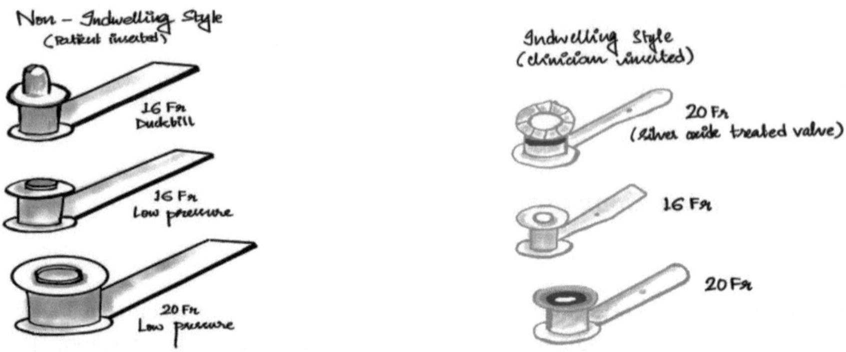

Fig. 6.12 Types of tracheoesophageal prosthesis

Parts of Instruments

Types of TEP (Table 6.1)

Indwelling and non-indwelling (Fig. 6.12).

Adverse Effects

- TEP fistula infection.
- TEP fistula bleeding.
- It can lead to deep neck (prevertebral) abscess.

- There is a high risk of development of granulation tissue at the inner side of the TEP fistula.
- There has been cases wherein the prosthesis invaginated and there has been leakage around the prosthesis [58].
- It can also lead to mediastinitis and abscess around paraesophageal region [59].

References

1. Liu L, Rodman C, Worobetz NE, Johnson J, Elmaraghy C, Chiang T. Topical biomaterials to prevent post-tonsillectomy hemorrhage. J Otolaryngol Head Neck Surg. 2019;48(1):45.
2. Spałek J, Ociepa P, Deptuła P, Piktel E, Daniluk T, Król G, et al. Biocompatible materials in otorhinolaryngology and their antibacterial properties. Int J Mol Sci. 2022;23(5):2575.
3. Bergmann CP, Stumpf A. Introduction. In: Dental ceramics [Internet]. Berlin/Heidelberg: Springer; 2013. p. 1–7. (Topics in mining, metallurgy and materials engineering). Available from: https://link.springer.com/10.1007/978-3-642-38224-6_1.
4. Sternberg K. Current requirements for polymeric biomaterials in otolaryngology. GMS Curr Top Otorhinolaryngol Head Neck Surg. 2009;8:Doc11. ISSN 1865-1011 [Internet]. Available from: http://www.egms.de/en/journals/cto/2011-8/cto000063.shtml
5. Kerr AG. Proplast and plastipore. Clin Otolaryngol. 1981;6(3):187–91.
6. Brown BL, Neel HB, Kern EB. Implants of supramid, proplast, plasti-pore, and silastic. Arch Otolaryngol Head Neck Surg. 1979;105(10):605–9.
7. Anderson TD, Sataloff RT. Complications of collagen injection of the vocal fold: report of several unusual cases and review of the literature. J Voice. 2004;18(3):392–7.
8. Bhat KV, Manjunath D. Atlas of instruments in otolaryngology: head and neck surgery [Internet]. Jaypee Brothers Medical Publishers (P) Ltd; 2012. Available from: https://www.jaypeedigital.com/book/9789350257135
9. Grote JJ. Biomaterials in otology: proceedings of the first international symposium "Biomaterials in otology". Leiden/Cham: Springer; 1983.
10. Lehnhardt E. Biocompatibility of cochlear implants. Eur Arch Otorhinolaryngol Suppl. 1992;1:223–33.
11. Stöver T, Lenarz T. Biomaterials in cochlear implants. GMS Curr Top Otorhinolaryngol Head Neck Surg. 2009;8:Doc10. ISSN 1865-1011 [Internet]. Available from: http://www.egms.de/en/journals/cto/2011-8/cto000062.shtml
12. Kronenberg J, Wolf M, Migirov L, Shapira Y, Aviel-Ronen S, Hildesheimer M. Foreign body reaction to cochlear implant. Oto-Rhino-Laryngol Nova. 2001;11(3–4):207–9.
13. Puri S, Dornhoffer JL, North PE. Contact dermatitis to silicone after cochlear implantation. Laryngoscope. 2005;115(10):1760–2.
14. Ratner BD, editor. Biomaterials science: an introduction to materials in medicine. 3rd ed. Amsterdam/Boston: Elsevier/Academic; 2013. p. 1519.
15. Brummer SB, Turner MJ. Electrical stimulation of the nervous system: the principle of safe charge injection with noble metal electrodes. Bioelectrochem Bioenerg. 1975;2(1):13–25.
16. Guex AA, Vachicouras N, Hight AE, Brown MC, Lee DJ, Lacour SP. Conducting polymer electrodes for auditory brainstem implants. J Mater Chem B. 2015;3(25):5021–7.
17. Dormer KJ, Gan RZ. Biomaterials for implantable middle ear hearing devices. Otolaryngol Clin N Am. 2001;34(2):289–97.
18. Sevy A, Arriaga M. The stapes prosthesis. Otolaryngol Clin N Am. 2018;51(2):393–404.
19. Fritsch MH. MRI scanners and the stapes prosthesis. Otol Neurotol. 2007;28(6):733–8.
20. Lippy WH, Burkey JM, Schuring AG, Berenholz LP. Comparison of titanium and robinson stainless steel stapes piston prostheses. Otol Neurotol. 2005;26(5):874–7.

21. Brar T, Passey JC, Agarwal AK. Comparison of hearing outcome using a Nitinol versus Teflon prosthesis in stapedotomy. Acta Otolaryngol (Stockh). 2012;132(11):1151–4.
22. Canu G, Lauretani F, Russo FY, Ferrary E, Lamas G, Sterkers O, et al. Early functional results using the nitibond prosthesis in stapes surgery. Acta Otolaryngol (Stockh). 2017;137(3):259–64.
23. Tenney J, Arriaga MA, Chen DA, Arriaga R. Enhanced hearing in heat-activated-crimping prosthesis stapedectomy. Otolaryngol Neck Surg. 2008;138(4):513–7.
24. Rajan GP, Diaz J, Blackham R, Eikelboom RH, Atlas MD, Shelton C, et al. Eliminating the limitations of manual crimping in stapes surgery: mid-term results of 90 patients in the nitinol stapes piston multicenter trial. Laryngoscope. 2007;117(7):1236–9.
25. Häusler R, Stieger C, Bernhard H, Kompis M. A novel implantable hearing system with direct acoustic cochlear stimulation. Audiol Neurotol. 2008;13(4):247–56.
26. Shimizu Y, Puria S, Goode RL. The floating mass transducer on the round window versus attachment to an ossicular replacement prosthesis. Otol Neurotol. 2011;32(1):98–103.
27. Neumann A, Jahnke K. Biomaterials for ossicular chain reconstruction. A review. Mater Werkst. 2003;34(12):1052–7.
28. Nouwen J, Hans S, De Mones E, Brasnu D, Crevier-Buchman L, Laccourreye O. Thyroplasty type I without arytenoid adduction in patients with unilateral laryngeal nerve paralysis: the montgomery implant versus the Gore-Tex implant. Acta Otolaryngol (Stockh). 2004;124(6):732–8.
29. Weitzel EK, Wormald PJ. A scientific review of middle meatal packing/stents. Am J Rhinol. 2008;22(3):302–7.
30. Desrosiers M, Evans GA, Keith PK, Wright ED, Kaplan A, Bouchard J, et al. Canadian clinical practice guidelines for acute and chronic rhinosinusitis. Executive summary. J Otolaryngol Head Neck Surg. 2011;40(Suppl 2):S91–8.
31. Maccabee MS, Trune DR, Hwang PH. Effects of topically applied biomaterials on paranasal sinus mucosal healing. Am J Rhinol. 2003;17(4):203–7.
32. Wang J, Cai C, Wang S. Merocel versus nasopore for nasal packing: a meta-analysis of randomized controlled trials. Lin SJ, editor. PLoS One. 2014;9(4):e93959.
33. Shah VN, Pasick LJ, Benito DA, Ghiam MK, D'Aguillo C. Complications associated with PROPEL mometasone furoate bioabsorbable drug-eluting sinus stents from 2012 to 2020. Am J Rhinol Allergy. 2022;36(2):185–90.
34. Toback J, Fayerman JW. Toxic shock syndrome following septorhinoplasty: implications for the head and neck surgeon. Arch Otolaryngol Head Neck Surg. 1983;109(9):627–9.
35. Wagner R, Toback JM. Toxic shock syndrome following septoplasty using plastic septal splints. Laryngoscope. 1986;96(6):609–10.
36. Lappin E, Ferguson AJ. Gram-positive toxic shock syndromes. Lancet Infect Dis. 2009;9(5):281–90.
37. Onwordi LN, Al Yaghchi C. Airway glottic insufficiency. In: StatPearls [Internet]. Treasure Island (FL): StatPearls Publishing; 2022. Available from: http://www.ncbi.nlm.nih.gov/books/NBK538207/.
38. Sittel C. Larynx: implants and stents. GMS Curr Top Otorhinolaryngol Head Neck Surg. 2009;8:Doc04. ISSN 1865-1011 [Internet]. Available from: http://www.egms.de/en/journals/cto/2011-8/cto000056.shtml
39. Friedrich G. Titanium vocal fold medializing implant: introducing a novel implant system for external vocal fold medialization. Ann Otol Rhinol Laryngol. 1999;108(1):79–86.
40. Schneider B, Denk DM, Bigenzahn W. Functional results after external vocal fold medialization thyroplasty with the titanium vocal fold medialization implant. Laryngoscope. 2003;113(4):628–34.
41. Romo T, Pearson JM. Nasal implants. Facial Plast Surg Clin N Am. 2008;16(1):123–32.
42. Wang TD. Multicenter evaluation of subcutaneous augmentation material implants. Arch Facial Plast Surg. 2003;5(2):153–4.
43. Arnold GE. Vocal rehabilitation of paralytic dysphonia: VIII. Phoniatric methods of vocal compensation. Arch Otolaryngol Head Neck Surg. 1962;76(1):76–83.

44. Ellis JC, McCaffrey TV, Desanto LW, Reiman HV. Migration of teflon after vocal cord injection. Otolaryngol Neck Surg. 1987;96(1):63–6.
45. Rubin HJ. Misadventures with injectable polytef (Teflon). Arch Otolaryngol Head Neck Surg. 1975;101(2):114–6.
46. Brandenburg JH, Kirkham W, Koschkee D. Vocal cord augmentation with autogenous fat. Laryngoscope. 1992;102(5):495–500.
47. Trebbi M, Villari D, Ruberto M. Materials for injection laryngoplasty: current application. In: Bergamini G, Presutti L, Molteni G, editors. Injection laryngoplasty [Internet]. Cham: Springer; 2015. p. 31–42. Available from: https://link.springer.com/10.1007/978-3-319-20143-6_4.
48. Chan RW, Titze IR. Viscosities of implantable biomaterials in vocal fold augmentation surgery. Laryngoscope. 1998;108(5):725–31.
49. Watanabe K, Hirano A, Honkura Y, Kashima K, Shirakura M, Katori Y. Complications of using Gore-Tex in medialization laryngoplasty: case series and literature review. Eur Arch Otorhinolaryngol. 2019;276(1):255–61.
50. Morris J, Thomas DM. Delayed airway extrusion of type 1 thyroplasty Gore-Tex implant. BMJ Case Rep. 2016;2016:bcr2016215704.
51. DeFatta RA, Chowdhury FR, Sataloff RT. Complications of injection laryngoplasty using calcium hydroxylapatite. J Voice. 2012;26(5):614–8.
52. Sanderson JD, Simpson CB. Laryngeal complications after lipoinjection for vocal fold augmentation. Laryngoscope. 2009;119(8):1652–7.
53. Tamura E, Fukuda H, Tabata Y, Nishimura M. Use of the buccal fad pad for vocal cord augmentation. Acta Otolaryngol (Stockh). 2008;128(2):219–24.
54. Laccourreye O, Papon JF, Kania R, Crevier-Buchman L, Brasnu D, Hans S. Intracordal injection of autologous fat in patients with unilateral laryngeal nerve paralysis: long-term results from the patient's perspective. Laryngoscope. 2003;113(3):541–5.
55. Kwon TK, Buckmire R. Injection laryngoplasty for management of unilateral vocal fold paralysis. Curr Opin Otolaryngol Head Neck Surg. 2004;12(6):538–42.
56. Hsiung MW, Woo P, Minasian A, Schaefer MJ. Fat augmentation for glottic insufficiency. Laryngoscope. 2000;110(6):1026–33.
57. Lewin JS, Baumgart LM, Barrow MP, Hutcheson KA. Device life of the tracheoesophageal voice prosthesis revisited. JAMA Otolaryngol Neck Surg. 2017;143(1):65–71.
58. Yan D, Zhang B, Li D, Li Z, Liu W, Xu Z. Larynx preservation and hypopharyngeal reconstruction in posterior hypopharyngeal wall sqamous cell carcinoma. Zhonghua Er Bi Yan Hou Tou Jing Wai Ke Za Zhi. 2014;49(7):548–52.
59. Imre A, Pınar E, Callı C, Sakarya EU, Oztürkcan S, Oncel S, et al. Complications of tracheoesophageal puncture and speech valves: retrospective analysis of 47 patients. Kulak Burun Bogaz Ihtis Derg KBB J Ear Nose Throat. 2013;23(1):15–20.

Chapter 7
Significant Risk Medical Devices – Gastroenterology and Urology

Manish Ranjan, Prerna Dabral, Namrata Khurana, and Nobel Bhasin

7.1 Gastroenterology Implants

The discipline of gastroenterology encompasses a diverse array of gastrointestinal (GI) disorders that can have a substantial impact on the well-being and overall quality of life of affected individuals. Gastrointestinal implants are medical devices specifically developed to manage complex diseases affecting the digestive tract and are recognized as playing a critical role in their treatment [1–5]. The utilization of these implants encompasses a range of applications, including, but not limited to, structural reinforcement and drug administration, and presents numerous potential benefits in the management of gastrointestinal disorders [1–4]. Nevertheless, it is of equal significance to meticulously assess their benefits and drawbacks in the context of clinical practice to guarantee the most favorable results for patients.

A range of gastrointestinal conditions, such as gastrointestinal cancers, inflammatory bowel disease (IBD), diverticulitis, achalasia, and fecal incontinence, among others, may necessitate the use of implants [1–4]. The aforementioned conditions have the potential to result in considerable morbidity and mortality, thereby impacting the physical, psychological, and social well-being of patients [6–9]. Gastrointestinal implants are frequently utilized in the management of these ailments, with the goal of reinstating typical anatomy and functionality, furnishing mechanical reinforcement, and mitigating symptoms [6–9]. Hence, it is imperative to have knowledge about prevalent gastrointestinal disorders that may require the implementation of implants to comprehend the clinical importance and influence of these instruments.

M. Ranjan · N. Khurana · N. Bhasin (✉)
Baylor College of Medicine, Houston, TX, USA
e-mail: nobel.bhasin@bcm.edu

P. Dabral
Vitalant Research Institute, San Francisco, CA, USA

P. S. Timiri Shanmugam et al. (eds.), *Significant and Nonsignificant Risk Medical Devices*, https://doi.org/10.1007/978-3-031-52838-5_7

Although gastrointestinal (GI) implants present potential advantages, such as enhanced functionality, minimally invasive techniques, personalized approaches, and extended symptom relief, they are not devoid of drawbacks and constraints. The drawbacks that require thorough evaluation include complications that may arise during implantation, foreign body reactions, financial burdens, complications that are specific to implants, and lifestyle modifications [6–9]. Hence, it is imperative to conduct a thorough evaluation of the benefits and drawbacks associated with gastrointestinal implants to facilitate informed clinical judgments and enhance patient results.

7.2 Types of Gastroenterology Implants

Gastrointestinal (GI) implants are medical devices that are surgically implanted within the gastrointestinal tract to address a range of medical conditions, such as obesity, gastroesophageal reflux disease (GERD), and gastrointestinal cancer. The following are prevalent categories of gastrointestinal implants:

1. Adjustable silicone bands are surgically placed around the upper portion of the stomach to create a small pouch, thereby limiting the amount of food intake. These are commonly referred to as gastric bands.
2. Gastric balloons are a type of temporary, inflatable medical device that is inserted into the stomach to induce satiety and subsequently decrease food consumption.
3. Gastric stimulators utilize electrical impulses to activate the stomach muscles, resulting in a potential amelioration of gastroparesis symptoms, a medical condition characterized by impaired gastric emptying.
4. Endoscopic suturing devices are utilized to generate folds or pleats within the stomach, resulting in a reduction in its size. This reduction in size can potentially restrict food intake.
5. Endoluminal sleeves refer to pliable tubes that are introduced into the stomach to establish a partition between ingested food and the stomach lining. This mechanism can potentially lead to a decrease in the assimilation of calories.
6. Enteral feeding tubes refer to medical devices that are inserted via the oral or nasal route and directed toward the stomach or small intestine. They are utilized to provide sustenance to individuals who are incapable of ingesting or swallowing food.

7.3 Gastric Bands

Gastric bands are gastrointestinal implants utilized to constrict the size of the stomach and curtail the quantity of food intake. A silicone ring is surgically implanted around the upper portion of the stomach to form a restricted pouch, commonly referred to as a gastric band. This intervention has the potential to decrease an individual's food intake, resulting in weight reduction [10, 11].

Advantages

- Non-surgical: Gastric bands are non-surgical and can be performed on an outpatient basis. This may be an attractive option for patients who are disinclined to undergo extensive surgical procedures [10, 11].
- The efficacy of gastric bands in achieving substantial weight loss has been demonstrated. According to research findings, the procedure has been associated with an average reduction in excess body weight ranging from 50% to 60% within a period of 2 years [10, 11].
- Gastric bands possess the attribute of reversibility, thereby allowing for the reversal of the procedure in the event of necessity [10, 11].
- According to sources, gastric bands exhibit a reduced incidence of complications in comparison to alternative weight loss interventions such as gastric bypass or sleeve gastrectomy [10, 11].

Disadvantages

- The achievement and maintenance of weight loss in patients with a gastric band necessitates strict adherence to dietary guidelines and lifestyle modifications [10, 11]. Non-compliance with these instructions may result in adverse outcomes, including complications and the restoration of previously lost weight [12].
- Regular adjustments may be necessary: Gastric bands may need to be regularly adjusted to ensure their proper functioning and to attain the most effective weight reduction. This may necessitate several consultations with a healthcare professional [10, 11].
- The utilization of gastric bands carries a potential risk of complications, such as band erosion, band slippage, and infection. Occasionally, it may be necessary to remove or replace the band [10, 11].
- The efficacy of gastric bands may not be universal among patients, particularly those with severe obesity or who have previously undergone weight loss surgery. Patients may also experience complications such as acid reflux or esophageal dilation [10, 11].

In general, gastric bands may represent a viable alternative for individuals seeking a non-invasive weight reduction approach that is both efficacious and reversible. It is imperative for patients to have a comprehensive understanding of the potential risks and limitations associated with the procedure prior to making an informed decision. Collaborating with a proficient healthcare practitioner is crucial in assessing the suitability of gastric bands for one's individual requirements and objectives. Furthermore, individuals who have undergone gastric band surgery are required to comply with rigorous dietary protocols and implement lifestyle modifications in order to attain and sustain weight reduction.

7.4 Gastric Balloons

Gastric balloons are gastrointestinal implants utilized for weight loss purposes. The gastric balloon is a pliable silicone device that is introduced into the stomach and subsequently inflated with either saline or air, as documented in sources 13 through 15. This procedure is temporary and is typically used for a period of 6 months to a year to help patients achieve weight loss.

Advantages

- Non-surgical weight loss interventions such as gastric balloons are available as an alternative to surgical procedures. This may be an attractive option for patients who are disinclined to undergo extensive surgical procedures [13–15].
- According to research, gastric balloons have been found to be a viable option for promoting substantial weight loss. The procedure has been reported to result in an average reduction of 30–40% of excess weight among patients within a 6-month period, according to sources [13–15].
- The reversibility of gastric balloons implies that the intervention can be reversed in the event of a need to do so, as indicated by sources [13–15].
- According to studies [13–15], gastric balloons exhibit a lower incidence of complications in comparison to alternative weight loss interventions, such as gastric bypass or sleeve gastrectomy.
- Gastric balloons have been found to be effective in initiating weight loss and promoting the development of healthy habits that can result in sustained success for patients [13–15].

Disadvantages

- Gastric balloons are designed to serve as a transitory measure for achieving weight loss. Following the removal of the balloon, it is imperative for patients to adhere to a nutritious diet and exercise regimen in order to sustain their weight loss, as indicated by sources [13–15].
- The success of weight loss in patients with a gastric balloon is contingent upon strict adherence to prescribed dietary guidelines and lifestyle modifications. Non-adherence to these instructions may result in complications and the recurrence of weight gain [13–15].
- The presence of the balloon in the stomach may result in discomfort or nausea for certain patients following the procedure, as reported in sources [13–15].
- Achieving weight loss goals may necessitate the implementation of multiple procedures for patients. The procedure's cost and time commitment may be elevated as indicated by previous research [13–15].

The utilization of gastric balloons carries a potential risk of complications such as stomach perforation, gastric ulcers, and bowel obstruction. Complications may necessitate the early removal of the balloon in exceptional circumstances [13–15].

In general, gastric balloons may represent a viable alternative for individuals seeking a non-invasive, transitory approach to address weight reduction. It is imperative that patients possess knowledge regarding the probable hazards and constraints associated with the procedure prior to arriving at a conclusion. Collaborating with a proficient healthcare practitioner is crucial in assessing the suitability of gastric balloons as a viable alternative for fulfilling one's individual requirements and objectives. Furthermore, individuals who undergo gastric balloon treatment are required to comply with rigorous dietary protocols and implement modifications to their lifestyle in order to attain and sustain weight reduction.

7.5 Gastric Stimulators

Gastric stimulators are a variety of gastrointestinal implants that are employed in the management of specific digestive disorders, including gastroparesis, and as a means of facilitating weight loss [16, 17]. The mechanism of action of these devices involves the administration of electrical stimuli to the muscles of the stomach, thereby modulating gastric motility and satiety.

Advantages

- Gastric stimulators have the potential to enhance gastric motility, the physiological mechanism responsible for propelling ingested food through the gastrointestinal tract. This can be particularly advantageous for individuals diagnosed with gastroparesis, a medical condition characterized by delayed gastric emptying resulting in symptoms such as nausea, vomiting, and bloating [16, 17].
- Gastric stimulators are classified as non-invasive medical devices, as they do not necessitate any surgical intervention. Rather than an alternative approach, they are inserted subcutaneously through a minimally invasive surgical intervention [16, 17].
- The adjustability of gastric stimulators enables healthcare professionals to tailor the degree of stimulation to the individualized requirements of the patient. The optimization of device effectiveness and minimization of potential side effects can be facilitated through this approach [16, 17].
- Gastric stimulators have the potential to aid in weight loss by decreasing appetite and enhancing satiety, as evidenced by studies [16, 17].

Disadvantages

- Insufficient evidence: Despite initial promising outcomes, the effectiveness of gastric stimulators for weight loss is not yet supported by sufficient evidence. The existing literature has presented inconclusive findings, indicating a need for further investigation to comprehensively comprehend the potential advantages and drawbacks of this particular technology [16, 17].
- Possible adverse reactions: Although gastric stimulators are generally deemed safe, there exist certain potential adverse reactions linked to their utilization, such as discomfort at the point of implantation, infection, and device malfunction [16, 17].

- The cost of gastric stimulators can be considerably high, varying from a few thousand dollars to tens of thousands of dollars. The affordability of healthcare can be challenging for certain patients [16, 17].
- Regular follow-up is necessary for gastric stimulators to ensure their proper functioning and make any required adjustments to the level of stimulation. This is stated in sources [16, 17].
- Gastric stimulators may not be universally applicable as they may pose risks for individuals with specific medical conditions or a prior history of gastrointestinal surgery [16, 17].

To sum up, gastric stimulators have the potential to serve as a viable treatment alternative for individuals diagnosed with gastroparesis and could also be explored as a means of promoting weight loss. Further research is required to comprehensively comprehend the efficacy and possible adverse effects of the aforementioned. Individuals contemplating the use of gastric stimulators are advised to collaborate closely with their healthcare provider in order to ascertain the suitability of this technology for their specific requirements and objectives. Furthermore, it is imperative for patients to be cognizant of the prospective expenses and prerequisites linked with gastric stimulator therapy, encompassing periodic monitoring and plausible adverse reactions.

7.6 Endoscopic Suturing

Endoscopic suturing devices are gastrointestinal implants utilized for the purpose of creating folds or pleats within the stomach, ultimately resulting in a reduction in size. This reduction in size can serve to restrict food intake. Various endoscopic suturing devices have been developed [18, 19].

Advantages

- Endoscopic suturing devices are a non-invasive option that can be executed on an outpatient basis. This may be an attractive option for patients who are disinclined to undergo extensive surgical procedures [18, 19].
- Endoscopic suturing devices possess the attribute of reversibility, thereby allowing for the possibility of undoing the procedure if deemed necessary [18, 19].
- The non-surgical nature of endoscopic suturing devices results in a comparatively shorter recovery time in contrast to alternative weight loss surgeries, as evidenced by sources [18, 19].

Disadvantages

- The efficacy of endoscopic suturing devices in achieving substantial weight loss is comparatively lower than other weight loss surgeries. In order to attain their desired outcomes, patients may need to integrate the aforementioned procedure with supplementary weight loss approaches, such as dietary modifications and physical activity [18, 19].

- The utilization of endoscopic suturing devices carries a potential risk of complications, such as bleeding, infection, and perforation of the stomach or other organs, as reported in literature sources [18, 19].
- The efficacy of endoscopic suturing devices may vary among patients, particularly those with severe obesity or who have previously undergone weight loss surgery, as suggested by sources [18, 19].

In general, endoscopic suturing devices may present a viable alternative for individuals seeking a non-invasive approach to weight loss that entails a comparatively abbreviated convalescence period. It is imperative for patients to have a comprehensive understanding of the potential risks and limitations associated with the procedure prior to making an informed decision. Collaborating with a proficient healthcare practitioner is crucial in assessing the suitability of endoscopic suturing devices for individual requirements and objectives.

7.7 Endoluminal Sleeves

Endoluminal sleeves refer to a category of gastrointestinal implants utilized for the purpose of lining the interior of the small intestine, thereby establishing a partition between ingested food and the intestinal wall. This phenomenon has the potential to decrease the assimilation of essential nutrients and consequently result in a reduction in body weight. Various categories of Endoluminal sleeves have been identified [20, 21].

Advantages

- Endoluminal sleeves are a non-invasive medical procedure that can be conducted on an outpatient basis. This may prove attractive to patients who are disinclined towards undergoing extensive surgical procedures [20, 21].
- The efficacy of endoluminal sleeves in inducing substantial weight loss has been demonstrated. Research has indicated that there is an average reduction in weight of 20–30% of excess body weight within a period of 6–12 months following the intervention [20, 21].
- Endoluminal sleeves are a reversible intervention, as indicated by sources, implying that the procedure can be reversed if deemed necessary [20, 21].
- The endoluminal sleeves have demonstrated a potential to enhance metabolic health by ameliorating blood sugar regulation in individuals diagnosed with type 2 diabetes [22].

Disadvantages

- Insufficient long-term data: The safety and efficacy of Endoluminal sleeves have not been fully established due to a lack of adequate long-term data. Although initial studies have demonstrated encouraging outcomes, further investigation is required to ascertain the enduring impacts of the intervention [20, 21].

- The utilization of Endoluminal sleeves carries a potential risk of complications, such as bleeding, infection, and device migration. In addition, there have been documented instances of inflammation and ulceration occurring within the intestinal tract [20, 21].
- The efficacy of endoluminal sleeves may vary among patients, specifically those with severe obesity or a prior history of weight loss surgery. Periodic adjustments or replacements of the device may be necessary to ensure its continued efficacy, as indicated by sources [20, 21].

In general, Endoluminal sleeves may serve as a viable alternative for individuals seeking a weight loss solution that is non-invasive, efficacious, and capable of being reversed. It is imperative for patients to possess knowledge regarding the probable hazards and constraints associated with the procedure prior to arriving at a conclusion. Collaborating with a proficient healthcare practitioner is crucial in assessing the suitability of Endoluminal sleeves as a viable alternative for fulfilling one's distinct requirements and objectives. Furthermore, it is recommended that patients adhere to a nutritious diet and engage in regular physical activity post-procedure in order to optimize the advantages of the implant.

7.8 Enteral Feeding Tubes

Enteral feeding tubes are gastrointestinal implants that are utilized for the purpose of administering nutrition directly to the stomach or small intestine via a tube that is inserted through the nose, mouth, or abdomen. Various forms of Enteral feeding tubes exist, such as nasogastric tubes, gastrostomy tubes, and jejunostomy tubes [22, 23].

Advantages

- Enhanced nourishment: Enteral feeding tubes have the capability to furnish patients with the essential nutrients required when they encounter difficulties in consuming or assimilating food in a regular manner. The significance of this matter is highlighted in the case of patients afflicted with ailments such as cancer, stroke, or severe burns, as evidenced by sources [22, 23].
- Enteral feeding tubes have been found to lower the likelihood of infection when compared to parenteral nutrition, a form of nutrition that is administered directly into the bloodstream [22, 23].
- Enteral feeding tubes have the potential to facilitate increased mobility and activity levels among patients, while simultaneously providing them with necessary nutrition. It is possible for patients to be discharged from the hospital and receive care at home while utilizing an Enteral feeding tube, as evidenced by sources [23, 24].
- Enteral feeding tubes are comparatively more cost-effective than parenteral nutrition, thereby providing a substantial benefit to patients requiring extended nutritional assistance [22, 23].

Disadvantages

- The use of Enteral feeding tubes carries a potential risk of complications, such as infection, blockage, and aspiration. It has been reported that patients may encounter sensations of discomfort or irritation in the area where the tube is introduced [23, 24].
- Enteral feeding tube patients necessitate meticulous monitoring to guarantee the appropriate quantity and quality of nutrition intake. Close monitoring by a healthcare professional and frequent modifications to the feeding schedule may be necessary, as indicated by sources [22, 23].
- The availability of food choices may be restricted due to the use of enteral feeding tubes, thereby limiting the dietary options for patients. According to sources, patients utilizing an Enteral feeding tube may also encounter symptoms such as nausea, vomiting, or diarrhea [22, 23].
- The suitability of enteral feeding tubes may be limited for certain patients, specifically those with particular medical conditions such as severe gastrointestinal bleeding or intestinal obstruction [22, 23].

In general, enteral feeding tubes may serve as a viable alternative for individuals who are incapable of consuming or metabolizing food in a typical manner and necessitate sustenance through nutritional means. It is imperative for patients to have a comprehensive understanding of the potential risks and limitations associated with the procedure prior to making an informed decision. Collaborating with a proficient healthcare practitioner is crucial in assessing the suitability of Enteral feeding tubes for an individual's distinct requirements and objectives. Furthermore, individuals who utilize Enteral feeding tubes necessitate meticulous supervision and adherence to an individualized dietary regimen to guarantee appropriate intake of nutrition in terms of quantity and quality.

7.9 Ongoing and Future Developments in Gastrointestinal Implants

The future of GI implants appears promising due to the continuous advancements in technology and research, despite the existing pros and cons associated with them. Research is being conducted to investigate the potential benefits of utilizing advancements in technology, such as 3D printing, biodegradable materials, and smart implants, to augment the safety, effectiveness, and individualization of gastrointestinal implants [24–26]. Furthermore, it is anticipated that continued clinical trials and extended follow-up studies will yield additional data regarding the advantages and drawbacks of gastrointestinal implants, ultimately resulting in enhanced patient selection, implant development, and clinical results [27, 28].

The realm of gastrointestinal (GI) implants has undergone swift advancements, and there exist numerous promising opportunities that could potentially enhance the diagnosis, treatment, and handling of GI ailments.

The following are potential areas of progress:

- The application of advanced imaging methods, including virtual colonoscopy, confocal laser endomicroscopy, and molecular imaging, exhibits promise for precise and non-invasive detection of gastrointestinal disorders [29, 30]. The aforementioned technologies have the capability to offer immediate visualization of the gastrointestinal (GI) tract on a cellular level, which can facilitate the timely identification and surveillance of ailments such as inflammatory bowel disease (IBD), colorectal cancer, and gastrointestinal bleeding [29, 30].
- The utilization of nanotechnology in the realm of gastrointestinal implants holds the promise of transformative advancements, as it facilitates precise drug delivery, heightened bioavailability, and superior therapeutic results [26]. Nanoparticles, nanostructured materials, and nanocomposites possess the capability to be engineered for targeted delivery of therapeutic agents to precise locations within the gastrointestinal tract. This enables the possibility of localized treatment of ailments such as Crohn's disease, ulcerative colitis, and gastrointestinal tumors. The aforementioned approach exhibits the capacity to mitigate systemic adverse effects and enhance the effectiveness of treatment [26, 31].
- The potential of bioengineered gastrointestinal implants, such as tissue-engineered constructs, 3D-printed implants, and biodegradable scaffolds, has been identified for the purpose of restoring or substituting damaged or diseased gastrointestinal tissues [32]. The implants possess the capability to replicate the configuration and operation of endogenous tissues, thereby expediting the process of tissue restoration and fostering recuperation. The utilization of bioengineered gastrointestinal implants presents a promising opportunity to transform the treatment of various medical conditions, including but not limited to esophageal strictures, gastric ulcers, and colorectal defects, as indicated by previous research studies [24, 32].
- The integration of smart and wireless technologies has the potential to revolutionize gastrointestinal implants, enabling them to function as intelligent devices capable of real-time monitoring, diagnosis, and treatment of gastrointestinal conditions [33, 34]. The utilization of wireless capsule endoscopy, a technique that entails ingestion of a miniature camera-equipped capsule capable of capturing images of the gastrointestinal tract, has significantly transformed the identification of small bowel ailments [33, 34]. Prospective developments may entail intelligent implants that have the capability to consistently monitor gastrointestinal functions, administer therapy, and transmit data through wireless means for prompt supervision and handling of ailments such as gastroesophageal reflux disease (GERD), motility disorders, and inflammatory bowel disease (IBD) [33, 34].
- The integration of robotic-assisted surgery and minimally invasive techniques has brought about significant advancements in the field of gastrointestinal surgery. These advancements have resulted in expedited recovery periods, decreased occurrence of complications, and enhanced patient outcomes. The potential for future developments in robotics, such as tele-robotics and autonomous robots,

has the capacity to improve the accuracy, safety, and results of gastrointestinal surgeries [35–37]. The potential adoption of robotic-assisted procedures, including but not limited to colorectal surgery, bariatric surgery, and hepatic surgery, may result in enhanced patient care [35–37].

- The utilization of genomics, proteomics, and other "omics" technologies has the potential to facilitate personalized medicine strategies for the treatment of gastrointestinal conditions, as evidenced by recent advancements in the field of personalized medicine and precision therapeutics. The utilization of personalized medicine techniques such as genetic profiling and molecular profiling may aid in the identification of patients who are susceptible to gastrointestinal conditions, assist in making informed treatment decisions, and enhance the efficacy of therapy response [37–39]. The utilization of precision therapeutics, such as targeted therapies, immunotherapies, and gene therapies, may increase in the management of gastrointestinal conditions, resulting in improved and personalized treatment approaches. This is suggested by recent research [38].

It is noteworthy that the potential benefits of gastrointestinal (GI) implants in the future are accompanied by several challenges and factors that must be taken into account. These include regulatory clearance, ethical implications, cost-effectiveness, and the long-term safety and effectiveness of such implants. The continuous progress in technology, imaging, diagnostics, materials science, and personalized medicine is expected to have a significant impact on the development of gastrointestinal implants and transform the approach to the diagnosis, treatment, and management of gastrointestinal disorders.

7.10 Urological Implants

The discipline of urology has undergone a revolution thanks to urological implants, which offer creative treatments for a range of urological issues that influence urinary tract function. These implants, which range from penile prostheses to urinary incontinence devices, offer encouraging results in terms of enhanced functioning, minimally invasive procedures, tailored solutions, and long-term management of urological problems. However, urological implants have inherent risks and potential side effects just like any medical procedure. In order to make wise therapeutic decisions, it is crucial to objectively assess the benefits and drawbacks of urological implants.

Restoration of regular urine function, advantages of minimally invasive treatments, solutions that are specially made for each patient's anatomy and demands, and long-lasting symptom relief are all benefits of urological implants. For instance, inflatable penile implants have been demonstrated to improve sexual health and quality of life in individuals with refractory erectile dysfunction by restoring erectile function [39]. Similar to artificial sphincters, artificial urinary sphincters have been shown to manage urine incontinence efficiently in individuals with neurogenic bladder dysfunction, enabling continence and enhanced social functioning [39–41].

On the other side, urological implants' drawbacks include surgical risks, sensitivities to foreign bodies, cost, potential problems, and lifestyle adjustments. Urological implantation operations may encounter difficulties due to implantation-related complications, including infection, erosion, and mechanical failure [42]. Certain types of urological implants may have foreign body reactions, including immunological responses and problems, which can have negative effects [42]. Patients and healthcare systems may be financially burdened by the price of urological implants and related costs, such as follow-up visits and device replacements [43, 44]. Even after a successful implantation, potential issues including device malfunction, discomfort, or more procedures could develop [44]. Patients' quality of life may also be impacted by lifestyle modifications such as routine adjustments, device maintenance, and restrictions on physical activity.

Careful assessment and patient selection are essential in clinical decision-making given the benefits and drawbacks of urological implants. Understanding the potential advantages and disadvantages of urological implants can help medical professionals make decisions that are based on the needs of specific patients. In addition, ongoing advancements in urology implant technology show promise for enhancing results and resolving current drawbacks. In order to give a thorough overview of this rapidly developing topic, we will examine the benefits and drawbacks of urological implants in detail in this part, combining pertinent data from clinical trials and actual case studies.

7.11 Types of Urological Implants

1. Men with erectile dysfunction (ED) who have not responded to traditional therapies including oral pills, injections, or suction devices may benefit from penile implants. Inflatable and semi-rigid penile implants are the two basic varieties. The most popular type of implant, known as an inflatable one, consists of two cylinders put in the penis and a pump in the scrotum [45, 46]. When the pump is turned on, fluid is poured into the cylinders from a reservoir, which results in an erection. Rods that are implanted in the penis and may be bent up for sex and down for comfort make up semi-rigid implants, which are easier to use [45, 46].
2. Men with urine incontinence are treated with an artificial urinary sphincter (AUS). The system includes a reservoir implanted in the abdomen, a pump inserted in the scrotum, and a cuff placed around the urethra. Squeezing the pump causes the cuff to deflate, releasing urine when the patient has to urinate. Once they have finished urinating, they let off of the pump, which causes the cuff to re-inflate and stop urine leakage [45–47].
3. Devices for suspending the bladder at the neck: These devices are used to treat female stress incontinence. The urethra is supported by these devices' sling, which also serves to stop urine leakage. Bladder neck suspension devices come

in a variety of forms, including artificial slings, ones created from the patient's own tissue, and minimally invasive slings that can be implanted without the need for significant surgery [45–47].

4. Similar to the AUS for men, the artificial urinary sphincter for women is a medical device used to treat female urine incontinence. The system comprises of a reservoir implanted in the abdomen, a pump implanted in the labia, and a cuff inserted around the urethra. Squeezing the pump causes the cuff to deflate, releasing urine when the patient has to urinate. Once they have finished urinating, they let off of the pump, which causes the cuff to re-inflate and stop urine leakage [45–47].
5. Overactive bladder (OAB), urge incontinence, and some forms of persistent pelvic pain are all treated with the sacral nerve stimulator (SNS). The system consists of a lead that is positioned close to the sacral nerves and a small generator that is inserted under the skin in the buttocks or abdomen. Electrical impulses are delivered to the nerves via the generator, which can assist control bladder function and lessen symptoms [45–47].
6. Urethral stents: The narrowing of the urethra known as urethral strictures, which can make it difficult to urinate, is treated with urethral stents. The urethra is kept open by the stent, a tiny tube, which facilitates easier urine flow [45–47].

7.12 Penile Implants

Men with erectile dysfunction (ED) who have not responded to alternative therapies including oral pills, injections, or suction devices may benefit from penile implants, one form of urological implant. Inflatable and semi-rigid penile implants are the two primary varieties [45–47].

7.12.1 Inflatable Penile Implants

Advantages

- Improved sexual function: Men with severe ED may benefit from inflatable penile implants, which help them develop and sustain an erection that is appropriate for sexual activity [45–47].
- Natural appearance: Inflatable implants, which allow for a more natural-looking erection that can be adjusted to the desired firmness, can look and feel more natural than semi-rigid implants [45–47].
- Greater patient satisfaction: According to studies, patients who have inflatable penile implants express more satisfaction with their sexual function and quality of life [45–47].
- Long-term efficacy: Studies have indicated that inflatable implants are beneficial over the long term, with a 10-year survival rate of up to 70% [45–47].

Disadvantages

- Higher risk of complications: Compared to semi-rigid implants, inflatable implants are more complicated and require longer surgery, increasing the risk of complications like infection, mechanical failure, and erosion [45–47].
- Lengthier recovery period: Patients who have inflatable penile implants may experience a lengthier recovery period than those who receive alternative therapies because they must wait for the wounds to heal and the device to fully inflate [45–47].
- Requirement for manual dexterity: Some patients may find it difficult to use the pump that inflates and deflates inflatable implants because it requires physical dexterity [45–47].

7.12.2 Semi-rigid Penile Implants

Advantages

- Easier procedure: Semi-rigid implants are implanted more quickly and easily than inflatable implants, lowering the risk of problems and accelerating recovery [45–47].
- Lower cost: Semi-rigid implants are typically more affordable than inflatable implants, making them an appealing choice for some patients [45–47].
- No requirement for manual dexterity: Patients with weak hands or poor hand-eye coordination may find semi-rigid implants to be a useful alternative because they don't require manual dexterity to operate [45–47].

Disadvantages

- Less natural appearance: Penises that are constantly or semi-permanently erect as a result of semi-rigid implants may appear less natural [45–47].
- Limited adjustability: Due to the lack of erection firmness changes possible with semi-rigid implants, it can be challenging to attain the ideal amount of firmness for sexual activity [45–47].
- Lower levels of patient satisfaction: Research has revealed that semi-rigid penile implant patients experience lower levels of sexual function and satisfaction than patients who receive inflatable implants [45–47].

In conclusion, men with severe ED who have not responded to previous therapies should consider penile implants. The choice of which implant to employ will depend on the patient's specific demands and medical history. Both inflatable and semi-rigid implants have specific benefits and drawbacks. To find the best course of treatment for their unique disease, people should discuss their options with their healthcare professional.

7.13 Artificial Urinary Sphincter

Men with urine incontinence may benefit from a urological implant called an artificial urinary sphincter (AUS). It comprises of an abdominal reservoir of fluid, a tiny pump inserted in the scrotum, and a cuff that is wrapped over the urethra. The pump controls the flow of urine by inflating and deflating the cuff [41, 42, 48]. AUS has several benefits, but it also has some drawbacks.

Advantages

- Improved quality of life: By minimizing or eliminating involuntary urine leakage, AUS can greatly enhance the quality of life for men with urinary incontinence [41, 42, 48].
- High success rate: Up to 90% of men who receive AUS report better urine continence [41, 42, 48].
- Long-lasting: AUS has been demonstrated to be long-lasting, with up to 80% of patients reporting an acceptable result after 10 years [41, 42, 48].
- Low risk of complications: When the device is implanted by a skilled surgeon, the risk of infection or erosion with AUS is low [41, 42, 48].
- Adjustable: The cuff's pressure can be changed to meet the needs of each patient, enabling the creation of a personalized treatment [41, 42, 48].

Disadvantages

- High price: AUS is relatively expensive, and for some patients, this can be a deterrent [41, 42, 48].
- Requires surgery: The implantation of AUS and any future changes require surgery, which entails risks such as infection, hemorrhage, and anesthesia-related issues [41, 42, 48].
- Prolonged recovery period: Patients who have AUS may need to take time off work to recover [41, 42, 48].
- Not appropriate for all patients: AUS is not appropriate for all patients, including those who have active infections or serious urethral or bladder damage [41, 42, 48].

Possible Complications

- Malfunction: Urinary incontinence or discomfort may result from AUS malfunction due to mechanical failure, tubing kinks, or fluid leaks [41, 42, 48].
- Erosion: The cuff or other AUS components may erode into nearby tissues or organs, which could result in an infection, discomfort, or the requirement to remove the device [41, 42, 48].
- Infection: AUS can get an infection, which can result in serious side effects such as sepsis or organ failure [41, 42, 48].
- Occlusion: According to some studies [41, 42, 48], AUS might result in occlusion of the bladder neck or urethra, which can induce urine retention.

In conclusion, AUS is a beneficial alternative for males with urine incontinence, especially those who have not responded to conventional therapies. With a high success rate and long lifespan, the device can greatly improve patients' quality of life [41, 42, 48]. The device is not appropriate for many people, and there are considerable disadvantages related to cost and the requirement for surgery. Patients thinking about AUS should talk to their doctor about the advantages and disadvantages to see if the device is right for them.

7.14 Bladder Neck Suspension Devices

Women with stress incontinence are treated with bladder neck suspension devices. When the muscles that support the bladder weaken, urine leaks during activities like coughing, sneezing, or exercise [49–52]. This condition is known as stress urinary incontinence. In order to stop urine leakage, bladder neck suspension devices support the bladder and urethra. They have certain benefits, but they also have some drawbacks [49–52].

Advantages

- High success rate: Up to 90% of patients who use bladder neck suspension devices for the treatment of stress urine incontinence report an improvement in their symptoms [49–52].
- Reliable: It has been demonstrated that bladder neck suspension devices are reliable, with up to 80% of patients reporting a satisfactory result after 10 years [49–52].
- Minimally invasive: Compared to open surgery, some bladder neck suspension devices can be implanted via a vaginal incision or laparoscopic surgery, which may cause less pain and hasten recovery [49–52].
- Better quality of life: Bladder neck suspension devices can enhance the quality of life for women with stress urinary incontinence by reducing or eliminating involuntary pee leakage [49–52].

Disadvantages

- Needs surgery: The insertion of bladder neck suspension devices requires surgery, which entails risks such as infection, hemorrhage, and anesthetic complications [49–52].
- Prolonged recovery time: Patients may need to take time off work to recover after using bladder neck suspension devices [49–52].
- Not suited for all patients: Patients with severe urine incontinence or those who have an active infection are not candidates for bladder neck suspension devices [49–52].

Potential Complications

- Failure: Over time, bladder neck suspension devices may stop working, which can cause recurrent incontinence [49–52].

- Infection: Bladder neck suspension devices have the potential to contract an infection, which can result in serious side effects such as sepsis or organ failure [49–52].
- Bladder neck suspension devices have the potential to block the urethra, which may result in urine retention or other complications [49–52].
- Erosion: The bladder neck suspension device may erode into nearby tissues or organs, resulting in discomfort or the requirement that the device be removed [49–52].

In conclusion, bladder neck suspension devices are a highly successful and long-lasting solution for treating stress urine incontinence in women. However, they come with some dangers and potential consequences, as well as requiring surgery. Patients who are interested in bladder neck suspension devices should talk to their doctor about the advantages and disadvantages to see if the device is right for them.

7.15 Artificial Urinary Sphincter

Women with stress urinary incontinence (SUI), a disorder that causes involuntary pee leakage during activities like coughing, sneezing, or exercise, have the option of receiving therapy with artificial urinary sphincters (AUS). AUS devices are made up of an inflatable cuff that is put around the urethra, a pump that regulates the flow of urine, and a reservoir that holds the fluid. AUS devices do offer some benefits for females with SUI, but they also have certain drawbacks [53–56].

Advantages

- High success rate: Up to 90% of patients who receive AUS devices for the treatment of SUI in women report improvements in their symptoms [53–56].
- Reliable: It has been demonstrated that AUS devices are reliable, with up to 80% of patients reporting an acceptable result after 10 years [53–56].
- Better quality of life: AUS devices can enhance the quality of life for women with SUI by minimizing or eliminating involuntary urine leakage [53–56].

Disadvantages

- Requires surgery: The insertion of AUS devices requires surgery, which entails risks such as infection, hemorrhage, and anesthetic complications [53–56].
- Prolonged recovery period: Patients who use AUS devices may need to take time off work to recover [53–56].
- Complications: AUS devices may result in issues such as urethral blockage, device failure, infection, and device erosion [53–56].
- Expensive: AUS devices are costly, and pre-approval may be necessary or insurance may not pay the cost [53–56].

Potential Complications

- Device malfunction: Over time, AUS devices may malfunction, which can cause recurrent incontinence [53–56].

- Infection: AUS devices may contract an infection, which can result in serious side effects such as sepsis or organ failure [53–56].
- Blockage: Urethral blockage brought on by AUS devices has been linked to urine retention and other complications [53–56].
- Erosion: The AUS device may erode into adjacent tissues or organs, resulting in discomfort or the requirement that the device be removed [53–56].

In conclusion, AUS devices are a highly successful and long-lasting therapy option for SUI in women. However, they come with some dangers and potential consequences, as well as requiring surgery. Patients thinking about AUS devices should talk to their doctor about the advantages and disadvantages to see if the device is right for them.

7.16 Sacral Nerve Stimulator

The sacral nerves, which are in charge of regulating the bladder, intestine, and pelvic floor muscles, are stimulated using a minimally invasive surgical procedure called sacral nerve stimulation (SNS). A small battery-operated generator that is implanted beneath the skin of the buttocks is the device, also known as a sacral nerve stimulator (SNS). SNS has shown encouraging outcomes for treating a variety of gastrointestinal and urological problems, but it may also have some drawbacks [57–59].

Advantages

- Effective: Studies have shown that SNS is a successful treatment for a number of diseases, including urine and fecal incontinence, overactive bladder, chronic pelvic discomfort, and constipation [57–59].
- Reversible: If the patient experiences unfavorable side effects or the treatment is no longer required, the process can be reversed and the device can be turned off or removed [57–59].
- Minimally invasive: Because SNS is an outpatient surgery with a low risk of complications, patients frequently go home the same day [57–59].
- Low risk of complications: Temporary pain or discomfort at the insertion site is the most frequent adverse effect of SNS and is generally considered to be low risk [57–59].

Disadvantages

- Expensive: SNS devices and surgical procedures can be costly, and insurance coverage varies or may not be available [57–59].
- Potential side effects: Although there is a chance of experiencing adverse effects such as pain or discomfort, infection, nerve injury, or device failure, SNS does carry some risk [57–59].
- Long-term efficacy: There are few long-term data on SNS's efficacy, and some studies have noted a deterioration in the treatment's efficacy with time [57–59].

- Demands battery replacement: Over time, the SNS device's battery will need to be changed, necessitating yet another surgical procedure [57–59].

Potential Complications

- Infection: SNS devices may contract an infection, which can result in serious side effects such as sepsis or organ failure [57–59].
- Nerve injury: The technique has the potential to harm the nerves, which may result in pain, numbness, or other sensory problems [57–59].
- Device malfunction: Over time, SNS devices may malfunction, resulting in recurring symptoms and the requirement for additional surgery [57–59].
- Lead migration or fracture: The wires connecting the device to the sacral nerves have the potential to move or break, which could result in device failure or malfunction [57–59].

In conclusion, SNS is a viable therapeutic option with a minimal risk of problems and possible advantages including increased quality of life for a number of urological and digestive diseases. There are some potential risks and side effects to take into account, though, and it is an expensive surgery. Patients thinking about SNS should talk to their doctor about the advantages and disadvantages to see if the procedure is right for them [57–59].

7.17 Urethral Stent

Small, flexible devices called urethral stents are inserted in the urethra to alleviate urine blockage. In addition to urethral strictures, benign prostatic hyperplasia (BPH), and urethral tumors, they are utilized to treat a number of other disorders. Urethral stents can be a successful therapeutic choice, but they may also have some potential benefits and drawbacks [60–62].

Advantages

- Minimally invasive: The implantation of a urethral stent is a minimally invasive operation that is frequently done as an outpatient, allowing patients to go home the same day [60–62].
- Enhances urine flow: In patients with urinary obstruction, urethral stents can enhance urine flow, alleviating symptoms such urinary retention, urgency, and frequency [60–62].
- Can be removed: If a patient develops unfavorable side effects or the treatment is no longer required, urethral stents may be removed [60–62].
- Non-surgical option: For individuals who are not candidates for more invasive procedures or who prefer not to undergo surgery, urethral stents are a good non-surgical option [60–62].

Disadvantages

- Potential side effects: Hematuria (blood in the urine) and urinary tract infections are the most frequent side effects of urethral stents [60–62].

- Migration: Urethral stents may move or migrate, which may cause issues such as recurring urinary obstruction [60–62].
- Encrustation: Urinary salts that adhere to urethral stents may irritate, inflame, and infect patients [60–62].
- Infection: Urethral stents may get an infection, which can result in serious side effects such as sepsis or organ failure [60–62].
- Difficult to remove: Urinary stent removal can be challenging and, in certain instances, necessitate a more intrusive procedure [60–62].
- Migration: As previously indicated, urethral stents have the potential to move out of place, which can result in consequences such as recurring urinary obstruction [60–62].
- Encrustation: If the stent accumulates urinary salts, this can irritate the surrounding tissue, lead to inflammation and infection, and make removal challenging [60–62].
- Infection: Urethral stents may get an infection, which can result in serious side effects such as sepsis or organ failure [60–62].
- Urine retention: The stent may occasionally result in urine retention, which may need the insertion of an interim urinary catheter [60–62].

In conclusion, urethral stents may be a useful choice for treating urinary blockage, particularly for patients who are not candidates for more invasive treatments. They could cause discomfort, infection, or migration, among other issues and side effects. Patients who are thinking about getting urethral stents should talk to their doctor about the advantages and disadvantages to see if the procedure is right for them [60–62].

The management of many urological disorders can benefit greatly from urological implants. For individuals with diseases like urine incontinence, erectile dysfunction, and urinary tract obstruction, they can significantly enhance sexual performance, quality of life, and urinary continence. For individuals who have tried conservative care but were unsuccessful, urological implants are frequently long-lasting and can provide permanent treatments. Additionally, urological implants are adaptable and may be made to fit each patient's specific needs, offering individualized therapy.

Urological implants do, however, have some drawbacks. Probable drawbacks of surgery include its risks, which include probable difficulties during implantation, reactions to foreign bodies, and the cost of the process. Urological implant-specific issues like infection, mechanical failure, erosion, and revision surgery can also be problematic. For patients with urological implants, a change in lifestyle may also be necessary, and the effect on daily life should be carefully examined.

The use of urological implants should only be chosen after careful assessment of the patient's overall health, medical history, and specific requirements. The patient's condition should be thoroughly assessed, taking into account factors like the seriousness of the urological problem, the patient's expectations, and potential risks and advantages. After addressing the potential benefits and drawbacks of urological implants, the patient should give their informed permission and be given a reasonable expectation of the procedure's results.

To ensure thorough assessment and management, a multidisciplinary approach involving urologists, surgeons, and other pertinent healthcare specialists should be used in the decision-making process. In order to identify and treat any potential side effects or complications related to urological implants, careful monitoring and follow-up are also necessary.

In conclusion, urological implants have the potential to be beneficial in treating a variety of urological problems, but they also have some risks and drawbacks. When contemplating urological implants as a therapeutic option, careful examination, patient selection, and detailed patient talks about potential risks and benefits are essential in clinical decision-making.

7.18 Ongoing and Future Developments in Urological Implants

Urological implants are a sector that is always changing, and there are a number of bright prospects for the future that could lead to more improvements in this discipline. Future possibilities for urological implants include:

1. The majority of urological implants are currently made of synthetic materials like silicone or polymeric materials. This is due to advancements in implant materials. The use of biocompatible and biodegradable materials, however, may help urological implants last longer and be safer, according to continuing research and development in the field of biomaterials [61, 63]. These developments may result in the creation of implants that are more resilient and feel more like native tissue, with less risks of problems like infection or responses to foreign bodies [61, 63].
2. The creation of urological implants with improved qualities is possible thanks to nanotechnology and drug-eluting implants. Drug-eluting implants, for instance, might release drugs locally to prevent issues like inflammation or tissue overgrowth. Nanotechnology-based coatings on implant surfaces, meanwhile, could help minimize bacterial adherence and prevent infection. These developments may enhance the effectiveness of urological implant procedures and lessen the need for follow-up procedures or revision surgeries [64–66].
3. Robotics and minimally invasive procedures: Robotics and minimally invasive procedures have been used more and more frequently in urological surgery, and this trend is expected to continue [67–69]. For patients having urological implant operations, robotic-assisted surgeries allow for more precision, control, and vision, which may enhance results and hasten recovery times. Additionally, ongoing improvements in minimally invasive procedures like laparoscopy and endoscopy may help to lessen the risks and consequences connected with urological implant surgery [67–69].
4. Implants that are personalized and patient-specific: With the advent of imaging technology and 3D printing, there is a growing interest in the creation of urologi-

cal implants that are both personalized and patient-specific. These implants enable more individualized and precise implantation because they may be made to fit a patient's particular anatomy and needs [70–72]. Better functional outcomes, lower complication risks, and more patient satisfaction could all result from patient-specific implants [70–72].

5. Integration of wireless and remote monitoring technologies is another potential development for urological implants in the future [73–75]. Real-time monitoring of implant function, patient symptoms, and consequences may be possible using these technologies, which could aid in the early identification and treatment of any problems. For instance, wireless sensors could be incorporated into urology implants to track variables like pressure, flow rate, or temperature and transmit the information to healthcare professionals for remote monitoring and prompt intervention [73–75].
6. Artificial intelligence and machine learning: By enhancing implant design, surgical planning, and patient selection, the use of artificial intelligence (AI) and machine learning algorithms in urological implants has the potential to completely transform the sector. To find patterns and trends that could guide therapeutic decision-making, AI systems could examine vast volumes of data, including patient demographics, implant attributes, and outcomes [76, 77]. Additionally, predictive models for patient outcomes, problems, or implant failures might be created using machine learning algorithms, which would help with risk assessment and customized treatment planning [76, 77].

In conclusion, urological implants are a sector that is continuously changing. There are a number of promising future prospects that promise to bring about more breakthroughs in this industry. Personalized and patient-specific implants, nanotechnology and drug-eluting implants, robotics and minimally invasive procedures, wireless and remote monitoring technologies, and the incorporation of artificial intelligence and machine learning are a few of these prospects. In the field of urological implants, continued research and innovation could potentially result in better results, lower risks of problems, and more patient satisfaction.

References

1. Cheng J, et al. Potential of electrical neuromodulation for inflammatory bowel disease. Inflamm Bowel Dis. 2020;26(8):1119–30.
2. Willms B, Arends J. Comparison of isolated (guar) and natural (Musli) dietary fiber in the treatment of type II diabetes. Med Klin (Munich). 1987;82(12–13):429–31.
3. Wang PM, et al. A wireless implantable system for facilitating gastrointestinal motility. Micromachines (Basel). 2019;10(8):525.
4. Lin Z, Chen JDZ. Developments in gastrointestinal electrical stimulation. Crit Rev Biomed Eng. 2017;45(1–6):263–301.
5. Dharmayanti C, et al. Drug-eluting biodegradable implants for the sustained release of bisphosphonates. Polymers (Basel). 2020;12(12):2930.

6. Jumbe S, Hamlet C, Meyrick J. Psychological aspects of bariatric surgery as a treatment for obesity. Curr Obes Rep. 2017;6(1):71–8.
7. Lloyd FMM, Hewison A, Efstathiou N. "It just made me feel so desolate": patients'narratives of weight gain following laparoscopic insertion of a gastric band. J Clin Nurs. 2018;27(3–4):732–42.
8. Delin CR, Anderson PG. A preliminary comparison of the psychological impact of laparoscopic gastric banding and gastric bypass surgery for morbid obesity. Obes Surg. 1999;9(2):155–60.
9. Pietrabissa G, et al. Psychological aspects of treatment with intragastric balloon for management of obesity: a systematic review of the literature. Obes Facts. 2022;15(1):1–18.
10. Flowers D, et al. Gastric bands: what the general radiologist should know. Clin Radiol. 2013;68(5):488–99.
11. Hopkins JC, et al. The use of adjustable gastric bands for management of severe and complex obesity. Br Med Bull. 2016;118(1):64–72.
12. Pacheco KA. Allergy to surgical implants. Clin Rev Allergy Immunol. 2019;56(1):72–85.
13. Lari E, et al. Intra-gastric balloons – the past, present and future. Ann Med Surg (Lond). 2021;63:102138.
14. Henry Z, et al. AGA clinical practice update on management of bleeding gastric varices: expert review. Clin Gastroenterol Hepatol. 2021;19(6):1098–1107e1.
15. Silva LB, Neto MG. Intragastric balloon. Minim Invasive Ther Allied Technol. 2022;31(4):505–14.
16. Camilleri M, Sanders KM. Gastroparesis. Gastroenterology. 2022;162(1):68–87e1.
17. Usai-Satta P, et al. Gastroparesis: new insights into an old disease. World J Gastroenterol. 2020;26(19):2333–48.
18. Committee AT, et al. Endoscopic closure devices. Gastrointest Endosc. 2012;76(2):244–51.
19. Kaan HL, Ho KY. Endoscopic robotic suturing: the way forward. Saudi J Gastroenterol. 2019;25(5):272–6.
20. Committee AT, et al. Endoluminal bariatric techniques. Gastrointest Endosc. 2012;76(1):1–7.
21. Cote GA, Edmundowicz SA. Emerging technology: endoluminal treatment of obesity. Gastrointest Endosc. 2009;70(5):991–9.
22. Williams NT. Medication administration through enteral feeding tubes. Am J Health Syst Pharm. 2008;65(24):2347–57.
23. Silva RME, et al. Immunosuppressives and enteral feeding tubes: an integrative review. J Clin Pharm Ther. 2020;45(3):408–18.
24. Jain P, Kathuria H, Dubey N. Advances in 3D bioprinting of tissues/organs for regenerative medicine and in-vitro models. Biomaterials. 2022;287:121639.
25. Lorenzo-Zuniga V, et al. Biodegradable stents in gastrointestinal endoscopy. World J Gastroenterol. 2014;20(9):2212–7.
26. Alici G. Towards soft robotic devices for site-specific drug delivery. Expert Rev Med Devices. 2015;12(6):703–15.
27. Goldstein DJ, et al. Association of clinical outcomes with left ventricular assist device use by bridge to transplant or destination therapy intent: the multicenter study of MagLev Technology in Patients Undergoing Mechanical Circulatory Support Therapy with HeartMate 3 (MOMENTUM 3) randomized clinical trial. JAMA Cardiol. 2020;5(4):411–9.
28. Li Y, et al. Optimal antiplatelet therapy for prevention of gastrointestinal injury evaluated by ANKON magnetically controlled capsule endoscopy: rationale and design of the OPT-PEACE trial. Am Heart J. 2020;228:8–16.
29. Rondonotti E, et al. Small-bowel capsule endoscopy and device-assisted enteroscopy for diagnosis and treatment of small-bowel disorders: European Society of Gastrointestinal Endoscopy (ESGE) technical review. Endoscopy. 2018;50(4):423–46.
30. Eslamy HK, Quon A. PET/CT imaging of gastrointestinal stromal tumor with calcified peritoneal implants after imatinib therapy. Clin Nucl Med. 2008;33(12):864–5.
31. Robinson DH, Mauger JW. Drug delivery systems. Am J Hosp Pharm. 1991;48(10 Suppl 1):S14–23.

32. O'Neill JD, et al. Gut bioengineering strategies for regenerative medicine. Am J Physiol Gastrointest Liver Physiol. 2021;320(1):G1–G11.
33. Kiourti A, Psathas KA, Nikita KS. Implantable and ingestible medical devices with wireless telemetry functionalities: a review of current status and challenges. Bioelectromagnetics. 2014;35(1):1–15.
34. Moore J, et al. Applications of wireless power transfer in medicine: state-of-the-art reviews. Ann Biomed Eng. 2019;47(1):22–38.
35. Damian DD. Regenerative robotics. Birth Defects Res. 2020;112(2):131–6.
36. Kinross JM, et al. Next-generation robotics in gastrointestinal surgery. Nat Rev Gastroenterol Hepatol. 2020;17(7):430–40.
37. Augustin AM, et al. Endovascular therapy of gastrointestinal bleeding. RöFo. 2019;191(12):1073–82.
38. Sharma S. Nanotheranostics in evidence based personalized medicine. Curr Drug Targets. 2014;15(10):915–30.
39. Quesada-Olarte J, et al. Penile implant instrument innovations. Curr Urol Rep. 2023;24(2):59–67.
40. Gotman I. Characteristics of metals used in implants. J Endourol. 1997;11(6):383–9.
41. Brant WO, Martins FE. Artificial urinary sphincter. Transl Androl Urol. 2017;6(4):682–94.
42. Khouri RK Jr, et al. Artificial urinary sphincter complications: risk factors, workup, and clinical approach. Curr Urol Rep. 2021;22(5):30.
43. Trost L. Future considerations in prosthetic urology. Asian J Androl. 2020;22(1):70–5.
44. Yang DY, Kohler TS. Damage control considerations during IPP surgery. Curr Urol Rep. 2019;20(2):10.
45. Verze P, et al. Two-piece inflatable and semi-rigid penile implants: an effective alternative? Int J Impot Res. 2020;32(1):24–9.
46. Le B, Burnett AL. Evolution of penile prosthetic devices. Korean J Urol. 2015;56(3):179–86.
47. Mulcahy JJ. The development of modern penile implants. Sex Med Rev. 2016;4(2):177–89.
48. Carson CC. Artificial urinary sphincter: current status and future directions. Asian J Androl. 2020;22(2):154–7.
49. Glazener CM, Cooper K, Mashayekhi A. Bladder neck needle suspension for urinary incontinence in women. Cochrane Database Syst Rev. 2017;7(7):CD003636.
50. Glazener CM, Cooper K. Bladder neck needle suspension for urinary incontinence in women. Cochrane Database Syst Rev. 2014;12:CD003636.
51. Gomelsky A, Dmochowski RR. Bladder neck pubovaginal slings. Expert Rev Med Devices. 2005;2(3):327–40.
52. Karram MM, Bhatia NN. Transvaginal needle bladder neck suspension procedures for stress urinary incontinence: a comprehensive review. Obstet Gynecol. 1989;73(5 Pt 2):906–14.
53. Barakat B, et al. A systematic review and meta-analysis of clinical and functional outcomes of artificial urinary sphincter implantation in women with stress urinary incontinence. Arab J Urol. 2020;18(2):78–87.
54. Sperling H, et al. Artificial urinary sphincter in women-too uncommon? Urologe A. 2017;56(12):1572–5.
55. Chartier-Kastler E, et al. Artificial urinary sphincter (AMS 800) implantation for women with intrinsic sphincter deficiency: a technique for insiders? BJU Int. 2011;107(10):1618–26.
56. Elliott DS, Barrett DM. The artificial urinary sphincter in the female: indications for use, surgical approach and results. Int Urogynecol J Pelvic Floor Dysfunct. 1998;9(6):409–15.
57. Malde S, et al. Sacral nerve stimulation for refractory OAB and idiopathic urinary retention: can phenotyping improve the outcome for patients: ICI-RS 2019? Neurourol Urodyn. 2020;39(Suppl 3):S96–S103.
58. Li LF, Ka-Kit Leung G, Lui WM. Sacral nerve stimulation for neurogenic bladder, vol. 90. World Neurosurg; 2016. p. 236–43.
59. Norderval S, et al. Sacral nerve stimulation. Tidsskr Nor Laegeforen. 2011;131(12):1190–3.

60. Peyton CC, Badlani GH. The management of prostatic obstruction with urethral stents. Can J Urol. 2015;22(Suppl 1):75–81.
61. Tammela TL, Talja M. Biodegradable urethral stents. BJU Int. 2003;92(8):843–50.
62. Kaplan SA. Prostatic urethral stents. Semin Urol. 1994;12(3):193–9.
63. Rowe CK, et al. Do the materials matter? A review of the literature and analysis of the materials properties of urethral stents for hypospadias repair. J Pediatr Urol. 2022;18(2):160–7.
64. Dai S, et al. Application of three-dimensional printing technology in renal diseases. Front Med (Lausanne). 2022;9:1088592.
65. Bailly GG, Carlson KV. The pubovaginal sling: reintroducing an old friend. Can Urol Assoc J. 2017;11(6Suppl2):S147–51.
66. Kogan P, Wald M. Male contraception: history and development. Urol Clin North Am. 2014;41(1):145–61.
67. Mittal S, Srinivasan A. Robotics in pediatric urology: evolution and the future. Urol Clin North Am. 2021;48(1):113–25.
68. Fuchs ME, DaJusta DG. Robotics in pediatric urology. Int Braz J Urol. 2020;46(3):322–7.
69. Hemal AK, Menon M. Robotics in urology. Curr Opin Urol. 2004;14(2):89–93.
70. Bilhim T, et al. Minimally invasive therapies for benign prostatic obstruction: a review of currently available techniques including prostatic artery embolization, water vapor thermal therapy, prostatic urethral lift, temporary implantable nitinol device and aquablation. Cardiovasc Intervent Radiol. 2022;45(4):415–24.
71. De Wachter S, et al. New technologies and applications in sacral neuromodulation: an update. Adv Ther. 2020;37(2):637–43.
72. Franco JV, et al. Minimally invasive treatments for lower urinary tract symptoms in men with benign prostatic hyperplasia: a network meta-analysis. Cochrane Database Syst Rev. 2021;7(7):CD013656.
73. Basu AS, et al. Is submucosal bladder pressure monitoring feasible? Proc Inst Mech Eng H. 2019;233(1):100–13.
74. Girtner F, Burger M, Mayr R. Sacral neuromodulation in under- and overactive detrusor-quo vadis?: principles and developments. Urologe A. 2019;58(6):634–9.
75. Sun B, et al. Wirelessly activated nanotherapeutics for in vivo programmable photodynamic-chemotherapy of orthotopic bladder cancer. Adv Sci (Weinh). 2022;9(16):e2200731.
76. Boussion N, et al. A machine-learning approach based on 409 treatments to predict optimal number of iodine-125 seeds in low-dose-rate prostate brachytherapy. J Contemp Brachytherapy. 2021;13(5):541–8.
77. Huang C, et al. Using deep learning in a monocentric study to characterize maternal immune environment for predicting pregnancy outcomes in the recurrent reproductive failure patients. Front Immunol. 2021;12:642167.

Chapter 8
Significant Risk Medical Devices – General and Plastic Surgery

Devi Prasad Mohapatra and Indumathy Jagadeeswaran

8.1 Introduction

While there are several general and plastic surgery devices (Fig. 8.1) intended for human use that are in commercial distribution, this chapter intends to highlight a few of these devices based on the significant and nonsignificant risks of medical devices. The difference in the risk status of the medical device is determined in accordance with investigational Device exemptions (IDE) regulation (21 CFR 812) [1].

8.1.1 Significant Risk Devices [1]

- Catheters for general hospital use—except for conventional long-term percutaneous, implanted, subcutaneous and intravascular
- Collagen implant material for use in plastic surgery
- Surgical lasers
- Tissue adhesives for use in general and plastic surgery
- Absorbable adhesion barrier devices and hemostatic agents
- Artificial skin and interactive wound and burn dressings
- Breast implants
- Injectable collagen
- Implantable craniofacial prostheses

D. P. Mohapatra (✉)
Department of Plastic Surgery, JIPMER, Puducherry, India

I. Jagadeeswaran
Pediatrics, UT Southwestern Medical Center, Dallas, TX, USA

P. S. Timiri Shanmugam et al. (eds.), *Significant and Nonsignificant Risk Medical Devices*, https://doi.org/10.1007/978-3-031-52838-5_8

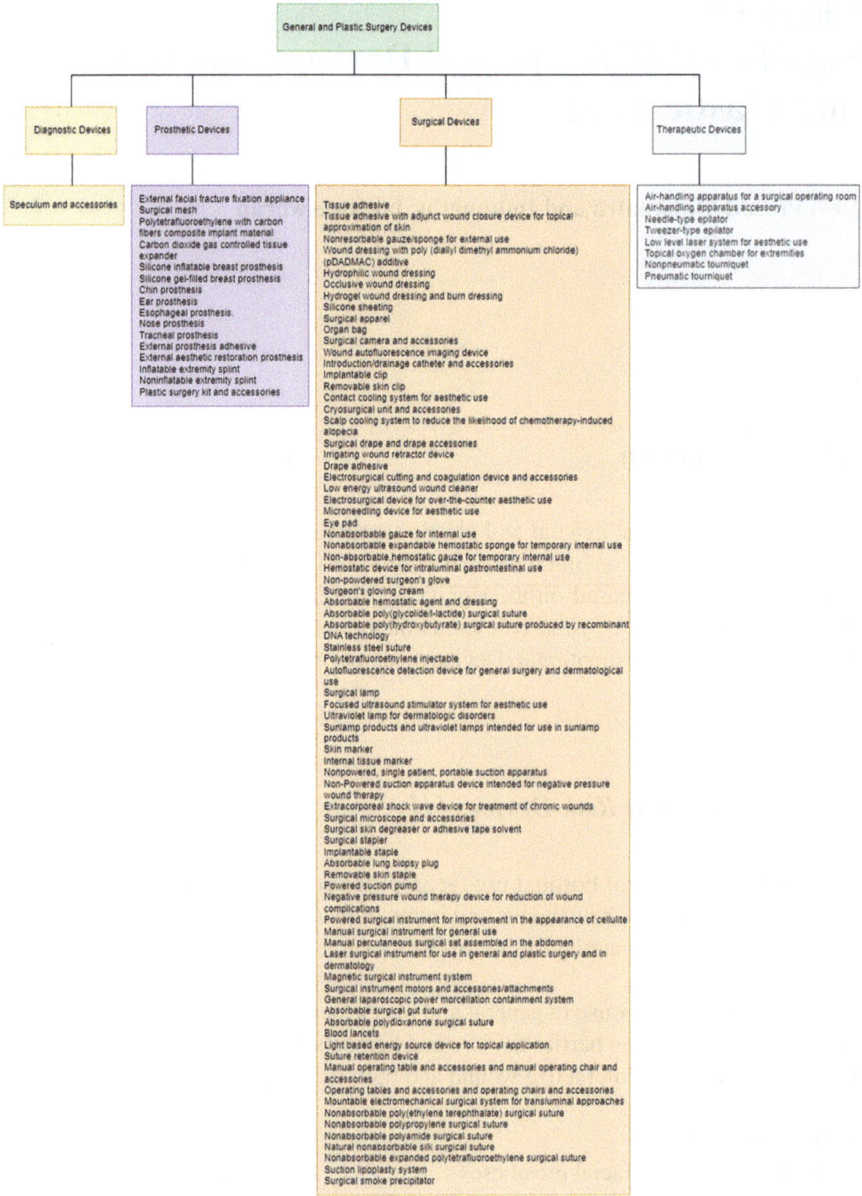

Fig. 8.1 List of approved devices used in plastic surgery

- Repeat access devices for surgical procedures
- Sutures

8.1.2 Nonsignificant Risk Devices [1]

Wound dressings, excluding absorbable hemostatic devices and dressings (also excluding interactive wound and burn dressings that aid or are intended to aid in the healing process).

8.2 Significant Risk Devices

1. *Catheters for General Hospital Use*

   1.1 These include.

   (a) *Foley Catheter*: A type of urinary catheter used to drain urine from the bladder. It's usually inserted through the urethra and into the bladder.
   (b) *Intermittent Catheter*: Also known as a short-term catheter, which is used to drain the bladder for short periods of time. It is often used for patients who cannot completely empty their bladders on their own.
   (c) *Suprapubic Catheter*: This is another type of urinary catheter. It's inserted into the bladder percutaneously. It might be used when the urethra is blocked or damaged [2].
   (d) *Nasogastric (NG) Tube*: While this is not a catheter in the traditional sense, it is a similar type of device. It is passed through the nose and down into the stomach and used to deliver food and medicine directly to the stomach or to remove substances from it [3].
   (e) *Endotracheal Tube*: This tube is inserted into the patient's windpipe (trachea) through the mouth or nose. It provides a passage for air and is used in situations where the patient cannot breathe on their own.
   (f) *Tracheostomy Tube*: This is a tube that's inserted into a hole made in the front of the neck and into the trachea, for use in patients who need help breathing [4].
   (g) *Arterial Catheters*: These are used to measure blood pressure in an artery or to take blood samples. Common insertion sites include the radial, femoral, or brachial arteries [5].

   1.2 Body contact:

   Foley catheter: It's a thin, flexible tube that is inserted through the urethra and into the bladder to allow urine to drain freely.
   Intermittent Catheter: They are inserted through the urethra and into the bladder, allowing urine to drain out. Once the bladder is empty, the catheter is removed.

Suprapubic catheter: it is inserted through the lower abdomen and into the bladder to drain urine, bypassing the urethra [2].
Nasogastric (NG) tube: It is inserted through the nose, down the esophagus, and into the stomach [3].
Endotracheal tube: It is inserted into the trachea (windpipe) through the mouth or nose to maintain an open airway or to deliver drugs or gases to the lungs.
Tracheostomy tube: It is a medical device that is inserted into a surgically created opening in the neck and trachea (windpipe) to help a patient breathe when the usual route for breathing is obstructed or needs to be bypassed [4].
Arterial catheter: Also known as an arterial line, it is a thin, sterile tube inserted into an artery, often in the wrist or groin [5].

1.3 Contact duration:

Foleys catheter: The duration of contact can range from a few hours to several weeks, depending on the patient's medical needs. However, it's important to note that long-term use of a Foley catheter can increase the risk of urinary tract infections (UTIs), so its use should be monitored and limited as much as possible.
Intermittent catheter: Intermittent catheters are typically used multiple times a day, depending on the individual's bladder capacity and fluid intake.
Suprapubic catheter: The duration of contact can range from a few days to several weeks or months, and even longer in certain cases, depending on the patient's medical condition and need [2].
Nasogastric (NG) tube: Duration of contact can vary widely based on the patient's specific needs, from a few hours to several weeks. However, long-term use is generally avoided [3].
Endotracheal tube: The duration of contact with an endotracheal tube can range from a few hours to several days, although long-term use is typically avoided when possible.
Tracheostomy tube: It can vary from a few weeks to several months or even indefinitely, depending on the patient's specific medical condition and breathing needs [4].
Arterial catheters: The duration of contact with an arterial catheter can vary from a few hours to several days, depending on the patient's condition and the need for continuous blood pressure monitoring or arterial blood sampling [5].

1.4 Single and/or multiple use:

Foley catheters: designed for single patient use.
Intermittent catheter: designed for single patient use. Each time a patient needs to empty their bladder, a new catheter is used and then discarded after use.

Suprapubic catheter: single-use devices, designed for a single patient. While the catheter itself can remain in place for a prolonged period in the same patient, it should not be used across multiple patients due to the risk of infection.

Nasogastric (NG) tube: designed for single-patient use. They can remain in place for several days or weeks but should be replaced regularly to prevent complications and should not be used on multiple patients to avoid the risk of infection transmission.

Endotracheal Tube: designed for single-patient use. They should not be used across multiple patients due to the high risk of infection transmission. Although the same tube can remain in place for several days in the same patient, the tube should be replaced regularly to prevent complications.

Tracheostomy tube: typically designed for single-patient use and should not be used on multiple patients due to the risk of infection transmission. A tracheostomy tube can be changed regularly (every few weeks to months) as per medical guidance to maintain hygiene and prevent complications.

Arterial catheter: designed for single-patient use and should not be used across multiple patients due to the risk of infection transmission. They are typically used for the duration of the surgical procedure or acute illness and then removed.

1.5 Benefits:

Foley catheters: They serve the intended use of draining urine from the bladder. This can be useful in situations where a patient is unable to urinate naturally due to medical conditions, during and after some surgical procedures, or when accurate urine output monitoring is required. The primary benefit of a Foley catheter is its ability to provide relief from urinary retention, enable accurate measurement of urinary output, and provide a means of urine collection in bedridden patients.

Intermittent catheters: They allow patients to maintain a regular bladder emptying schedule, which can help prevent UTIs and bladder or kidney damage. They also give patients more freedom and control over their urinary routine, which can significantly improve their quality of life.

Suprapubic catheters: They allow the bladder to drain urine when other methods are not possible or desirable. They also offer more comfort for long-term use and reduce the risk of damage to the urethra. They can also make personal hygiene and sexual activity easier compared to urethral catheters [2].

Nasogastric (NG) tube: Main benefits of NG tubes are their ability to ensure the patient receives adequate nutrition and hydration when they cannot eat or drink sufficiently. This method of feeding can support their overall health and recovery. They are also beneficial in managing conditions that cause a buildup of fluid or air in the stomach [3].

Endotracheal tube: The main benefit of an endotracheal tube is the ability to maintain an open airway in a patient who is unconscious or unable to breathe independently. It allows for controlled and efficient delivery of oxygen, anesthetic gases, or medications directly to the lungs.
Tracheostomy tube: It provides a direct and reliable access to the lower airway for oxygen delivery and facilitates respiratory secretion management. This can significantly improve breathing in patients who have difficulty maintaining an open and clear airway. It also tends to be more comfortable for long-term use compared to an endotracheal tube [4].
Arterial catheters: The main benefit of an arterial catheter is the ability to continuously monitor the blood pressure and quickly adjust medical or surgical treatment as necessary. It also facilitates regular blood gas analysis without the need for repeated arterial puncture [5].

1.6 Side/adverse/toxicological effects on the patient:

Foley catheter: The most common complication is UTIs. Other potential risks include damage to the urethra during insertion or removal, bladder stones, and kidney damage in severe cases of long-term use.
Intermittent catheter: This can include discomfort during insertion and removal, risk of UTIs, and potential injury to the urethra if not inserted correctly. Regular use can also cause urethral inflammation in some individuals.
Suprapubic catheter: This can include the risk of infection at the insertion site or in the urinary tract, bladder spasms, and accidental dislodgement of the catheter. In rare cases, injury to the bowel during catheter placement may occur [2].
Nasogastric (NG) tube: This can include discomfort or pain during insertion, nose and throat irritation, nasal injuries, and the risk of aspiration if the tube is incorrectly placed in the trachea instead of the esophagus. Necrosis on nostril skin is a common but avoidable complication [3].
Endotracheal tube: This can include discomfort or injury during insertion, damage to the teeth or larynx, infection, aspiration (breathing food or stomach acid into the lungs), and the development of vocal cord lesions or tracheal stenosis with prolonged use.
Tracheostomy tube: This include infection, bleeding, damage to the trachea or surrounding tissues, difficulty swallowing, changes in voice, and tracheal stenosis or tracheomalacia with prolonged use [4].
Arterial catheter: This can include pain at the insertion site, bleeding, infection, arterial injury, and in rare cases, formation of a blood clot that could reduce blood flow to the limb and possible loss of digits [5].

1.7 Use error:

Foley catheter: This might include improper sterilization leading to infection, incorrect insertion causing discomfort or injury, or mismanagement such as failing to empty the urine collection bag, leading to backflow of urine and potential infection. It's crucial to have appropriate medical training and follow precise procedures when inserting, maintaining, and removing Foley catheters to minimize these risks.

Intermittent catheter: This might include not properly sterilizing the area before catheterization, inserting the catheter incorrectly, or not using a new catheter each time. Proper training on how to use an intermittent catheter is essential to minimize these risks and ensure the most effective use of the device.

Suprapubic catheter: This could include improper site care leading to infection, incorrect catheter size causing discomfort or blockage, or not securing the catheter properly, which can cause it to dislodge. Therefore, healthcare providers should follow all relevant procedures and guidelines when inserting, maintaining, and removing suprapubic catheters. Proper patient education is also vital to prevent complications at home [2].

Nasogastric (NG) tube: This could include incorrect placement of the tube, improper securing of the tube leading to displacement, or inadequate monitoring and care of the insertion site. Proper training for healthcare providers is crucial to minimizing these risks, and placement should always be confirmed, ideally using an x-ray or routinely by detecting sound during air insufflation into the stomach before the tube is used for feeding or medication administration [3].

Endotracheal tube: This can include incorrect tube placement (such as into the esophagus instead of the trachea), overinflation of the tube cuff causing tracheal injury, or inadequate securing of the tube leading to accidental dislodgement. ET tube insertion should always be done by properly trained individuals.

Tracheostomy tube: This can include incorrect tube size selection, improper placement, or tube change, not adequately securing the tube, which may lead to accidental decannulation and failure to keep the stoma clean and dry. Both healthcare providers and caregivers must be adequately trained in the correct techniques for tracheostomy care and tube changes to minimize these risks [4].

Arterial catheters: This might include incorrect placement, failure to maintain sterility leading to infection, or failure to monitor and respond to changes in the patient's condition. Proper training and following established procedures are essential to minimize these risks. Staff must be adept in the insertion technique, catheter care and maintenance, and the interpretation of data derived from the catheter to ensure patient safety [5].

2. *Collagen Implant Material for Use in Plastic Surgery* [6]

2.1 The device: Collagen implant material is a medical device used in plastic surgery procedures to enhance soft tissue augmentation. It is usually derived from bovine or human sources and is processed to be biocompatible and resorbable.

2.2 Body contact: It comes into direct contact with soft tissues during the procedure. Once implanted, it interacts with the surrounding tissues under the skin.
2.3 Contact duration: The duration of contact is typically long-term. While collagen is eventually resorbed by the body, the timeline can vary from a few months to several years depending on the specific product, procedure, and individual patient factors.
2.4 Single and/or multiple use: It is intended for single-patient use. Each package or unit should not be used across multiple patients due to the risk of infection transmission.
2.5 Benefits: The primary benefit of collagen implant material is its ability to augment soft tissue and improve aesthetic appearance in areas like the face, lips, or other body parts. It is a popular choice in plastic and reconstructive surgeries due to its biocompatibility, resorbability, and the natural-looking results it can provide.
2.6 Side/adverse/toxicological effects on the patient: These include local reactions at the implantation site such as redness, swelling, pain, or bruising. In rare cases, patients may have an allergic reaction to the material, especially if it's derived from bovine sources. Over time, the implant may also migrate from the original placement site or cause lumpiness or asymmetry in the appearance of the skin.
2.7 Use error: These might include injecting the material too superficially or too deeply, not using the correct amount, or not following aseptic techniques, leading to an infection. Proper training in handling and injecting the material, as well as a thorough understanding of the anatomy of the area to be augmented, are essential to preventing these issues.

3. *Surgical Lasers* [7]

3.1 Surgical lasers are medical devices that deliver high-intensity light to treat or cut through body tissues. Various types of lasers are used in surgery, including carbon dioxide (CO_2), argon, and neodymium-doped yttrium aluminum garnet (Nd:YAG) (Fig. 8.2), among others, each with different applications and properties.
3.2 Body contact: Surgical lasers interact with the body's tissues. They do not physically touch the patient, but the light they emit is absorbed by the tissue, resulting in various therapeutic effects.
3.3 Contact duration: The duration of contact with a surgical laser is typically short-term and only for the duration of the surgical procedure.
3.4 Single and/or multiple use: Surgical lasers themselves are multiple-use devices, but they should be maintained properly between uses to ensure optimal function and safety. Accessories or attachments that come in direct contact with patients, like some types of laser fibers, are typically single use.
3.5 Benefits: They provide precise control, which allows surgeons to limit damage to surrounding tissues. Certain types of lasers can coagulate blood

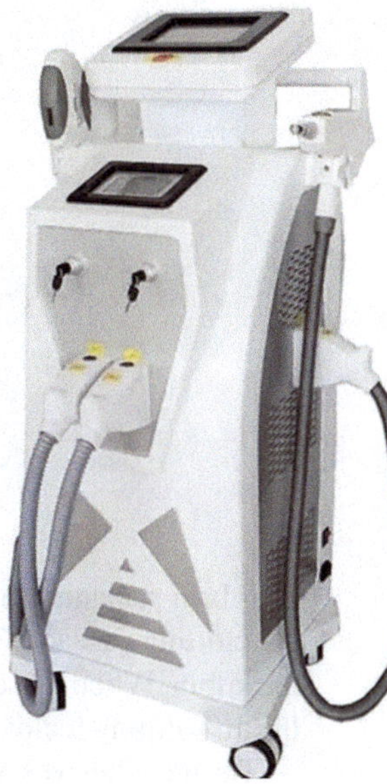

Fig. 8.2 Hair removal laser device

vessels while cutting, thereby reducing blood loss. Furthermore, some surgical procedures are only possible with lasers.

3.6 Side/adverse/toxicological effects on the patient: These include potential burns, unwanted tissue damage if the laser is misdirected, and temporary or permanent vision damage if eye protection is not used during procedures involving the face or eyes. Lasers can also create surgical smoke that may be harmful if inhaled.

3.7 Use error: These include inappropriate settings for a particular procedure, improper targeting of the laser, or lack of proper eye protection. It is essential that all users of surgical lasers receive appropriate training on their use and safety. Furthermore, surgical lasers should be regularly serviced and maintained to ensure their safe operation.

4. *Tissue Adhesives for Use in General and Plastic Surgery* [8]

4.1 Tissue adhesives used in general and plastic surgery come in various forms, with the most common ones being cyanoacrylate-based adhesives and fibrin sealants. Cyanoacrylate-based adhesives are synthetic compounds that quickly bond to the skin surface. On the other hand, fibrin sealants are derived from human blood products and mimic the final stages of the body's natural clotting process. Cyanoacrylate-based adhesives work by rapidly polymerizing in the presence of water or blood, forming a strong and flexible film that binds the edges of a wound together. They provide immediate wound closure and a microbial barrier, aiding in infection prevention.

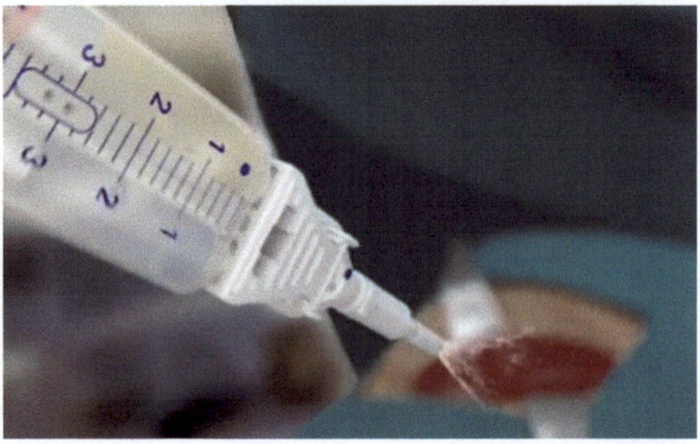

Fig. 8.3 Fibrin sealants

Fibrin sealants (Fig. 8.3), on the other hand, replicate the final stage of the clotting cascade. They consist of two main components: fibrinogen and thrombin. When these two are combined, thrombin converts fibrinogen into fibrin, forming a clot that seals the wound and aids tissue healing.

Tissue adhesives are primarily used for wound closure following surgical procedures in both general and plastic surgery. They can replace or supplement sutures, staples, and tapes for closing skin incisions or lacerations. Fibrin sealants can also be used for hemostasis, to control bleeding in surgical procedures, or to adhere tissue layers together.

4.2 Body contact: These adhesives come into direct contact with the skin or internal tissues. They are usually applied topically onto surgical incision or wound for external use and can also be applied internally in the case of minimally invasive or laparoscopic procedures, nerve repair, and vascular repair.

4.3 Contact duration: On average, tissue adhesives begin to peel off naturally after 5–10 days for external use, while internal applications dissolve over time as part of the healing process.

4.4 Single and/or multiple use: Typically designed for single-patient use to reduce the risk of cross-contamination. Some products may be used multiple times on the same patient, but they should be discarded after the treatment of one patient.

4.5 Benefits: The benefits of tissue adhesives include quick application, immediate wound closure, excellent cosmetic outcomes, and reduced pain compared to traditional sutures. They also offer increased convenience for patients, as there's no need for suture or staple removal. The microbial barrier provided by some adhesives helps reduce the risk of wound infections. Internally used adhesives have the advantage of aiding the healing of tissues like nerve and vessels where they are used.

4.6 Side/adverse/toxicological effects on the patient: These include allergic reactions, particularly in those with a sensitivity to cyanoacrylates. There may be a mild inflammatory reaction, and if used improperly, they can cause wound dehiscence (separation of wound edges). In rare cases, the adhesive may also cause irritation or a foreign body reaction.

4.7 Use error: The most common use error involves applying the adhesive inappropriately, either too much or too little, which can affect the wound healing process. Furthermore, they should not be used on wounds under high tension or on mucosal surfaces. Lastly, care must be taken to avoid contact with the eyes, as these adhesives can cause serious eye irritation.

5. *Absorbable Adhesion Barrier Devices and Hemostatic Agents* [9, 10]

5.1 Absorbable adhesion barrier devices and hemostatic agents (Fig. 8.4) are frequently used in various surgical procedures. Absorbable adhesion barrier devices are used to prevent post-surgical adhesions. Hemostatic agents are used to control bleeding during surgery. Absorbable adhesion barrier devices are bioresorbable membranes that physically separate the internal tissues during the healing process to prevent the formation of surgical adhesions. Their mechanism of action relies on providing a physical barrier that keeps the tissues separated during the initial, critical phase of healing.

Hemostatic agents, on the other hand, work by accelerating the clotting process to control surgical bleeding. These agents typically contain substances that encourage the aggregation of platelets, thereby forming a clot and helping stop bleeding. Absorbable adhesion barrier devices are used primarily in surgeries where there's a high risk of postoperative adhesions, like pelvic, abdominal, and cardiac surgery. Hemostatic agents are used in various surgical procedures to control bleeding, particularly when conventional methods to control bleeding are ineffective.

5.2 Body contact: These devices come into direct contact with the internal tissues, specifically, the surgical site. Absorbable adhesion barrier devices are placed over the surgical site to prevent the tissues from adhering to each other during healing. Hemostatic agents are applied directly to the site of bleeding to aid in the clotting and cessation of bleeding.

5.3 Contact duration: These devices are designed to remain in the body post-surgery. Over time, they are absorbed by the body—the duration varies

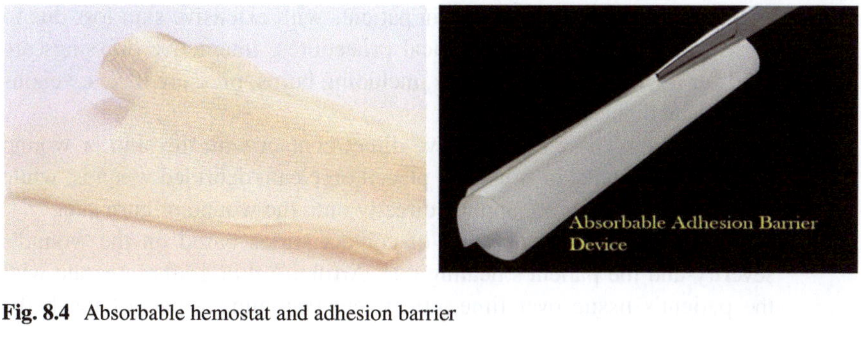

Fig. 8.4 Absorbable hemostat and adhesion barrier

depending on the specific product and the individual's healing rate. Generally, absorption happens over a period of weeks to months.

5.4 Single and/or multiple use: Both absorbable adhesion barrier devices and hemostatic agents are designed for single use and should be discarded after the procedure. They are typically provided in sterile packaging that is opened just before use to maintain sterility.

5.5 Benefits: Absorbable adhesion barrier devices can reduce the incidence of post-surgical adhesions, thereby reducing associated complications such as chronic pain, infertility, and bowel obstruction. Hemostatic agents can quickly control bleeding, potentially reducing surgical time and improving patient outcomes.

5.6 Side/adverse/toxicological effects on the patient: Potential side effects from absorbable adhesion barrier devices may include an inflammatory reaction or the formation of a seroma (a pocket of clear serous fluid). Hemostatic agents may occasionally cause a mild inflammatory response. In rare instances, they may also cause an allergic reaction.

5.7 Use error: Potential errors with the use of these devices could include inappropriate placement of the adhesion barrier, leading to ineffective prevention of adhesions. For hemostatic agents, use error could include an incorrect amount or placement of the agent, leading to ineffective hemostasis or, in rare cases, thrombosis (the formation of a blood clot within a blood vessel). It's also crucial to ensure that the patient does not have known allergies to the components of these devices before their use.

6. *Artificial Skin and Interactive Wound and Burn Dressings* [11]

6.1 Artificial skin products along with interactive wound and burn dressings (Fig. 8.5) are innovative medical devices utilized in the treatment of various types of wounds and burns. Artificial skin, often known as a skin substitute, consists of a synthetic epidermis and a dermis layer, which is typically collagen-based. This device promotes the regeneration of skin and can serve as a permanent solution for skin loss.

Interactive wound and burn dressings are advanced dressings that work by maintaining a moist environment, which promotes healing, and some also have antimicrobial properties. These dressings interact with the wound to aid in the removal of dead tissue and promote the formation of new tissue. Artificial skin is primarily used in patients with extensive skin loss due to burns, chronic wounds, or surgical procedures. Interactive dressings are used for a wide range of wounds, including burns, pressure ulcers, venous ulcers, and diabetic foot ulcers.

6.2 Body contact: These devices have direct contact with the skin or wound surface. Artificial skin is usually placed on clean, debrided wounds, while interactive dressings are applied directly onto the wound or burn site.

6.3 Contact duration: The duration of contact varies based on the wound's severity and the patient's healing rate. Artificial skin grafts integrate with the patient's tissue over time, effectively becoming a part of the body.

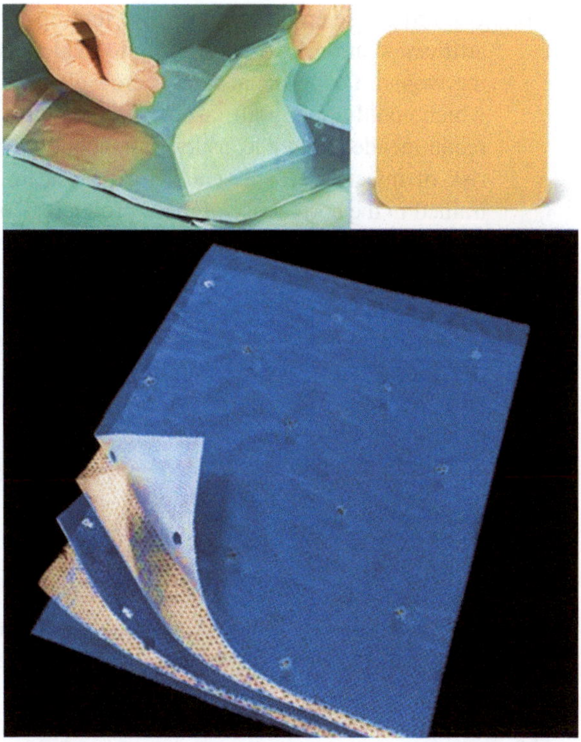

Fig. 8.5 Interactive wound dressings picture

Interactive dressings are typically changed every few days or as per the healthcare provider's instructions.

6.4 Single and/or multiple use: Artificial skin is a single-use product, created for one-time application to a patient's wound. Interactive dressings are also typically designed for single use, but the treatment course often requires multiple dressing changes, thus multiple dressings.

6.5 Benefits: Artificial skin offers a life-saving treatment for patients with extensive skin loss when traditional grafts aren't possible. It can accelerate healing, reduce scarring, and improve the functional and aesthetic outcome. Interactive dressings maintain a moist wound environment, which accelerates healing, reduces pain, and allows for less frequent dressing changes. Some interactive dressings also contain antimicrobial agents, which help prevent infection.

6.6 Side/adverse/toxicological effects on the patient: The use of artificial skin and interactive dressings is generally safe, but some potential adverse effects can occur. These include infection, inflammation, allergic reaction, and, in the case of artificial skin, graft failure. Care must be taken to monitor the wound for signs of infection, including increased pain, redness, swelling, or pus discharge.

6.7 Use error: Use errors could involve improper placement or securing of the artificial skin or dressing, leading to inadequate coverage or protection of the wound. For artificial skin, there's also a risk of incorrect preparation, which could affect graft take. Dressings must also be changed as per the recommended schedule, as leaving them on for too long could increase the risk of infection. It's essential for healthcare providers to be thoroughly trained in the correct use of these products to minimize such errors.

7. *Breast Implants* [12]

7.1 Breast implants (Fig. 8.6) are medical devices used in procedures such as breast augmentation, reconstruction after mastectomy, or correction of developmental deformities. There are two main types: saline-filled implants and silicone gel-filled implants. Saline-filled implants have a silicone outer shell filled with sterile saline solution, while silicone gel-filled implants have a silicone outer shell filled with silicone gel. They work by increasing the size and enhancing the shape of the breasts.

7.2 Body contact: Breast implants are surgically inserted into the body, coming into contact with internal tissues. They are placed either under the breast tissue (subglandular placement) or under the chest muscle (submuscular placement). Breast implants are used in cosmetic breast augmentation procedures to increase breast size or improve breast shape. They're also used in reconstructive surgeries to restore a breast that's been removed due to cancer or other diseases, and in corrective surgeries to amend congenital or developmental deformities.

7.3 Contact duration: Breast implants are intended to be permanent devices, and they remain in the body indefinitely unless complications arise that necessitate their removal or replacement.

7.4 Single and/or multiple use: Breast implants are designed for single use. If removed, they should not be reimplanted. While they are intended to be permanent, some patients may need or choose to have the implants replaced or removed at some point.

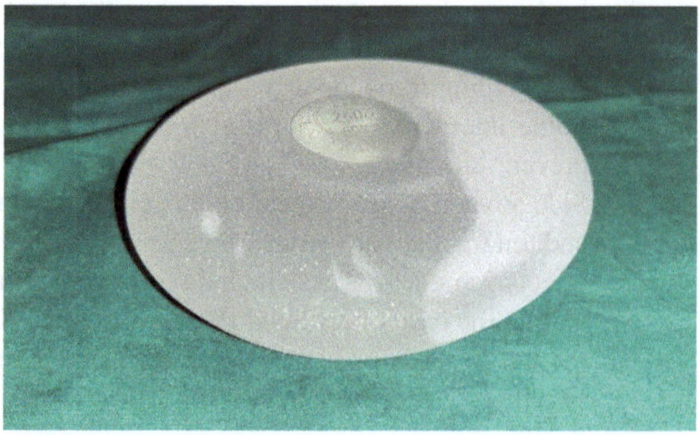

Fig. 8.6 Round silicone filled 260 cc breast implant with textured surface

7.5 Benefits: The benefits of breast implants include improved breast size and shape, which can lead to increased self-confidence and body image. In the case of reconstructive surgeries, implants can help restore a natural appearance after a mastectomy, aiding in physical and emotional recovery.

7.6 Side/adverse/toxicological effects on the patient: Potential side effects and complications of breast implants include capsular contracture (hardening of the area around the implant), rupture or deflation of the implant, asymmetry, infection, and changes in nipple or breast sensation. There has been some concern about a rare type of cancer known as Breast Implant-Associated Anaplastic Large Cell Lymphoma (BIA-ALCL) associated with textured implants. Silicone implants also carry the risk of silicone leakage, which can cause local reactions and potentially impact systemic health.

7.7 Use error: Use error with breast implants could involve incorrect implant placement during surgery, leading to asymmetry or complications. Additionally, failure to follow proper post-operative care instructions could lead to complications such as infection or displacement of the implant. Regular follow-ups and imaging tests are important to ensure that the implants are functioning correctly and to monitor for complications.

8. *Injectable Collagen* [6]

8.1 Injectable collagen, is a type of dermal filler used in cosmetic and reconstructive procedures. Injectable collagen is primarily derived from bovine or human sources. It works by adding volume to the skin's dermal layer, effectively plumping up the skin to smooth out wrinkles or enhance certain features, such as lips. The injected collagen integrates with the body's natural collagen and other skin structures, providing a natural-looking result. Injectable collagen is used for smoothing facial lines, wrinkles, and scars, enhancing the volume of lips, improving contours of the face, and treating deeper wrinkles and folds. It can also be used in reconstructive procedures to fill in areas where tissue loss has occurred.

8.2 Body contact: Injectable collagen comes into direct contact with the skin and underlying soft tissues. It is injected beneath the dermal layer, in the area of wrinkles, fine lines, or other areas that need volume enhancement.

8.3 Contact duration: The effects of injectable collagen typically last for several months, up to a year, depending upon the individual's metabolism, the area of treatment, and the specific product used. After this period, the collagen is safely absorbed by the body, and the procedure can be repeated if desired.

8.4 Single and/or multiple use: Injectable collagen is a single-use device, with each syringe meant for one treatment session for a single patient. However, multiple sessions might be required over time to maintain the desired effect, as the body gradually absorbs the injected collagen.

8.5 Benefits: The benefits of injectable collagen include minimally invasive treatment with immediate results, improved facial volume and contour, reduction of wrinkles and fine lines, and enhanced self-confidence and appearance. Additionally, the procedure involves minimal downtime, allowing individuals to resume their usual activities quickly.

8.6 Side/adverse/toxicological effects on the patient: Potential side effects from injectable collagen include allergic reactions, particularly in those allergic to bovine products. Other possible effects include bruising, redness, swelling at the injection site, or unevenness or lumps under the skin. In rare cases, collagen may migrate from the injection site, causing unintended effects.

8.7 Use error: Use error with injectable collagen could involve incorrect injection technique or injection into an inappropriate site, leading to lumps, unevenness, or migration of the filler. Overfilling is also a potential risk. It's crucial that injectable collagen is administered by a trained healthcare provider familiar with the anatomy of the treated area to minimize these risks. Additionally, a skin test should be conducted prior to treatment to identify patients with hypersensitivity to bovine collagen.

9. *Implantable Craniofacial Prostheses* [13]

9.1 Implantable craniofacial prostheses, also known as craniofacial implants, are medical devices used to reconstruct or replace parts of the face and skull, typically after surgery, trauma, or congenital abnormalities. Common types include dental implants, orbital implants, nasal prostheses, and cranial plates. Implantable craniofacial prostheses are typically constructed from biocompatible materials such as titanium, medical-grade silicone, or ceramic. They are designed to mimic the shape, structure, and function of the missing or damaged craniofacial part, restoring both form and function. Craniofacial implants are used for various applications, including facial reconstruction after trauma or cancer surgery, correction of congenital anomalies, or replacement of dental structures. They are intended to restore both the aesthetic appearance and functional capacity of the craniofacial region.

9.2 Body contact: These devices come into direct contact with bone and soft tissue within the craniofacial region. The prosthesis is usually secured to the underlying bone structure through surgical placement.

9.3 Contact duration: Implantable craniofacial prostheses are intended to be permanent fixtures within the body, often remaining in place for the duration of the patient's life, unless complications arise that require removal or replacement.

9.4 Single and/or multiple use: Craniofacial prostheses are designed for single use and intended to be a permanent solution for the patient. If complications arise or the prosthesis fails, a new device is fabricated and installed.

9.5 Benefits: Craniofacial prostheses can significantly improve a patient's quality of life by restoring the aesthetics and function of the face and skull. This can enhance self-esteem, facilitate social interaction, and improve physical capabilities such as speech and mastication.

9.6 Side/adverse/toxicological effects on the patient: Potential adverse effects of craniofacial prostheses include infection, inflammation, rejection, or mechanical failure of the implant. There may also be aesthetic dissatisfaction if the implant doesn't adequately match the surrounding tissue in color or shape. In some cases, additional surgeries may be required to adjust or replace the prosthesis.

9.7 Use error: Use errors with craniofacial prostheses could include improper sizing, positioning, or securing of the implant, which could lead to inadequate function, discomfort, or aesthetic dissatisfaction. There could also be an error in the choice of material, leading to an allergic reaction or intolerance. It's critical for healthcare providers to have specialized training in the placement of these prostheses to minimize these risks.

10. *Repeat Access Devices for Surgical Procedures*

10.1 Repeat access devices for surgical procedures, such as trocars or surgical ports, are medical tools that allow for repeated access to the same area during minimally invasive surgery. They are most commonly used in laparoscopic or endoscopic procedures. Trocars are devices that include a cannula (a hollow tube) and a sharp, pointed obturator. The obturator is used to puncture the body cavity, after which it's withdrawn, leaving the cannula in place as a conduit for other surgical instruments. Surgical ports, on the other hand, are implanted under the skin and provide access for repeated injections or withdrawals of fluid without the need for repeated skin puncture. Repeat access devices are used in a variety of surgical procedures where repeated access to a body cavity is necessary. This includes laparoscopic procedures such as cholecystectomy (gallbladder removal), appendectomy, and bariatric surgery. Repeat access devices are used in breast surgery (Fig. 8.7) to create a pocket for implants. Surgical ports are often used in chemotherapy treatments, allowing for repeated administration of medication directly into the bloodstream.

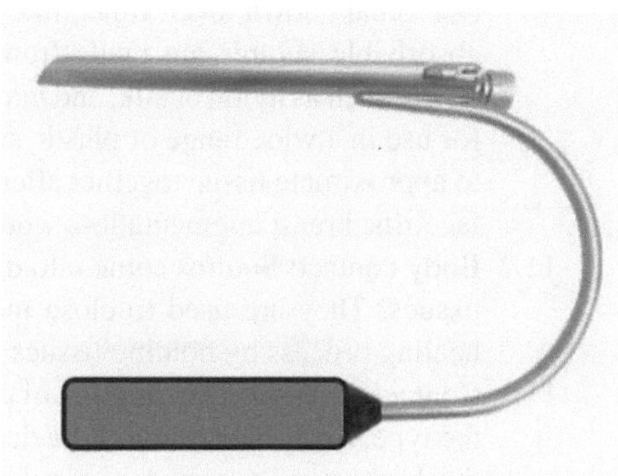

Fig. 8.7 Repeat access device for breast surgery

10.2 Body contact: These devices come into contact with the skin and internal tissues. They're inserted through a small incision in the skin, creating a passageway into the body.

10.3 Contact duration: Repeat access devices remain in the body for the duration of the surgical procedure. They're removed once the procedure is complete, and therefore don't stay in contact with the body for an extended period of time.

10.4 Single and/or multiple use: Most repeat access devices are single-use items that are discarded after surgery. However, certain devices, like implanted surgical ports, can facilitate multiple uses over a longer time span, such as during ongoing chemotherapy treatments.

10.5 Benefits: Repeat access devices allow for minimally invasive procedures, which can minimize patient discomfort and reduce recovery time. By providing a safe and efficient way to access body cavities or blood vessels, these devices reduce the risk of complications related to repeated punctures, such as infection or tissue damage.

10.6 Side/adverse/toxicological effects on the patient: Potential complications of repeat access devices include infection, hemorrhage, damage to adjacent organs or tissues, and, in rare cases, device malfunction or failure. In the case of surgical ports, there may be additional risks related to long-term implantation, such as thrombosis or embolism.

10.7 Use error: Use error with repeat access devices could involve incorrect placement or removal of the device, potentially leading to injury to the patient. It's crucial that healthcare providers follow established protocols for the use of these devices to minimize potential risks. Proper sterilization is also critical to preventing infection.

11. *Sutures* [14]

11.1 Sutures are crucial for use in plastic surgery for different procedures. They come in a variety of materials, both absorbable and non-absorbable, such as polydioxanone (PDS), vicryl, monocryl, nylon, and silk (Fig. 8.8). Sutures are thread-like materials used to close wounds or surgical incisions. Absorbable sutures are typically made from materials that the body can break down over time, like catgut or synthetic polymers. Non-absorbable sutures are made from materials that the body can't break down, such as nylon or silk, and may require removal. Sutures are intended for use in a wide range of plastic surgery procedures to close incisions or to approximate tissue together after surgery. This includes procedures like facelifts, breast augmentation, wound repair, and many others.

11.2 Body contact: Sutures come into direct contact with the skin and internal tissues. They are used to close incisions or wounds, thus aiding in the healing process by holding tissues together.

11.3 Contact duration: The duration of contact for sutures varies depending on the type. Absorbable sutures are designed to be gradually broken down by the body over a period of weeks to months, whereas non-absorbable sutures remain in the body indefinitely unless removed.

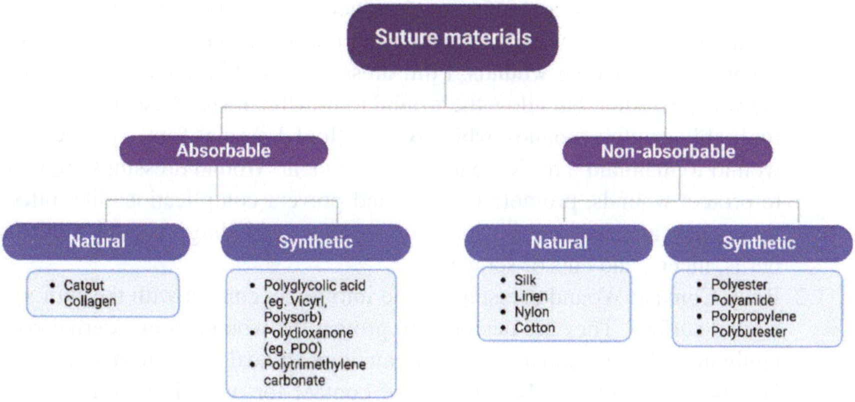

Fig. 8.8 Suture Material and types used in healthcare

11.4 Single and/or multiple use: Sutures are single-use devices. Each suture is used for one patient and one procedure only and then discarded to prevent cross-contamination.

11.5 Benefits: Sutures aid in the healing process by keeping the wound closed, reducing the risk of infection, minimizing scar appearance, and speeding up recovery. They also help to maintain the aesthetic outcomes of the surgery by ensuring the precise approximation of tissues.

11.6 Side/adverse/toxicological effects on the patient: Potential adverse effects of sutures include allergic reactions, wound dehiscence (separation of the wound edges), infection, or suture granulomas (small lumps of inflamed tissue). Certain types of sutures can also cause more noticeable scarring, and there's a risk of stitch abscesses with non-absorbable sutures left in too long.

11.7 Use error: Use errors with sutures could involve choosing an inappropriate suture material or technique for the wound type or location, which could affect the healing process or aesthetic outcome. It also includes not removing non-absorbable sutures in a timely manner, leading to potential complications. The surgeon's skill and experience in suture placement greatly affect the final outcome.

8.3 Nonsignificant Risk Devices

1. *Wound dressings, excluding absorbable hemostatic devices and dressings (also excluding interactive wound and burn dressings that aid or are intended to aid in the healing process)*

   1.1 Wound dressings, such as gauze, film dressings, foam dressings, and hydrocolloid dressings, are non-invasive medical devices designed to cover a wound and promote a suitable environment for healing. Wound dressings come in various forms and serve different purposes. Gauze is often used for cleaning and packing wounds. Film dressings are thin, flexible sheets that are water-resistant but allow the wound to breathe. Foam dressings are used for highly exuding wounds, while hydrocolloid dressings form a gel over the wound to maintain a moist healing environment. Wound dressings are used to protect wounds, promote healing, and prevent complications like infection. They are used on a variety of wounds, including surgical wounds, burns, ulcers, and cuts or scrapes.

1.2 Body Contact: Wound dressings come into direct contact with the skin and wound surface. They are designed to protect the wound from external contaminants, absorb exudate, and maintain a moist healing environment.

1.3 Duration of Contact: The duration of contact for wound dressings varies depending on the wound's severity and the type of dressing used. Dressings are typically changed regularly, often daily or every few days, until the wound is fully healed.

1.4 Single and/or Multiple Use: Wound dressings are typically single-use products. Each dressing is disposed of at the time of change, and a new dressing is applied.

1.5 Benefits: The benefits of wound dressings include providing a barrier against bacteria and other contaminants, absorbing wound exudate, maintaining a moist healing environment, reducing pain, and aiding the healing process. In turn, this can prevent complications, minimize scarring, and improve patient comfort.

1.6 Any Side/Adverse/Toxicological Effects on the Patient: Potential adverse effects of wound dressings include allergic reactions or skin irritation. Improper application could potentially cause further injury to the wound site. There's also a risk of infection if the dressing is not changed regularly or if the dressing becomes contaminated.

1.7 Use Error: Use errors with wound dressings could include improper selection of dressing type for the wound condition, incorrect application or removal of the dressing, or not changing the dressing regularly, which could inhibit the healing process or lead to infection. It's crucial for healthcare providers and caregivers to understand how to choose and use the appropriate dressing to minimize these risks.

References

1. U.S. Food and Drug Administration. Updated January 2006, https://www.fda.gov/regulatory-information/search-fda-guidance-documents/significant-risk-and-nonsignificant-risk-medical-device-studies. Accessed 26 Jun 2023.
2. Sigmon DF, An J. Nasogastric tube. [Updated 2022 Oct 31]. In: StatPearls [Internet]. Treasure Island: StatPearls Publishing; 2023. Available from: https://www.ncbi.nlm.nih.gov/books/NBK556063/.

3. Ahluwalia RS, Johal N, Kouriefs C, Kooiman G, Montgomery BS, Plail RO. The surgical risk of suprapubic catheter insertion and long-term sequelae. Ann R Coll Surg Engl. 2006;88(2):210–3. https://doi.org/10.1308/003588406X95101.
4. Mehta AK, Chamyal PC. Tracheostomy complications and their management. Med J Armed Forces India. 1999;55(3):197–200. https://doi.org/10.1016/S0377-1237(17)30440-9.
5. Pierre L, Pasrija D, Keenaghan M. Arterial lines. [Updated 2023 May 17]. In: StatPearls [Internet]. Treasure Island: StatPearls Publishing; 2023. Available from: https://www.ncbi.nlm.nih.gov/books/NBK499989/.
6. Cho KH, Uthaman S, Park IK, Cho CS. Injectable biomaterials in plastic and reconstructive surgery: a review of the current status. Tissue Eng Regen Med. 2018;15(5):559–74. https://doi.org/10.1007/s13770-018-0158-2.
7. Wu EC, Wong BJF. Lasers and optical technologies in facial plastic surgery. Arch Facial Plast Surg. 2008:381–90. https://doi.org/10.1001/archfaci.10.6.381.
8. Sanders L, Nagatomi J. Clinical applications of surgical adhesives and sealants. Crit Rev Biomed Eng. 2014;42(3-4):271–92. https://doi.org/10.1615/critrevbiomedeng.2014011676.
9. Waldron MG, Judge C, Farina L, O'Shaughnessy A, O'Halloran M. Barrier materials for prevention of surgical adhesions: systematic review. BJS Open. 2022;6(3):zrac075. https://doi.org/10.1093/bjsopen/zrac075.
10. Bloom JA, Erlichman Z, Foroutanjazi S, Beqiraj Z, Jonczyk MM, Persing SM, Chatterjee A. The use of hemostatic agents to decrease bleeding complications in general plastic surgery procedures. Plast Reconstr Surg Glob Open. 2021;9(8):e3744. https://doi.org/10.1097/GOX.0000000000003744.
11. Przekora A. A concise review on tissue engineered artificial skin grafts for chronic wound treatment: can we reconstruct functional skin tissue in vitro? Cells. 2020;9(7):1622. https://doi.org/10.3390/cells9071622.
12. U.S. Food and Drug Administration. Risks and complications of breast implants. Retrieved from https://www.fda.gov/medical-devices/breast-implants/risks-and-complications-breast-implants. Accessed on 6th June 2023.
13. Alberga J, Eggels I, Visser A, et al. Outcome of implants placed to retain craniofacial prostheses – a retrospective cohort study with a follow-up of up to 30 years. Clin Implant Dent Relat Res. 2022;24(5):643–54. https://doi.org/10.1111/cid.13106.
14. Byrne M, Aly A. The surgical suture. Aesthet Surg J. 2019;39(Supplement_2):S67–72. https://doi.org/10.1093/asj/sjz036.

Chapter 9
Significant Risk Medical Devices – General Hospital

Thamizharasan Sampath, Sandhiya Thamizharasan, V. Krithaksha, Pralhad Wangikar, and Prakash Srinivasan Timiri Shanmugam

Abbreviations

AAMI	Association of the Advancement of Medical Instrumentation
ANSI	American National Standards Institute
ASTM	American Society for Testing and Materials
CDRH	Center for Devices and Radiological Health
CFR	Code of Federal Regulations
CGM	Continuous Glucose Monitoring Sensors
CTSI	Clinical and Translational Sciences Institute
DMIST	Digital Mammographic Imaging Screening Trial
ECG	Electrocardiogram
ECT	Electroconvulsive Therapy
EMA	European Medicines Agency
ETO	Ethylene Oxide
EU	European Union
FDA	Food and Drug Administration
FDC	Federal Food, Drug, and Cosmetic Act
FFC	Filtering Face piece Respirators

T. Sampath (✉)
Department of Pharmacology & Toxicology, VMCGH, DR.YSR University of Health Sciences, Kurnool, AP, India

S. Thamizharasan
The Tooth Doctor - Advanced Implant Centre, Vellapanchavadi, Chennai, India

V. Krithaksha
Georgian National University SEU, Tbilisi, Georgia

P. Wangikar
PRADO, Preclinical Research and Development Organization, Pvt. Ltd., Pune, Maharashtra, India

P. S. Timiri Shanmugam
Global Product Safety & Toxicology, Avanos Medical Inc., Alpharetta, GA, USA

P. S. Timiri Shanmugam et al. (eds.), *Significant and Nonsignificant Risk Medical Devices*, https://doi.org/10.1007/978-3-031-52838-5_9

GMP	Good Manufacturing Practices
HCP	Healthcare personnel
HPPP	Human Research Protection Program
ICU	intensive care units
IRB	Institutional Review Board
ISO	International Organization for Standardization
IV	Intravenous
IVD	In Vitro Diagnostics
IVDR	In Vitro Diagnostic Medical Device Regulation
LED	Light emitting diodes
LLLT	Low Level Laser Therapy
LCG	liquid chemical germicide
MDR	Medical Device Regulation
MDDS	Medical Device Data Systems
MRI	Magnetic Resonance Imaging
NEMA	The National Electrical Manufacturers Association
NSR	Non-Significant Risk
NAS	Network Attached Storage
NCIT	Non-contact infrared thermometers
NPPTL	National Personal Protective Technology Laboratory
NIOSH	National Institute for Occupational Safety and Health
OSHA	Occupational Safety and Health Administration
PORP	Partial Ossicular Replacement Prosthesis
PCA	Patient-controlled analgesia
PPE	Personal protective equipment
PVC	Polyvinyl chloride
RF	Radiofrequency
SAN	Storage area network
SOP	Standard Operating Procedures
SR	Significant Risk
TMJ	Temporomandibular Joint
TSS	Toxic Shock Syndrome
UTI	Urinary Tract Infections

9.1 Introduction

In the realm of healthcare, medical devices play a vital role in diagnosing, treating, and managing a wide range of medical conditions. These devices range from simple tools like thermometers and stethoscopes to complex machinery such as MRI scanners and surgical robots. With advancements in technology, the medical device industry has witnessed remarkable growth, leading to an ever-expanding array of devices available for use in general hospitals. The safety and efficacy of medical devices are of paramount importance to ensure optimal patient care. To regulate the use of these devices, various regulatory bodies, such as the Food and Drug Administration (FDA) in the United States and the European Medicines Agency (EMA) in Europe, have established comprehensive guidelines and classifications to categorize devices based on their associated risks.

This chapter aims to delve into the classification of medical devices as significant and non-significant risk devices in the context of general hospital settings. By understanding the differences between these categories and the associated regulatory requirements, healthcare professionals, administrators, and device manufacturers can make informed decisions regarding device selection, implementation, and ongoing monitoring. Furthermore, we will examine case studies and real-life examples to illustrate the impact of significant and non-significant risk devices on patient outcomes and healthcare organizations. By highlighting these examples, we aim to enhance the understanding of the challenges and considerations involved in managing and implementing medical devices in general hospital settings. Ultimately, this chapter intends to provide healthcare professionals with a comprehensive overview of significant and non-significant risk devices, equipping them with the knowledge necessary to make informed decisions and ensure the highest standards of patient care and safety in general hospital settings.

The US Food and Drug Administration (FDA) regulates a wide range of medical devices used in general hospital settings to ensure their safety and effectiveness. The FDA classifies medical devices into different categories based on the level of risk they pose to patients. Here are some examples of general hospital devices regulated by the FDA:

9.2 Classification

Class I Devices These devices pose the lowest risk to patients and are subject to general controls, such as labeling requirements and adherence to good manufacturing practices. Examples of Class I devices used in hospitals include:

(a) Non-powered wheelchairs
(b) Elastic bandages
(c) Surgical instruments (non-cutting)

Class II Devices Class II devices are considered to have a moderate risk to patients, and they require specific performance standards, special labeling, and post-market surveillance. Examples of Class II devices used in hospitals include:

(a) Infusion pumps
(b) Pulse oximeters
(c) X-ray machines
(d) Powered wheelchairs

(e) Electrocardiography (ECG) monitors

Class III Devices These devices pose the highest risk to patients and require a rigorous pre-market approval process. Class III devices generally support or sustain human life, are implanted, or present potential risks that cannot be controlled through general or special controls. Examples of Class III devices used in hospitals include:

(a) Implantable pacemakers
(b) Artificial heart valves
(c) Ventricular assist devices
(d) Automated external defibrillators (AEDs)

The FDA's role in regulating these devices involves reviewing pre-market submissions, conducting inspections and quality system audits, and overseeing post-market surveillance. The FDA ensures that these devices meet safety and effectiveness standards, undergo proper testing, and are properly labeled and marketed.

9.3 General Hospital Devices

- *Non-contact Infrared Thermometers*

Measuring a person's temperature can be done in several ways. One method to measure a person's surface temperature is with the use of non-contact infrared thermometers (NCITs). NCITs may be used to reduce cross-contamination risk and minimize the risk of spreading disease. While typically 98.6 °F (37.0 °C) is considered a "normal" temperature, some studies have shown that "normal" body temperature can be within a wide range, from 97 °F (36.1 °C) to 99 °F (37.2 °C). Before NCITs are used, it is important to understand the benefits, limitations, and proper use of these thermometers. Improper use of NCITs may lead to inaccurate measurements of temperature [1].

Benefits of NCITs

- Non-contact approach may reduce the risk of spreading disease between people being evaluated.
- Easy to use.
- Easy to clean and disinfect.
- Measures temperature and displays a reading rapidly.
- Provides the ability to retake a temperature quickly.

Limitations of NCITs

- How and where the NCIT is used may affect the measurement (for example, head covers, environment, positioning on forehead).
- The close distance required to properly take a person's temperature represents a risk of spreading disease between the person using the device and the person being evaluated.

Proper Use of NCITs

The person using the device should strictly follow the manufacturer's guidelines and instructions for use for the specific NCIT being used. The manufacturer's instructions for use typically include the following information and recommendations for proper use.

Preparing the Environment and NCIT

The use environment may impact the performance of the NCIT. Instructions will typically include recommendations for optimal use, such as the following:

- Use in a draft-free space and out of direct sun or near radiant heat sources.
- Determine if conditions are optimal for use. Typically, the environmental temperature should be between 60.8–104 °F (16–40 °C) and relative humidity below 85 percent.
- Place the NCIT in the testing environment or room for 10–30 minutes prior to use to allow the NCIT to adjust to the environment.
- For cleaning NCITs between uses, most NCITs should never be immersed in water or other liquids.

Preparing the Person Being Evaluated

In preparation for taking a temperature measurement with an NCIT, the person using the NCIT should typically ensure that:

- The test area of the forehead is clean, dry, and not blocked during measurement.
- The person's body temperature or temperature at the forehead test area has not been increased or decreased by wearing excessive clothing or head covers (e.g., headbands and bandanas) or by using facial cleansing products (for example cosmetic wipes).

Using the NCIT

As previously noted, the person using the device should strictly follow the manufacturer's guidelines and instructions for use for the specific NCIT being used. In particular, the following are typical instructions for NCIT usage [2].

- Hold the NCIT sensing area perpendicular to the forehead and instruct the person to remain stationary during measurement(s).
- The distance between the NCIT and the forehead is specific to each NCIT. Consult the manufacturer's instructions for correct measurement distances.
- Do not touch the sensing area of the NCIT and keep the sensor clean and dry.

- *IV Stands*

IV stands, also known as intravenous stands or infusion stands, are essential general hospital devices used to support intravenous (IV) therapy. They are designed to hold and suspend IV fluid bags, medication containers, and infusion pumps, allowing for the safe and efficient administration of fluids, medications, and nutrients to patients.

Height adjustability: IV stands are typically height-adjustable to accommodate patients of different heights and ensure that IV bags are positioned at the appropriate level. This feature helps prevent excessive tension on the IV tubing and ensures a smooth flow of fluids.

Stable base: IV stands have a stable base with multiple legs or wheels to provide support and prevent tipping or toppling. The base is designed to maintain stability, even when the stand is extended to its maximum height or when multiple IV bags are hung.

Hanging hooks and poles: IV stands are equipped with hooks or poles to hang IV bags, medication containers, and other infusion-related items. These hooks or poles are adjustable and allow for easy attachment and removal of IV bags and related accessories.

Maneuverability: Many IV stands come with wheels or casters, allowing for easy mobility and transport within the hospital setting. This feature enables healthcare providers to move the IV stand alongside the patient or between different treatment areas, ensuring convenience and flexibility in IV therapy administration.

Cable and cord management: IV stands often have built-in features or clips to manage the IV tubing and electrical cords from infusion pumps or monitoring devices. This helps prevent tangling and tripping hazards, ensuring a safer environment for patients and healthcare providers.

Stability and load capacity: IV stands are designed to support the weight of IV bags, medication containers, and infusion pumps. They have a specific load capacity and are engineered to maintain stability even with heavy or multiple bags hanging from them.

Collapsible or folding design: Some IV stands feature a collapsible or folding design, allowing for easy storage and transportation when not in use. This is particularly beneficial in situations where space is limited or when IV stands need to be transported to different hospital areas.

IV stands are used in various healthcare settings, including hospital wards, emergency departments, intensive care units (ICUs), outpatient clinics, and home healthcare settings. They are an integral part of IV therapy, ensuring the safe and effective delivery of fluids and medications to patients. IV stands may vary in design and features depending on the specific manufacturer and model. Some advanced IV stands may include additional features such as built-in infusion pumps, adjustable telescopic poles, or integrated power outlets for electrical devices [3].

- *Mobility Aids*

Devices such as wheelchairs, walkers, and crutches assist patients with mobility limitations, enabling them to move around the hospital safely. Wheelchairs are widely used general hospital devices that provide mobility and support for individuals with limited or impaired physical mobility. They are designed to assist patients in moving within healthcare facilities, performing daily activities, and improving their overall independence. Here are some key features and functions of wheelchairs:

Seat and frame: Wheelchairs typically consist of a seat made of padded material for comfort and a sturdy frame to provide stability and support. The frame is usually constructed from materials such as steel, aluminum, or titanium, depending on the specific wheelchair model.

Wheels: Wheelchairs have two large rear wheels and two smaller front wheels. The rear wheels are usually equipped with hand rims, allowing users to propel themselves forward by pushing the rims. Some wheelchairs may have additional wheels, such as anti-tip wheels or casters, for enhanced stability and maneuverability.

Brakes: Wheelchairs are equipped with brakes that can be operated by the user or a caregiver. The brakes are used to lock the wheels, preventing the wheelchair from moving or rolling unintentionally.

Footrests and leg supports: Many wheelchairs feature detachable or swing-away footrests to support the user's legs and provide comfort. Some models also offer adjustable leg supports or elevating leg rests for individuals with specific needs, such as those with leg injuries or circulation problems.

Armrests: Wheelchairs come with armrests to provide support and comfort to the user's arms. Armrests can be fixed or removable, depending on the wheelchair model, and they may be padded or contoured for added comfort.

Seat adjustability: Some wheelchairs offer seat adjustability features to accommodate users of different sizes and preferences. These features may include adjustable seat height, seat angle, or seat width, ensuring optimal comfort and proper posture for the user.

Folding or transportable design: Many wheelchairs have a folding mechanism that allows for easy transportation and storage. This feature is particularly useful for individuals who need to travel or for healthcare facilities with limited space.

Accessories and customizations: Wheelchairs can be customized with various accessories to meet individual needs. These accessories may include cushions for pressure relief, anti-tipping devices, oxygen tank holders, or trays for activities and mealtime.

It's important to note that wheelchairs come in different types and variations to suit specific user requirements. For example, manual wheelchairs are propelled by the user or a caregiver, while electric-powered wheelchairs utilize a motorized system for mobility.

In healthcare facilities, wheelchairs are used in various departments, including rehabilitation centers, outpatient clinics, emergency departments, and inpatient wards. They provide individuals with mobility limitations the opportunity to move around independently, participate in activities, and access healthcare services.

- *Sterilization for Medical Devices*

Medical devices are sterilized in a variety of ways including using moist heat (steam), dry heat, radiation, ethylene oxide gas, vaporized hydrogen peroxide, and other sterilization methods (e.g., chlorine dioxide gas, vaporized peracetic acid, and nitrogen dioxide).

Sterilization methods: There are several common methods used for sterilizing medical devices:

(a) Steam sterilization (autoclaving): This method uses high-pressure steam to kill microorganisms. It is widely used and effective for a wide range of medical devices that can withstand heat and moisture.
(b) Ethylene oxide (ETO) sterilization: ETO gas is commonly used for temperature-sensitive devices that cannot withstand steam sterilization. It penetrates packaging materials to reach all surfaces of the device and effectively kills microorganisms.

    For many medical devices, sterilization with ethylene oxide may be the only method that effectively sterilizes and does not damage the device during the sterilization process. Medical devices made from certain polymers (plastic or resin), metals, or glass, or that have multiple layers of packaging or hard-to-reach places (e.g., catheters) are likely to be sterilized with ethylene oxide. Literature shows that about fifty percent of all sterile medical devices in the United States are sterilized with ethylene oxide. The types of devices that are sterilized with ethylene oxide range from devices used in general healthcare practices (e.g., wound dressings) to more specialized devices used to treat specific areas of the body (e.g., stents).
(c) Hydrogen peroxide gas plasma sterilization: This low-temperature sterilization method utilizes hydrogen peroxide gas plasma to eliminate microorganisms. It is suitable for heat-sensitive devices and delicate instruments.
(d) Dry heat sterilization: Dry heat sterilization involves the use of hot air or flames to kill microorganisms. It is typically used for items that are sensitive to moisture, such as powders, oils, and sharp instruments.
(e) Chemical sterilization: Chemical agents, such as glutaraldehyde and peracetic acid, can be used to sterilize medical devices. These methods are often utilized for heat-sensitive and delicate equipment, but they may require longer exposure times.

Liquid Chemical Sterilization
Although the terms are similar, "liquid chemical sterilization" is different from thermal and gas/vapor/plasma low temperature "sterilization." As explained on this webpage, the FDA believes that sterilization with liquid chemical sterilants does not convey the same sterility assurance as sterilization using thermal or gas/vapor/plasma low temperature sterilization methods.

Traditional Sterilization
"Sterilization," as defined in the FDA's Liquid Chemical Sterilants/High Level Disinfectants guidance document, is a validated process used to render a product free of all forms of viable microorganisms. In many cases, thermal methods, such as steam, are used to achieve sterilization. Thermal sterilization methods have been studied and characterized extensively. In addition, the survival kinetics for gas/vapor/plasma low temperature sterilization methods have also been well characterized.

Liquid chemical sterilization involves a two-part process:

(i) Devices are treated with a liquid chemical germicide (LCG).
(ii) The processed devices are rinsed with water to remove the chemical residues.

There are several limitations to liquid chemical sterilization. Although the rinse water is treated to minimize any bioburden, it is not sterile. Because the rinse water is not sterile, devices rinsed with this water cannot be assured to be sterile. Furthermore, devices cannot be wrapped or adequately contained during processing in a liquid chemical sterilant. This means that there is no way to maintain sterility once devices have been processed.

Recommendations: For the reasons stated above, FDA recommends that the use of liquid chemical sterilants be limited to reprocessing only critical devices that are heat-sensitive and incompatible with sterilization methods such as steam and gas/vapor/plasma low temperature processes.

Biological and chemical indicators for liquid chemical sterilization: Biological Indicators are not appropriate or required for monitoring liquid chemical sterilization process. They are generally used for monitoring traditional sterilization processes where a SAL 10-6 is achieved. FDA has not cleared any biological indicators for monitoring liquid chemical sterilization process.

Chemical indicators are appropriate and are required for monitoring the minimum required concentration of most liquid chemical sterilants. The FDA has cleared many chemical indicators for monitoring the concentration of liquid chemical sterilant. Refer to the manufacturer's instructions for a compatible chemical indicator that is cleared by the FDA for use with the liquid chemical sterilant.

- *Patient Lifts*

Patient lifts are designed to lift and transfer patients from one place to another (e.g., from bed to bath, chair to stretcher). These should not be confused with stairway chair lifts or elevators. Patient lifts may be operated using a power source or manually. The powered models generally require the use of a rechargeable battery and the manual models are operated using hydraulics. While the design of patient lifts will vary based on the manufacturer, basic components may include a mast (the vertical bar that fits into the base), a boom (a bar that extends over the patient), a spreader bar (which hangs from the boom), a sling (attached to the spreader bar, designed to hold the patient), and a number of clips or latches (which secure the sling).

These medical devices provide many benefits, including reduced risk of injury to patients and caregivers when properly used. However, improper use of patient lifts can pose significant public health risks. Patient falls from these devices have resulted in severe patient injuries including head traumas, fractures, and deaths.

The FDA has compiled a list a best practices that, when followed, can help mitigate the risks associated with patient lifts. Users of patient lifts should:

- Receive training and understand how to operate the lift.
- Match the sling to the specific lift and the weight of the patient. A sling must be approved for use by the patient lift manufacturer. No sling is suitable for use with all patient lifts.
- Inspect the sling fabric and straps to make sure they are not frayed or stressed at the seams or otherwise damaged. If there are signs of wear, do not use it.

- Keep all clips, latches, and hanger bars securely fastened during operation.
- Keep the base (legs) of the patient lift in the maximum open position and situate the lift to provide stability.
- Position the patient's arms inside the sling straps.
- Make sure that the patient is not restless or agitated.
- Lock the wheels on any device that will receive the patient such as a wheelchair, stretcher, bed, or chair.
- Make sure that the weight limitations for the lift and sling are not exceeded.
- Follow the instructions for washing and maintaining the sling.
- Create and follow a maintenance safety inspection checklist to detect worn or damaged parts that need immediate replacement.

In addition to following these best practices, users of patient lifts must read all instructions provided by the manufacturer in order to safely operate the device.

Safe patient handling laws mandating the use of patient lifts to transfer patients have been passed in several states. Due to the passage of these laws, and the clinical community's goal of reducing patient and caregiver injury during patient transfers, it is expected that the use of patient lifts will increase. The best practices listed above are designed to help reduce the risks while enhancing the benefits of these medical devices.

- *Surgical Staplers and Staples*

Surgical staplers for internal use are used to deliver staples to tissues inside the body during surgery for:

- Removing part of an organ (resection)
- Cutting through and sealing organs and tissues (transection)
- Creating connections between structures (anastomoses)

Surgical staplers and staples for external use are used outside the body to close large wounds or surgical cuts on a patient's skin or scalp.

Advantages of surgical staplers and staples include:

- Quick placement
- Minimal tissue reaction
- Low risk of infection
- Strong wound closure

The FDA is identifying the special controls for surgical staplers for internal use that the Agency believes are necessary to provide a reasonable assurance of the safety and effectiveness of the device. FDA is issuing this reclassification on its own initiative based on new information. As part of this reclassification, the FDA is also amending the existing classification for "manual surgical instrument for general use" to remove staplers and to create a separate classification regulation for surgical staplers that distinguishes between surgical staplers for internal use and external use.

Risks

The FDA Panel recommended removing the increased risk of cancer recurrence and adverse tissue reaction from the risks to health presented at the Panel meeting. While surgical stapler malfunctions have resulted in complications such as anastomotic leaks, which have been associated with an increased risk of cancer recurrence, the FDA agrees that there is limited evidence directly linking surgical stapler failure or malfunction with an increased risk of cancer recurrence. Therefore, due to the limited evidence directly linking surgical stapler failure or malfunction with an increased risk of cancer recurrence, the FDA agrees with removing the increased risk of cancer recurrence from the list of complications associated with device failure/malfunction. FDA does not agree that adverse tissue reaction should be removed as a risk to health, as staplers for internal use contain patient-contacting materials that contact internal tissues, and these patient-contacting device materials may pose a risk of adverse tissue reaction if not adequately demonstrated to be biocompatible. The demonstration of biocompatibility for these devices is consistent with our approach for other devices with similar types and duration of contact; therefore, the FDA has not removed the applicable special control regarding biocompatibility.

- *Infusion Pumps*

An infusion pump is a medical device that delivers fluids, such as nutrients and medications, into a patient's body in controlled amounts. Infusion pumps are in widespread use in clinical settings such as hospitals, nursing homes, and in the home.

In general, an infusion pump is operated by a trained user, who programs the rate and duration of fluid delivery through a built-in software interface. Infusion pumps offer significant advantages over manual administration of fluids, including the ability to deliver fluids in very small volumes, and the ability to deliver fluids at precisely programmed rates or automated intervals. They can deliver nutrients or medications, such as insulin or other hormones, antibiotics, chemotherapy drugs, and pain relievers.

There are many types of infusion pumps, including large volume, patient-controlled analgesia (PCA), elastomeric, syringe, enteral, and insulin pumps. Some are designed mainly for stationary use at a patient's bedside. Others, called ambulatory infusion pumps, are designed to be portable or wearable [4].

Because infusion pumps are frequently used to administer critical fluids, including high-risk medications, pump failures can have significant implications for patient safety. Many infusion pumps are equipped with safety features, such as alarms or other operator alerts that are intended to activate in the event of a problem. For example, some pumps are designed to alert users when air or another blockage is detected in the tubing that delivers fluid to the patient. Some newer infusion pumps, often called smart pumps, are designed to alert the user when there is a risk of an adverse drug interaction, or when the user sets the pump's parameters outside of specified safety limits.

Over the past several years, significant safety issues related to infusion pumps have come to FDA's attention. These issues can compromise the safe use of external infusion pumps and lead to over- or under-infusion, missed treatments, or delayed therapy.

From 2005 through 2009, FDA received approximately 56,000 reports of adverse events associated with the use of infusion pumps, including numerous injuries and deaths. During this time period, manufacturers conducted 87 infusion pump recalls to address identified safety concerns. Seventy of these recalls were designated as Class II, a category that applies when the use of the recalled device may cause temporary or medically reversible adverse health consequences, or when the probability of serious adverse health consequences is remote. Fourteen recalls were Class I – situations in which there is a reasonable probability that use of the recalled device will cause serious adverse health consequences or death. These adverse event reports and device recalls have not been isolated to a specific manufacturer, type of infusion pump, or use environment; rather, they have occurred across the board.

Although some adverse events may be the result of user error, many of the reported events are related to deficiencies in device design and engineering, which can either create problems themselves or contribute to user error. The most common types of reported problems have been associated with software defects, user interface issues, and mechanical or electrical failures.

Examples of Reported Infusion Pump Problems

Software problems: A software error message is displayed, stating that the pump is inoperable. This occurs in the absence of an identifiable problem. The infusion pump interprets a single keystroke as multiple keystrokes (a problem called a "key bounce"). For example, the user programs an infusion rate of 10 mL/hour, but the device registers an infusion rate of 100 mL/hour.

Alarm errors: The infusion pump fails to generate an audible alarm for a critical problem, such as an occlusion (e.g., clamped tubing) or the presence of air in the infusion tubing. The infusion pump generates an occlusion alarm in the absence of an occlusion.

Inadequate user interface design ("human factors" issues): The design of the infusion pump screen confuses the user, or the infusion pump does not respond as it should (i.e., with a warning or alarm) when inappropriate data is entered. The infusion pump screen doesn't make clear which units of measurement the user is expected to enter. For example, the user may enter weight in pounds when the infusion pump requires it in kilograms.

Pump labels or components become damaged under routine use. For example, cleaning the pump, as the user-maintainer believes is acceptable practice, may damage the pump, making it unreliable for clinical use. Users with long fingernails may damage the print on the pump keys, making them unreadable. User instructions or cues for mechanical set-up are not specific or clear enough. For example, an instruction to attach a tubing set in all required tube holder-clips before closing the pump's access door may be unclear, resulting in clamped tubing and under-infusion.

Inadequately designed alarm functions and settings cause users to miss problems or respond late. For example, an alarm indicating low battery charge may not be displayed in time for a user to prevent pump shut-off during a critical infusion while a patient is in transport. False ("nuisance") alarms may decrease users' sensitivity to all alarms.

The infusion pump screen design is clunky or confusing to users, causing a delay in therapy. For example, the "Start Infusion" key may be located next to the "Power" key, and a user may turn off the infusion pump instead of initiating infusion. In some cases, programmed settings are lost when a user turns the pump off, and the infusion settings have to be re-entered after the pump restarts.

Warnings are displayed so often that users come to ignore them (similar to "nuisance alarms"), are not detailed enough to prevent misuse, or represent values in ways that are unfamiliar to the user.

Warning messages are unclear. In the example below, it is unclear if the user is confirming the warning message or the infusion settings.

User manuals are confusing, inadequate, outdated, or unavailable. This is particularly of concern for home-based users.

When communicating the critical aspects of the pump's operational, default, or "piggyback" status, the system does not use user-friendly language or does not give enough information to guide users through appropriate actions.

Broken components: The infusion pump may have been dropped or damaged during use, which may result in an over-infusion or an under-infusion if the pump continues to be used without being repaired. The plastic casing of an insulin pump, although promoted as waterproof, is prone to cracking, allowing water to enter the case and to cause the pump to malfunction. Slight misalignment of tubing places stress on the pump door, resulting in eventual cracking of pump case.

Battery failures: A design issue causes over-heating of the battery and leads to premature battery failure. A patient returns from ambulating and forgets to plug in the infusion pump. The infusion pump alarms with a low battery message, but the speaker volume is set too low, and the alarm goes unnoticed. The infusion pump powers off after the battery is depleted. The battery is not replaced during the recommended end of life routine maintenance.

Fire, sparks, charring, or shocks: The user plugs in or unplugs the device from an electrical outlet and receives a shock, and/or sparks are seen. A burning smell or flames are noted on the infusion pump [5].

- *Personal Protective Equipment*

Personal protective equipment (PPE) refers to protective clothing, helmets, gloves, face shields, goggles, facemasks and/or respirators or other equipment designed to protect the wearer from injury or the spread of infection or illness.

PPE is commonly used in healthcare settings such as hospitals, doctor's offices, and clinical labs. When used properly, PPE acts as a barrier between infectious materials such as viral and bacterial contaminants and your skin, mouth, nose, or eyes (mucous membranes). The barrier has the potential to block transmission of contaminants from blood, body fluids, or respiratory secretions. PPE may also protect patients who are at high risk for contracting infections through a surgical procedure or who have a medical condition, such as an immunodeficiency, from being exposed to substances or potentially infectious material brought in by visitors and healthcare workers. When used properly and with other infection control practices such as hand-washing, using alcohol-based hand sanitizers, and covering coughs and sneezes, it minimizes the spread of infection from one person to another. Effective use of PPE includes properly removing and disposing of contaminated PPE to prevent exposing both the wearer and other people to infection [6].

The FDA's role in regulating personal protective equipment:

All personal protective equipment (PPE) that is intended for use as a medical device must follow The FDA's regulations and should meet applicable voluntary consensus standards for protection. This includes surgical masks, N95 respirators, medical gloves, and gowns. The consensus standards and the FDA's requirements vary depending on the specific type of PPE. When these standards and regulations are followed, they provide reasonable assurance that the device is safe and effective.

Some PPEs are reviewed by the FDA before they can be legally sold in the United States. In this review, known as Premarket Notification or 510(k) clearance, the manufacturers have to show they meet specific criteria for performance, labeling, and intended use to demonstrate substantial equivalence. One-way substantial equivalence may be demonstrated, in part, is by conforming to consensus standards for barrier performance and resistance to tears and snags. Voluntary consensus standards may also be used to demonstrate sterility (when applicable), biocompatibility, fluid resistance, and flammability. Manufacturers must validate the methods used to test conformance to standards and support each product with appropriate performance test data.

- *N95 Respirators and Surgical Masks*

N95 respirators and surgical masks are examples of personal protective equipment that are used to protect the wearer from particles or from liquid contaminating the face. The Centers for Disease Control and Prevention (CDC) National Institute for Occupational Safety and Health (NIOSH) also regulates N95 respirators. The Department of Labor's Occupational Safety and Health Administration (OSHA) regulates entities for compliance with worker safety rules and OSHA standards, including, for example, the proper use of respirators in different work environments.

It is important to recognize that the optimal way to prevent transmission of microorganisms, such as viruses, is to use a combination of interventions from across the hierarchy of controls, not just PPE alone.

N95 Respirators: An N95 respirator is a respiratory protective device designed to achieve a very close facial fit and very efficient filtration of airborne particles. Note that the edges of the respirator are designed to form a seal around the nose and mouth. Surgical N95 Respirators are commonly used in healthcare settings and are a subset of N95 Filtering Facepiece Respirators (FFRs), often referred to as N95s.

Comparing Surgical Masks and Surgical N95 Respirators

The FDA regulates surgical masks and surgical N95 respirators differently based on their intended use.

A surgical mask is a loose-fitting, disposable device that creates a physical barrier between the mouth and nose of the wearer and potential contaminants in the immediate environment. These are often referred to as face masks, although not all face masks are regulated as surgical masks. Note that the edges of the mask are not designed to form a seal around the nose and mouth.

An N95 respirator is a respiratory protective device designed to achieve a very close facial fit and very efficient filtration of airborne particles. Note that the edges of the respirator are designed to form a seal around the nose and mouth.

Surgical N95 Respirators are commonly used in healthcare settings and are a subset of N95 Filtering Facepiece Respirators (FFRs), often referred to as N95s.

General N95 Respirator Precautions

- People with chronic respiratory, cardiac, or other medical conditions that make breathing difficult should check with their healthcare provider before using an N95 respirator because the N95 respirator can make it more difficult for the wearer to breathe.
- Some models have exhalation valves that can make breathing out easier and help reduce heat build-up. Note that N95 respirators with exhalation valves should not be used when sterile conditions are needed.
- All FDA-cleared N95 respirators are labeled as "single-use," disposable devices. If your respirator is damaged or soiled, or if breathing becomes difficult, you should remove the respirator, discard it properly, and replace it with a new one. To safely discard your N95 respirator, place it in a plastic bag and put it in the trash. Wash your hands after handling the used respirator.
- N95 respirators are not designed for children or people with facial hair. Because a proper fit cannot be achieved on children and people with facial hair, the N95 respirator may not provide full protection.

N95 Respirators in Industrial and Healthcare Settings

Most N95 respirators are manufactured for use in construction and other industrial-type jobs that expose workers to dust and small particles. They are regulated by the National Personal Protective Technology Laboratory (NPPTL) in the National Institute for Occupational Safety and Health (NIOSH), which is part of the Centers for Disease Control and Prevention (CDC).

However, some N95 respirators are intended for use in a healthcare setting. Specifically, single-use, disposable respiratory protective devices are used and worn by healthcare personnel during procedures to protect both the patient and healthcare personnel from the transfer of microorganisms, body fluids, and particulate material. These surgical N95 respirators are class II devices regulated by the FDA, under 21 CFR 878.4040, and CDC NIOSH under 42 CFR Part 84.

N95 respirators regulated under product code MSH are class II medical devices exempt from 510(k) premarket notification, unless:

- The respirator is intended to prevent specific diseases or infections.

- The respirator is labeled or otherwise represented as filtering surgical smoke or plumes, filtering specific amounts of viruses or bacteria, reducing the amount of and/or killing viruses, bacteria, or fungi, or affecting allergenicity.
- The respirator contains coating technologies unrelated to filtration (e.g., to reduce and or kill microorganisms).

- *Face Masks*

A face mask is a product that covers the wearer's nose and mouth. Face masks are for use as source control by the general public and healthcare personnel (HCP) in accordance with CDC recommendations and are not personal protective equipment. Face masks may or may not meet any fluid barrier or filtration efficiency levels; therefore, they are not a substitute for N95 respirators or other Filtering Facepiece Respirators (FFRs), which provide respiratory protection to the wearer, or for surgical masks, which provide fluid barrier protection to the wearer.

Barrier Face Coverings

A barrier face covering, as described in ASTM F3502-21, is a product worn on the face specifically covering at least the wearer's nose and mouth, with the primary purpose of providing source control and to provide a degree of particulate filtration to reduce the amount of inhaled particulate material. Barrier face coverings are not a substitute for N95 respirators and other Filtering Facepiece Respirators (FFRs), which provide respiratory protection to the wearer, or for surgical masks, which provide fluid barrier and particulate material protection to the wearer.

Barrier face coverings may be made from a variety of materials that are not flammable. By definition, a barrier face covering should meet the particulate filtration efficiency, airflow resistance, and leakage assessment recommendations as described in ASTM F3502-21.

- *Surgical Masks*

A surgical mask is a loose-fitting, disposable device that creates a physical barrier between the mouth and nose of the wearer and potential contaminants in the immediate environment. Surgical masks are regulated under 21 CFR 878.4040. Surgical masks are not to be shared and may be labeled as surgical, isolation, dental, or medical procedure masks. They may come with or without a face shield. These are sometimes referred to as face masks, as described above, although not all face masks are regulated as surgical masks.

Surgical masks are made in different thicknesses and with different abilities to protect you from contact with liquids. These properties may also affect how easily you can breathe through the face mask and how well the surgical mask protects you. If worn properly, a surgical mask is meant to help block large-particle droplets, splashes, sprays, or splatter that may contain germs (viruses and bacteria), keeping it from reaching your mouth and nose. Surgical masks may also help reduce exposure of your saliva and respiratory secretions to others.

While a surgical mask may be effective in blocking splashes and large-particle droplets, a face mask, by design, does not filter or block very small particles in the air that may be transmitted by coughs, sneezes, or certain medical procedures. Surgical masks also do not provide complete protection from germs and other contaminants because of the loose fit between the surface of the mask and your face.

Surgical masks are not intended to be used more than once. If your surgical mask is damaged or soiled, or if breathing through the mask becomes difficult, you should remove it, discard it safely, and replace it with a new one. To safely discard your surgical mask, place it in a plastic bag and put it in the trash. Wash your hands after handling the used mask.

- *Medical Gloves*

Medical gloves are examples of personal protective equipment used to protect the wearer and/or the patient from the spread of micro-organisms that may potentially cause infection or illness during medical procedures and examinations. Medical gloves are one part of an infection-control strategy. Medical gloves are disposable and include examination gloves, surgical gloves, and medical gloves for handling chemotherapy agents (chemotherapy gloves). Medical gloves are regulated by the FDA as Class I reserved medical devices that require a 510(k) premarket notification. Generally, the FDA reviews these devices to ensure that performance criteria, such as leak resistance, certain physical properties, and biocompatibility, are met.

When to Use Medical Gloves

Use medical gloves when your hands may touch someone else's body fluids (such as blood, respiratory secretions, vomit, urine, or feces), certain hazardous drugs, or some potentially contaminated items.

Instruction to use:

- Wash your hands before putting on sterile medical gloves.
- Make sure your medical gloves fit properly for you to wear them comfortably during all patient care activities.
- Some people are allergic to the natural rubber latex used in some medical gloves. The FDA requires manufacturers to identify on the package labeling the materials used to make the medical gloves. If you are or your patient is allergic to natural rubber latex, you should choose medical gloves made from other synthetic materials (such as polyvinyl chloride (PVC), nitrile, or polyurethane).
- Be aware that sharp objects can puncture medical gloves.
- Always change your medical gloves if they rip or tear.
- After removing medical gloves, wash your hands thoroughly with soap and water or alcohol-based hand rub.
- Never reuse medical gloves.
- Never wash or disinfect medical gloves.
- Never share medical gloves with other users.
- Ban on powdered gloves

On December 19, 2016, the FDA published a final rule banning powdered gloves (Powdered Surgeon's Gloves, Powdered Patient Examination Gloves, and

Absorbable Powder for Lubricating a Surgeon's Gloves) based on the unreasonable and substantial risk of illness or injury to individuals exposed to the powdered gloves. The risks to both patients and healthcare providers when internal body tissue is exposed to the powder include severe airway inflammation and hypersensitivity reactions. Powder particles may also trigger the body's immune response, causing tissue to form around the particles (granulomas) or scar tissue formation (adhesions) which can lead to surgical complications. For a detailed description of the risks that the FDA identified, please refer to the final rule.

- *Medical Gowns*

Medical gowns are examples of personal protective equipment used in healthcare settings. They are used to protect the wearer from the spread of disease-causing microorganisms if the wearer comes in contact with potentially infectious liquid or solid material. They may also be used to help prevent the wearer from transferring microorganisms that could harm vulnerable patients, such as those with compromised immune systems. Gowns are intended to provide broad barrier protection. At this time, the FDA has not cleared, approved, or authorized any gowns for specific protection or prevention against the virus that causes COVID-19. Gowns are one part of an overall infection-control strategy.

Many names are used to refer to gowns intended for use in healthcare settings, including, surgical gowns, isolation gowns, surgical isolation gowns, non-surgical gowns, procedural gowns, and operating room gowns.

Levels of Medical Gowns

When choosing gowns, look for product labeling that describes intended use with the desired level of protection based on the risk levels described below:

The FDA recognizes the consensus standard American National Standards Institute/Association of the Advancement of Medical Instrumentation (ANSI/AAMI) PB70, "Liquid barrier performance and classification of protective apparel and drapes intended for use in healthcare facilities." This standard establishes a system of classification for protective apparel and drapes used in healthcare facilities based on their liquid barrier performance and specifies related labeling requirements and standardized test methods for determining compliance. The standard is intended to ultimately assist end-users in determining the type(s) of protective product most appropriate for a particular task or situation:

Level 1: Minimal risk, to be used, for example, during basic care, standard isolation, cover gown for visitors, or in a standard medical unit

Level 2: Low risk, to be used, for example, during blood draw, suturing, in the intensive care unit (ICU), or a pathology lab

Level 3: Moderate risk, to be used, for example, during arterial blood draw, inserting an intravenous (IV) line, in the emergency room, or for trauma cases

Level 4: High risk, to be used, for example, during long, fluid intense procedures, surgery, when pathogen resistance is needed or infectious diseases are suspected (non-airborne)

Surgical Gowns

A surgical gown is regulated by the FDA as a Class II medical device that requires a 510(k) premarket notification. A surgical gown is a personal protective garment intended to be worn by healthcare personnel during surgical procedures to protect both the patient and healthcare personnel from the transfer of microorganisms, body fluids, and particulate matter. Because of the controlled nature of surgical procedures, critical zones of protection have been described by national standards. All surgical gowns must be provided sterile and labeled as a surgical gown [7].

Surgical Isolation Gowns

Surgical isolation gowns are used when there is a medium to high risk of contamination and a need for larger critical zones than traditional surgical gowns. Surgical isolation gowns, like surgical gowns, are regulated by the FDA as a Class II medical device that requires a 510(k) premarket notification. All areas of the surgical isolation gown except bindings, cuffs, and hems are considered critical zones of protection and must meet the highest liquid barrier protection level for which the gown is rated. All seams must have the same liquid barrier protection as the rest of the gown. Additionally, the fabric of the surgical isolation gown should cover as much of the body as is appropriate for the intended use.

Non-surgical Isolation Gowns

Non-surgical isolation gowns are Class I devices (exempt from premarket review) intended to protect the wearer from the transfer of microorganisms and body fluids in low or minimal risk patient isolation situations. Non-surgical gowns are not worn during surgical procedures, invasive procedures, or when there is a medium to high risk of contamination.

Like surgical isolation gowns, non-surgical gowns should also cover as much of the body as is appropriate to the task. All areas of the non-surgical gown except bindings, cuffs, and hems are considered critical zones of protection and must meet the highest liquid barrier protection level for which the gown is rated. All seams must have the same liquid barrier protection as the rest of the gown.

Non-surgical Non-isolation Gown

Non-sterile, non-isolation gowns are intended to be worn by healthcare personnel to provide moderate or high barrier protection in non-sterile and non-patient isolation situations. These gowns are regulated by the FDA as a Class II medical device that requires a 510(k) premarket notification. Cloth gowns that will not be used in a sterile field, such as surgery, can be reused if they are laundered in enzymatic detergent or per the hospital's standard operating procedures.

Standards for Gowns

The FDA recognizes consensus standards for gowns as listed in the FDA's Recognized Consensus Standards database.

Sterility Information for Gowns

For a device sold sterile, the FDA recommends sponsors provide the following information as detailed in the final guidance entitled Submission and Review of

Sterility Information in Premarket Notification (510(k)) Submissions for Devices Labeled as Sterile. This information may include:

- Sterilization method that will be used.
- A description of the method that will be used to validate the sterilization cycle, but not the validation data itself (for established sterilization methods).
- Reference to a standard method (e.g., AAMI Radiation Standard) usually is sufficient for established sterilization methods with FDA-recognized standards.
- The sterility assurance level (SAL) for the device which the firm intends to meet. An SAL of 10-6 is required for surgical drapes and surgical gowns which are to be used during surgical procedures.
- A description of the packaging's ability to maintain the device's sterility.
- If sterilization involves ethylene oxide (EtO), the maximum levels of residues of ethylene oxide and ethylene chlorohydrin remain on the device. The levels should be consistent with the FDA-recognized consensus standards for ethylene oxide.
- In the case of radiation sterilization, the radiation dose.

Biocompatibility Information for Gowns

Medical gowns are devices that are considered surface-contacting devices with intact skin with a contact duration of ≤24 hours. The FDA recommends that cytotoxicity (ISO 10993-5), sensitization (ISO 10993-10), and irritation or intracutaneous reactivity (ISO 10993-10) be evaluated for a device.

- *Hospital Beds*

Hospital beds are essential medical devices used in general hospitals to provide comfort and support to patients during their stay. Here are some key features and functions of hospital beds:

Adjustable positions: Hospital beds can be adjusted to various positions to accommodate the needs of patients. They typically have adjustable backrests, leg rests, and overall height. These adjustments help improve patient comfort and facilitate medical procedures, such as examinations and surgeries [8].

Side rails: Hospital beds often come equipped with side rails that can be raised or lowered. These rails help prevent patients from falling out of bed, especially those who are elderly, have limited mobility, or are at risk of wandering or confusion.

Mattresses: Hospital beds are designed to work with specific mattresses that offer pressure relief and support for patients. These mattresses are typically made of foam or air-filled cells to prevent the development of bedsores or pressure ulcers.

Mobility and maneuverability: Hospital beds are equipped with wheels, allowing for easy movement within the healthcare facility. They can be locked in place to ensure stability when needed.

Patient monitoring: Some advanced hospital beds include integrated monitoring systems that can measure vital signs, such as heart rate, blood pressure, and oxygen saturation. These beds can transmit real-time data to nursing stations, improving patient care and reducing the need for additional monitoring equipment.

Patient safety features: Hospital beds may have additional safety features, such as built-in alarms that alert healthcare providers when a patient tries to get out of bed without assistance. These features help prevent falls and ensure patient safety.

Accessibility: Hospital beds are designed to be accessible for healthcare providers. They often have adjustable height settings, allowing medical professionals to position the bed at a comfortable working level for tasks like patient assessment, wound care, or administering medication.

Hospital beds can vary in features and functionalities depending on the specific model and manufacturer. Different specialized areas within a hospital, such as intensive care units or maternity wards, may have beds with additional features tailored to the specific needs of those departments. FDA received 901 incidents of patients caught, trapped, entangled, or strangled in hospital beds. The reports included 531 deaths, 151 nonfatal injuries, and 220 cases where staff needed to intervene to prevent injuries. Most patients were frail, elderly, or confused.

The efforts of the FDA and the Hospital Bed Safety Workgroup have culminated in the FDA's release of Hospital Bed System Dimensional and Assessment Guidance to Reduce Entrapment. This guidance provides recommendations for manufacturers of new hospital beds and for facilities with existing beds (including hospitals, nursing homes, and private residences).

Healthcare facilities developing comprehensive bed safety programs should consider the Clinical Guidance for the Assessment and Implementation of Bed Rails to assess an individual patient's needs when using a side rail and consulting with the hospital bed manufacturer and their facilities' risk managers. The HBSW has developed a Bed Safety Entrapment Kit containing information and tools that can be used to assess the risk of entrapment in hospital beds. Additional information about the kit is also provided in the FDA Guidance.

Covers for Hospital Bed Mattresses

Hospital beds refer to a variety of medical devices that are classified as beds. FDA regulations classify hospital beds as Class I and Class II devices. These devices are used for patients in acute care, long-term care, or home care settings. A hospital bed system encompasses the bed frame and its components, including the bedside rails, head and footboard, the mattress, and any accessories added to the bed, such as a detachable mattress cover.

A hospital bed mattress cover provides outer protection to a mattress by preventing blood and other body fluids from entering the inside (inner core) of the mattress. Such covers may be coated with or contain an antimicrobial solution that kills germs (viruses or bacteria) or prevents bacterial growth. There are multiple terms used to describe hospital bed mattress covers: water-resistant (keeps liquid away from the material), water-proof (prevents liquid from entering inside the material), or water-repellent (keeps liquid away from the material and prevents liquid from entering inside the material). Covers are usually detachable from the mattress or the mattress lining, meaning that they can be removed or replaced.

Safety Concerns

Over time, hospital bed mattress covers can wear out and allow blood and body fluids to penetrate and get trapped inside mattresses. If blood or body fluids from one patient penetrate and get absorbed in a mattress, the fluids can leak out the next time the mattress is used. Coming into contact with these fluids poses a risk of infection to patients using the bed. The FDA issued a safety communication External Link Disclaimer in 2013 alerting healthcare providers, healthcare facility staff, and caregivers to these safety concerns. From 2011 through 2016, the FDA has received over 700 reports of a hospital bed mattress cover failing to prevent blood or body fluids from leaking into the mattress.

Recommendations

These recommendations are based on guidelines for environmental infection control in healthcare facilities issued by the Centers for Disease Control and Prevention (CDC). They are intended to help healthcare providers, healthcare facility staff, and caregivers ensure hospital bed mattress covers are safe for use in healthcare settings.

Develop an Inspection Plan

- Create an inspection plan for all hospital bed mattresses and mattress covers in your facility.
- Check the manufacturers' guidelines for an expected lifetime on the hospital bed mattress and mattress covers, and follow any additional recommendations listed there.
- Contact the mattress cover manufacturer for any additional questions not covered here.
- Regularly check each hospital bed mattress cover for any visible signs of damage or wear such as cuts, tears, cracks, pinholes, snags, or stains.
- Routinely remove the hospital bed mattress cover and check its inside surface. Once the mattress cover is removed, inspect the mattress for wet spots, staining, or signs of damage or wear. Check all sides and the bottom of the mattress.
- Be aware that it may be difficult to identify damaged or soiled mattresses without removing the mattress covers first. Mattress covers tend to be dark in color, making it hard to see what lies underneath.

Remove and Replace

Remove any damaged, worn, or visibly stained hospital bed mattress according to the healthcare facility's procedures and manufacturer's instructions.

Immediately replace any hospital bed mattress cover with visible signs of stains, damage, or wear to reduce the risk of infection to patients.

Clean and disinfect undamaged hospital bed mattress covers according to the manufacturer's guidelines.

DO NOT stick needles into a hospital bed mattress through the mattress cover.

- *Medical Device Connectors*

Patients in healthcare settings receive food, medication, and other therapies through a variety devices or delivery systems, such as syringes, catheters, and tubing sets that connect to each other. Connectors are the parts of devices that attach tubing, catheters, and syringes to other medical devices. Medical devices are often packaged together in tubing sets or co-packaged with another device (e.g., feeding set and enteral feeding tube). These sets comprise all the parts needed to use the tubing for its intended purpose, including the connectors that attach tubes to the other parts of the set or to other devices.

In a typical hospital setting, several different types of medical devices, each with their own connections, may be in use at the same time on a single patient. In specialized settings, such as intensive care units (ICUs), cardiac care units, or emergency departments, patients may require dozens of different devices at once. Devices that need to connect to each other are also used in the home setting and other environments beyond professional healthcare facilities. Patients may have to use these devices for the duration of an illness, recuperation, long-term care, or throughout their lives.

Medical Device Misconnections

Medical device misconnections may occur when one type of medical device is mistakenly attached to another type of medical device that performs a different function. Because these connectors are easy to use and may be compatible with different medical devices, users can mistakenly connect unrelated systems to one another. This may cause medication or other substances to be delivered through the wrong tubing into the incorrect area of the body. These errors are sometimes called tubing misconnections, wrong route errors, catheter misconnections, or Luer misconnections and can result in patient injury or death. For examples of potential misconnections, see the Examples of Medical Device Misconnections page, which includes case studies of errors and how to correct or prevent them [9].

Device misconnections can occur for many reasons, including: The similar design of many connectors and widespread use of connectors with similar sizes and shapes. Human error arising from conditions such as multiple connections on one patient, poor lighting, lack of training, time pressure, fatigue, or high-stress environments.

Reducing Risks Associated with Medical Device Misconnections: Manufacturers and healthcare facilities have tried many methods to prevent device misconnections, including color-coding, labels, tags, and training. However, these methods alone have not effectively solved the misconnection problem because they are not consistently applied, nor do these methods physically prevent the misconnections. Soon, the way devices connect to each other may be changing, greatly reducing the risk for misconnections. New design standards are being developed for tubing connectors for high-risk medical applications (e.g., enteral, respiratory, neuraxial), so that unrelated devices cannot connect with each other.

The FDA's Role

The FDA is working with standards organizations, federal partners, professional societies, advocacy groups, patients, and other stakeholders to reduce the chance of medical device misconnections. The FDA continues to address this issue through:

- Participating in international consensus standards development
- Developing draft and final guidance
- Evaluating medical device adverse event reports about medical device misconnections
- Communicating with patients and other stakeholders to understand the impact of new connector designs

Reducing Risks Through Standards Development for Medical Device Connectors

The FDA, the standards community, notably the International Organization for Standardization (ISO), and the medical device industry are taking actions to reduce the likelihood of medical device misconnections. These actions include the development of standardized connector designs for specific medical applications intended to physically prevent connections with devices used for other medical applications. The FDA participates in standards development for connector designs because of the direct association between the connections and other medical devices.

Standards for Small-Bore Connectors

Many different types of medical devices incorporate small-bore connectors. Small-bore connectors are parts used to connect medical devices such as tubing, syringes, and other accessories that deliver fluids and gases for patient care. Small-bore refers to the small size of the diameter opening (less than 8.5 millimeters) of the connector.

Two standards provide overarching recommendations for small-bore connectors:

- ISO 80369-1:2016External Link Disclaimer, "Small-bore connectors for liquids and gases in healthcare applications" – Part 1: General requirements – specifies general provisions for small-bore connectors and methodology for assessing design characteristics to reduce the risk of misconnections between medical devices or accessories.
- ISO 80369-20:2015External Link Disclaimer, "Small-bore connectors for liquids and gases in healthcare applications" – Part 20: Common test methods – specifies test methods to support small-bore connectors' functional requirements. The FDA currently recognizes this standard.

Examples of Medical Devices that Use Connectors

The FDA anticipates recognizing additional standards for specific small-bore connector applications as they are developed. These standards are expected to include specific dimensions and performance requirements for the respective medical device or medical application. Listed below is a description of the medical device or medical application and status information about the relevant international design standard:

- Blood pressure cuffs and other non-invasive blood pressure devices are used to inflate the blood pressure cuff to test a patient's blood pressure. IEC 80369-5:2016External Link Disclaimer was published in March 2016 to provide specifications for the small-bore connectors used with blood pressure cuffs.

- Breathing or respiratory systems such as anesthesia machines and ventilators are used to facilitate a patient's breathing. Work on the international standard for breathing and respiratory systems is still underway. The FDA anticipates recognizing this standard once it is finalized.
- Enteral devices deliver liquid nutrients or medicine to the stomach or intestines in patients who are unable to eat or drink by mouth or need supplemental nutrition. Feeding tubes are often inserted into the patient's abdomen. Patients use pre-packaged food purchased from nutrition manufacturers or blend their own diets at home. ISO 80369-3:2016 External Link Disclaimer was published in July 2016, and the FDA recognizes this standard. This standard provides specifications for connectors intended for enteral applications. Manufacturers are now transitioning to the new standard to address the misconnection issue and reduce the risk of misconnections between enteral and non-enteral devices.
- Intravascular or hypodermic devices, such as arterial or intravenous (IV) lines, are generally used to deliver medications or fluids through a patient's neck, chest, or veins in the arm. ISO 80369-7:2021External Link Disclaimer is being finalized and will provide manufacturers with specifications for intravenous and hypodermic applications.
- Neuraxial devices, such as epidural catheters, are used to deliver medicines or anesthesia to neuraxial sites, such as the epidural space, or are used to monitor or remove cerebral-spinal fluid for therapeutic or diagnostic purposes. ISO 80369-6:2016External Link Disclaimer was published in March 2016 to provide specifications for designing connectors for use with neuraxial devices. The FDA recognizes this standard.
- Limb tourniquet cuffs are compression devices used to apply pressure, such as to the radial artery, to help stop bleeding after a procedure. IEC 80369-5: 2016External Link Disclaimer "Small-Bore Connectors for Liquids and Gases in Healthcare Applications - Part 5: Connectors for Limb Cuff Inflation Applications" was published in March 2016 to specify dimensions and requirements for the design and functional performance of connectors used in limb cuff inflation.

Examples of Medical Device Misconnections

The FDA has recognized voluntary consensus standards for certain medical device connector designs intended to enhance device safety such as the ISO 80369 series. Many such connectors that rely on the ISO 80369 standards are currently available and are also incorporated into product-specific standards. Connectors that conform to these standards are designed to help reduce the risk of medical device misconnections. As new connector designs for high-risk delivery systems become more widely available, the likelihood for medical device misconnections is expected to decrease. In addition to these standards, standards development organizations are developing standards applicable to connectors for other medical applications.

If the connectors you use do not conform to one of these standards, or the connectors do not yet have an applicable standard, misconnections can still occur. The below case studies describe device misconnections that have been reported to the FDA. This information offers patients and healthcare providers safety tips and recommendations to reduce device misconnections. The potential for harm designations: High, Medium, and Low refer to the severity of patient harm that could result from the depicted type of misconnection.

Case Study 1:Epidural Tubing Erroneously Connected to IV Tubing

An anesthetist and a midwife mistakenly connected an epidural set to the patient's IV tubing.

The epidural medicine was delivered to the IV.

The patient died.

Potential for harm: High.

Safety advice: For certain high-risk catheters (e.g., epidural, intrathecal, arterial), label the catheter and do not use catheters that have injection ports.

Case Study 2: IV Tubing Erroneously Connected to Trach Cuff

A child in a pediatric intensive care unit had both an IV line and a trach tube.

The IV tubing was mistakenly connected to the trach cuff port.

The IV fluid over-expanded the trach cuff to the point of breaking and continuous IV fluids entered the child's lungs.

The child died.

Potential for harm: High.

Safety advice: Emphasize the risk of tubing misconnections in orientation and training.

Case Study 3: IV Tubing Erroneously Connected to Nebulizer

During a nebulizer treatment, the patient's oxygen tubing fell off the nebulizer and the patient's IV tubing was inadvertently attached to the nebulizer.

When the patient inhaled, a moderate amount of IV fluids was aspirated into the patient's lungs.

The misconnection was identified by the respiratory therapist and the patient survived.

Potential for harm: High.

Safety advice: Do not purchase non-intravenous equipment that is equipped with connectors that can physically mate or attach with a female Luer IV line connector.

Case Study 4: Oxygen Tubing Erroneously Connected to a Needleless IV Port

A patient's oxygen tubing became disconnected from his nebulizer and was accidentally reattached to his IV tubing Y-site by a staff member who was completing a double shift.

The patient died from an air embolism, even though the connection was broken within seconds.

Potential for harm: High.

Safety advice: Identify and manage conditions and practices that may contribute to healthcare worker fatigue, and take appropriate action.

Case Study 5: Blood Pressure Tubing Erroneously Connected to IV Catheter

An ER patient had an IV heparin lock but no IV fluids had been started. The patient also had a noninvasive automatic BP cuff placed for continuous monitoring.

The BP cuff tubing was disconnected when the patient went to the bathroom.

When she returned, her spouse mistakenly connected the BP cuff tubing to the IV catheter and approximately 15 mL of air was delivered to the IV catheter.

The patient died from a fatal air embolus, despite resuscitation efforts.

Potential for harm: High.

Safety advice: Inform non-clinical staff, patients, and their families that they must get help from clinical staff whenever there is a real or perceived need to connect or disconnect devices or infusions.

Case Study 6: IV Tubing Erroneously Connected to Nasal Cannula

A nurse's aide inadvertently connected a patient's IV tubing to the nasal oxygen cannula upon transfer to the step-down unit.

The misconnection was not noted until four hours later, when the patient complained of chest tightness and difficulty breathing.

The patient was treated for congestive heart failure and survived.

Potential for harm: High.

Safety advice: Recheck connections and trace all patient tubes and catheters to their sources upon the patient's arrival in a new setting or service as part of the handoff process. Standardize this "line reconciliation" process.

Case Study 7: Syringe Erroneously Connected to Trach Cuff

The patient had both a central line with three ports and a trach tube.

Medicine intended for the central line was inadvertently injected into the trach cuff.

The trach cuff was damaged and the medicine entered the patient's lungs.

A new trach tube was inserted and the patient survived.

Potential for harm: High.

Safety advice: Always trace a tube or catheter from the patient to the point of origin before connecting any new device or infusion.

Case Study 8: Enteral Feeding Tube Erroneously Connected to Ventilator In-Line Suction Catheter

A patient's feeding tube was inadvertently connected to the instillation port on the ventilator in-line suction catheter.

Tube feeding was delivered into the patient's lungs.

The patient died.

Potential for harm: High.

Safety advice: Emphasize the risk of tubing misconnections in orientation and training.

Case Study 9: Pulsatile Anti-embolism Stocking Erroneously Connected to IV Heparin Lock

A patient admitted for stroke had a pulsatile anti-embolism stocking (PAS) on the left lower extremity and an IV heparin lock in the right ankle.

The patient was alert and oriented on admission but shortly after was found unresponsive and cyanotic.

The PAS pump tubing was found connected to the IV heparin lock in the patient's right ankle.

The patient died of a massive air embolus.

Potential for harm: High.

Safety advice: Manufacturers should implement "designed incompatibility" as appropriate to prevent dangerous misconnections of tubes and catheters.

Case Study 10: IV Tubing Erroneously Connected to Enteral Feeding Tube

A child had both a gastric feeding tube for nutrition and an IV for medicine and hydration.

When the child's gown was changed, a family member inadvertently attached the IV tubing to the gastric feeding tube.

The medicine was delivered through the feeding tube into the stomach.

There was no patient harm since the event was noted in a timely manner.

Potential for harm: Moderate.

Safety advice: Inform non-clinical staff, patients, and their families that they must get help from clinical staff whenever there is a real or perceived need to connect or disconnect devices or infusions.

Case Study 11: Foley Catheter Erroneously Connected to NG Tube

A patient was found with her Foley catheter disconnected from its drainage bag. One end of the catheter was still in her bladder and the other end was connected to her nasogastric (NG) tube.

Urine was noted to be flowing into her NG tube.

The NG tube was connected to suction and more than 300 mL of urine drained.

The patient's vital signs were stable and her laboratory results were within normal limits.

Potential for harm: Low.

Safety advice: Inform non-clinical staff, patients, and their families that they must get help from clinical staff whenever there is a real or perceived need to connect or disconnect devices or infusions.

Case Study 11: Air-Filled Syringe for Limb Tourniquet Cuff Erroneously Connected to Introducer SSheath

Misconnection: Air injected into introducer sheath

- Before removing the arterial sheath from the radial artery, a clinician mistakenly injected air into the arterial sheath instead of into the air inflation port for the tourniquet cuff.

Air was injected directly into the patient's radial artery.

The patient suffered a stroke from the air embolism.

Potential for harm: High.

Safety advice: Emphasize the risk of catheter misconnections in orientation and training, and confirm during every procedure that the correct port is being used for an air injection.

Case Study 12: Incorrect Dialysate Canister Mix-Up During Hemodialysis Therapy

A patient was receiving hemodialysis (HD) therapy when it was noted:

The citric acid disinfectant canister was erroneously connected to the dialysate line instead of the intended bicarbonate dialysate.

The bicarbonate dialysate was erroneously connected to the disinfectant solution canister instead of the citric acid disinfectant canister.

The mix-up of these treatment canisters has the potential to result in electrolyte and/or acid-base imbalance, both of which may lead to serious patient injury or death.

During treatment, the patient became unstable with low blood pressure and general distress. Despite discontinuing the treatment, the patient's status deteriorated, and the patient died before the end of the treatment.

Potential for harm: High.

Safety advice: To help prevent or minimize cross-connections or use of mismatched concentrates. Follow and conduct a safety check for every treatment, every time. Develop and use a system of labeling connector types and matching containers. This includes checking to make sure the proper connections of dialysates are completed prior to starting therapy. Measure conductivity and pH to ensure delivery of the proper composition of the dialysate.

Case Study 13: Air Inflation Line from Non-invasive Vascular Diagnostic System Erroneously Connected to IV Catheter

A technician in training inadvertently connected the air inflation line from a peripheral vascular diagnostic system to a patient's intravenous (IV) saline lock, instead of to the blood pressure cuff.

Pressurized air was injected directly into the patient's IV.

The error was recognized when the blood pressure cuff was not inflating.

The patient suffered a cardiac arrest and died from an air embolus.

The connector for the blood pressure cuff was compatible with the Luer connector used on the IV tubing. This permitted a connection between the air inflation tubing and the IV tubing. The international voluntary consensus standard IEC 80369-5:2016External Link Disclaimer was published in March 2016 to provide specifications for the connectors used with blood pressure cuffs so that they do not misconnect with other connectors to prevent this type of event.

Potential for harm: High

Safety advice: Emphasize the risk of IV tubing misconnections in orientation and training. Do not purchase non-intravenous equipment with connectors that can physically connect to or attach with a female IV line connector. Manufacturers should implement "designed incompatibility" as appropriate, to prevent dangerous misconnections of tubes and catheters.

- *Medical Device Data Systems*

Medical Device Data Systems (MDDS) are hardware or software products intended to transfer, store, convert formats, and display medical device data. An MDDS does not modify the data or modify the display of the data, and it does not by itself control the functions or parameters of any other medical device. MDDS may or may not be intended for active patient monitoring [10].

A medical device data system (MDDS) is a device that is intended to provide one or more of the following uses, without controlling or altering the functions or parameters of any connected medical devices:

- The electronic transfer of medical device data
- The electronic storage of medical device data
- The electronic conversion of medical device data from one format to another format in accordance with a preset specification
- The electronic display of medical device data

An MDDS may include software, electronic or electrical hardware such as a physical communications medium (including wireless hardware), modems, interfaces, and a communications protocol. This identification does not include devices intended to be used in connection with active patient monitoring.

In practice, a medical device data system (MDDS) is a medical device intended to provide one or more of the following functions:

- The electronic transfer or exchange of medical device data from a medical device, without altering the function or parameters of any connected devices. For example, this would include software that collects output from a ventilator about a patient's CO2 level and transmits the information to a central patient data repository.
- The electronic storage and retrieval of medical device data, without altering the function or parameters of connected devices. For example, software that stores historical blood pressure information for later review by a healthcare provider.
- The electronic conversion of medical device data from one format to another in accordance with a preset specification. For example, software that converts digital data generated by a pulse oximeter into a digital format that can be printed.
- The electronic display of medical device data, without altering the function or parameters of connected devices. For example, software that displays the previously stored electrocardiogram for a particular patient.

MDDS include the following, provided the intended use is consistent with the MDDS regulation:

- Any assemblage or arrangement of network components that includes specialized software or hardware expressly created for a purpose consistent with the intended use in the MDDS regulation.
- Products specifically labeled (per 21CFR 801) by the manufacturer as an MDDS, provided such products do not provide additional functionality.
- Custom software that is written by entities other than the original medical device manufacturer (for example, hospitals, third-party vendors) that directly connects to a medical device, to obtain medical device information.
- Modified portions of software or hardware that are part of an IT infrastructure created and/or modified for specific MDDS functionality. For example, when

modifying software (writing and compiling software source code), the modified portion is considered MDDS.

What Is Not an MDDS?

General-purpose IT infrastructure used in healthcare facilities that is not altered or reconfigured outside of its manufactured specifications. Modifications within the off-the-shelf parameters of operation are still considered general IT infrastructure and not MDDS. For example, components with the following functions by themselves are NOT considered MDDS if they are used as part of general IT infrastructure even though they may transfer, store, display, or convert medical device data, in addition to other information:

- The electronic transfer of medical device data
- Network router
- Network hub
- Wireless access point
- The electronic storage of medical device data
- Network-attached storage (NAS)
- Storage area network (SAN)

The electronic conversion of medical device data from one format to another in accordance with a preset specification:

- Virtualization System (ex: VM Ware)
- PDF software
- The electronic display of medical device data.
- Computer monitor
- Big screen display
- Networks used to maintain medical devices to see which systems are running or malfunctioning, or other similar uses that do not meet the definition of medical device under 201(h) of the FD&C Act.
- Standard IT software that is not specifically sold by the manufacturer as an MDDS, which may have MDDS functionality such as reading serial numbers, barcodes, UDI, or other data from a medical device, but is not used in providing patient care.
- Off-the-shelf passive network sniffing software that is generally used to monitor any network performance by reading TCP/IP packets on a network if this software is not intended to connect directly to a medical device.

Per section 520(o)(1)(D) of the Federal Food, Drug, and Cosmetic Act:

Software functions that are solely intended to transfer, store, convert formats, and display medical device data or medical imaging data, are not devices and are not subject to FDA regulatory requirements applicable to devices. The FDA describes these software functions as “Non-Device-MDDS.”

Hardware functions that are solely intended to transfer, store, convert formats, and display medical device data or results are “Device-MDDS.”

Examples of Non-Device-MDDS include software functions that:

- Store patient data, such as blood pressure readings, for review at a later time
- Convert digital data generated by a pulse oximeter into a format that can be printed
- Display a previously stored electrocardiogram for a particular patient

Devices that were not in commercial distribution prior to May 28, 1976, are generally referred to as post-amendment devices and are classified by operation of law under section 513(f) of the Food Drug and Cosmetic Act (21 U.S.C. 360c(f)) as Class III devices. CDRH evaluates such post-amendment devices to establish the appropriate degree of regulatory controls needed to provide reasonable assurance of their safety and effectiveness. CDRH may decide to classify such a device as Class I (requiring general controls), Class II (requiring special controls), or Class III (requiring premarket approval).

Risks associated with MDDS include the potential for inaccurate, incomplete, or untimely data transfer, storage, conversion, or display of medical device data. In some cases, this can lead to incorrect patient diagnosis or treatment. Based on the evaluation of these risks, the FDA has determined that general controls such as the Quality System Regulation (21 CFR part 820), will provide a reasonable assurance of safety and effectiveness. Therefore, special controls and premarket approval are not necessary.

The risks associated with general hospital devices are significant and require careful consideration. These risks can encompass a wide range of factors, including technological limitations, human error, cybersecurity vulnerabilities, and regulatory compliance issues. It is crucial for healthcare institutions and device manufacturers to acknowledge these risks and implement robust mitigation strategies to ensure patient safety and the integrity of healthcare operations.

Firstly, technological limitations pose inherent risks to general hospital devices. Complex medical equipment may suffer from hardware or software failures, leading to malfunctions or inaccurate readings. Inadequate maintenance, outdated firmware, or lack of interoperability can further compromise device performance and compromise patient care. It is essential for healthcare providers to conduct regular maintenance, implement software updates, and ensure compatibility with other hospital systems to minimize these risks.

Human error also represents a significant risk factor. Improper operation, insufficient training, or inadequate supervision can lead to errors in device usage, resulting in adverse events or compromised patient outcomes. Healthcare institutions must prioritize comprehensive training programs for staff members and establish clear protocols and standard operating procedures to mitigate the potential for human error.

Cybersecurity vulnerabilities present another critical risk. With the increasing digitization and connectivity of hospital devices, the threat of unauthorized access, data breaches, or malicious attacks becomes more prominent. Cybercriminals may exploit vulnerabilities in device software, network infrastructure, or wireless communication, compromising patient privacy, tampering with medical data, or even gaining control over critical medical equipment. Healthcare organizations should adopt robust cybersecurity measures, including strong access controls, encryption protocols, regular security assessments, and prompt patching of known vulnerabilities.

Moreover, regulatory compliance plays a vital role in managing risks associated with general hospital devices. Compliance with relevant laws, standards, and regulations ensures the safety, effectiveness, and quality of medical devices. Failure to meet regulatory requirements may result in legal consequences, loss of reputation, and compromised patient trust. Healthcare providers and device manufacturers must stay updated with evolving regulations and proactively adhere to compliance standards to mitigate risks and ensure patient safety.

In conclusion, the risks associated with general hospital devices require a comprehensive approach that addresses technological limitations, human error, cybersecurity vulnerabilities, and regulatory compliance. By implementing proactive measures such as regular maintenance, staff training, robust cybersecurity protocols, and adherence to regulatory standards, healthcare institutions and device manufacturers can mitigate these risks and promote the safe and effective use of medical devices, ultimately improving patient outcomes and enhancing the quality of healthcare delivery.

References

1. Guidance on the Content of Premarket Notification [510(K)] Submissions for Clinical Electronic Thermometers, March 1993.
2. U.S. National Library of Medicine: MedlinePlus. (2019 February). Body Temperature Norms. https://medlineplus.gov/ency/article/001982.htm
3. Centers for Disease Control and Prevention-Safe Patient Handling https://www.cdc.gov/niosh/topics/safepatient/
4. https://www.fda.gov/medical-devices/general-hospital-devices-and-supplies/infusion-pumps
5. Import Alert 89-04 "Detention without physical examination of devices without approved PMA's or IDE's and other devices not substantially equivalent or without a 510(k)".
6. Medical glove guidance manual - guidance for industry and FDA Staff.
7. Guidance on Premarket Notification [510(k)] submissions for surgical gowns and surgical drapes.
8. Guidance: use of International Standard ISO 10993-1, Biological evaluation of medical devices - Part 1: Evaluation and testing within a risk management process.
9. Clinical guidance for the assessment and implementation of Bed Rails in Hospitals, long term care facilities and home care settings.
10. https://www.fda.gov/medical-devices/general-hospital-devices-and-supplies/medical-device-connectors

Chapter 10
Significant Risk Medical Devices – Neurology

Krishnapriya Neelambaran, Anil Kumar Sharma, Hemasri Velmurugan, and Pugazhenthan Thangaraju

10.1 Neurology: Medical Devices

Significant risk devices under neurology can be broadly classified into the following:

I. Electroconvulsive therapy (ECT) devices
II. Hydrocephalus shunts [1]
III. Implanted intracerebral/subcortical stimulators [2, 3]
IV. Implanted intracranial pressure monitors [4–8]
V. Implanted spinal cord and nerve stimulators and electrodes [9–13]
VI. Neurological catheters (e.g., cerebrovascular, occlusion balloon) [14–16]
VII. Transcutaneous electric nerve stimulation (TENS) devices for treatment of chest pain/angina [17–19]

10.2 Electroconvulsive Therapy (ECT) Devices

In a patient who is sedated or is under general anesthesia, electroconvulsive therapy (ECT) creates a generalized cerebral seizure using an electric current to create a generalized cerebral seizure. It is used to treat patients suffering from severe psychiatric disorders such as treatment-resistant depression, schizophrenia, bipolar

K. Neelambaran · H. Velmurugan · P. Thangaraju (✉)
Department of Pharmacology, All India Institute of Medical Sciences (AIIMS), Raipur, Chhattisgarh, India

A. K. Sharma
Department of Neurosurgery, All India Institute of Medical Sciences (AIIMS), Raipur, Chhattisgarh, India

P. S. Timiri Shanmugam et al. (eds.), *Significant and Nonsignificant Risk Medical Devices*, https://doi.org/10.1007/978-3-031-52838-5_10

disorder, catatonia, and neuroleptic malignant syndrome. However, its use also has a stigma attached to it due to misinformation regarding procedural methodology.

Equipment

ECT equipment and recovery areas should be in accordance with ASA guidelines. The equipment needed in general are as follows:

- Stethoscope
- BP monitor with multiple blood pressure cuffs
- ECG monitor
- Pulse oximeter
- Suction apparatus
- Oxygen delivery system
- Anesthetic induction supplies and medications
- Ventilatory and resuscitatory equipment
- Oxygen masks
- Nerve stimulator
- EMG and EEG leads

Team Comprises of an anesthesiologist, psychiatrist, and a nurse.

Preparation

- Proper history and clinical examination should be done to rule out significant risk factors like cardiac and intracranial pathologies.
- Rule out history of use of herbal medications that can interfere with ECT like *Ginkgo biloba*, *Ginseng*, St. John's wort, valerian, and kava.

Drug Interactions

- Beta-blockers: reduce ECT-related hypertension and tachycardia and can affect ECT efficacy as the shorten the seizure duration.
- Cardiac medications (aspirin, clopidogrel, statins, antihypertensive agents, anti-anginal medications can be continued on the day of the procedure).
- As ECT treatments raise blood glucose levels, serum glucose levels should be monitored both preoperatively and in the recovery room.
- In patients with defibrillator: detection mode should be turned off and external defibrillation should be available at the patient's bedside.
- Pregnant patients: fetal monitoring should be done (non-invasive/non-stress test).

Anesthesia

- General anesthesia is used.
- Induction agents: barbiturates (thiopental, methohexital) and nonbarbiturates (propofol, etomidate).
- Methohexital: most used induction agent due to its quick onset, effectiveness, low cost, and minimal effect on seizure duration.

Technique

- Patient should be NPO.
- All vitals should be monitored, including EMG.

- Preoxygenation of the patient via nasal cannula or face mask.
- Anesthetic induction and paralysis.
- Once the patient is rendered unconscious, muscle relaxant is administered, along with bag valve mask ventilation with 100% oxygen.
- Following induction, a bite block should be placed to protect the patient's tongue and teeth.
- Skeletal muscle relaxation (using succinylcholine) during ECT to minimize a motor seizure and avoid musculoskeletal injury.
- If a patient is experiencing a prolonged seizure for more than 2 min, to suppress seizure activity and avoid neurologic injury induction agents like propofol or methohexital is given at half dose or a benzodiazepine is given.

Complications

- Mild memory loss over long term.
- Transient cognitive impairment, seen more with bilateral than unilateral ECT.
- Brady arrhythmias during the tonic phase of seizure.
- Tachycardia and hypertension during the clonic phase of seizure.
- ECT therapy increases both cerebral blood flow and intra cranial pressure.

In 2018, FDA has reclassified ECT devices from class III to class II (special controls) for use in treating catatonia or a severe major depressive episode (MDE) associated with major depressive disorder (MDD) or bipolar disorder (BPD) in patients13 years of age and older who are treatment-resistant or who require a rapid response due to the severity of their psychiatric or medical condition.

The three categories of devices are class I (general controls), class II (special controls), and class III (premarket approval).

While the reclassification is a positive step forward, ECT devices will then remain to a more restrictive category (i.e., class III) for patients who are diagnosed with schizophrenia, schizophreniform disorder, schizoaffective disorder, bipolar mania, catatonia, or mixed states and for patients less than 18 years of age.

ECT is a relatively safe and low-risk procedure that requires interprofessional care and coordination. The antidepressant effect is seen quickly and last up to a few years. The mortality rate is also very low.

10.3 Hydrocephalus Shunts

The most common treatment for hydrocephalus is the placement of a shunt, which is a flexible tube-like catheter that is placed in the lateral ventricle of brain where CSF is produced. The catheter is then passed under the skin to another region of the body which can naturally absorb the excess CSF produced.

A shunt consists of three parts (Fig. 10.1):

- Inflow/proximal catheter
- A valve mechanism
- An outflow/distal catheter

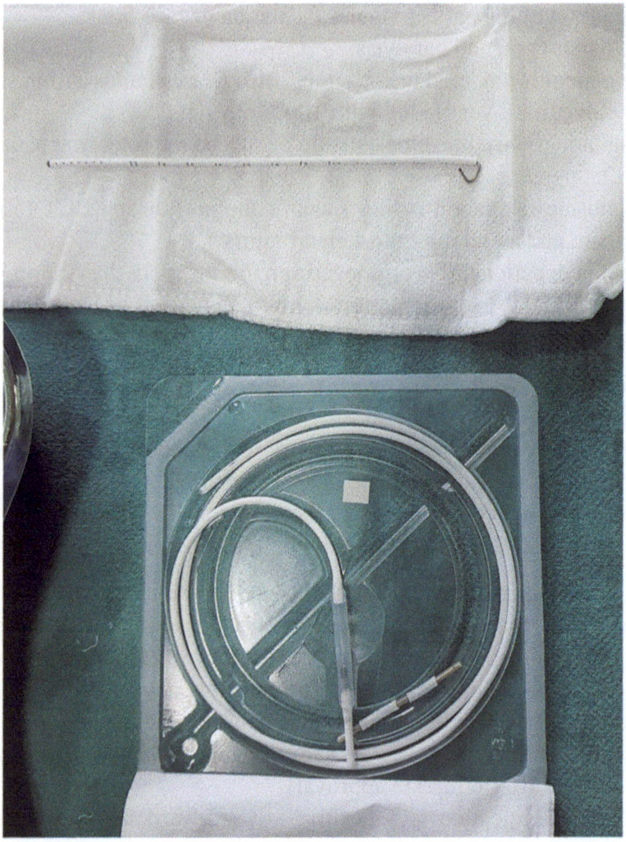

Fig. 10.1 Hydrocephalus shunt

Inflow/Proximal Catheter

This catheter drains CSF from the lateral ventricles and then runs for a short distance under the skin by leaving the brain through a small hole drilled in the skull.

Valve Mechanism

This regulates the intracranial pressure by controlling fluid flow. This device is connected to the proximal catheter and lies, usually on top or the back of head or behind the ear, between the skin and the skull and they operate within a specific pressure range.

Outflow/Distal Catheter

It runs under the skin and directs the CSF from valve to the drainage site.

Types of Shunt Systems

1. *Ventriculoperitoneal (VP) shunts*: *They* divert CSF from the ventricles of the brain into peritoneal cavity. The tip of the distal catheter rests near the intestinal loops. The CSF is reabsorbed into the bloodstream and excreted through normal urination.

2. *Ventriculoatrial (VA) shunts*: *They* divert CSF into the right atrium of the heart and the CSF passes directly into bloodstream and is excreted through normal urination.
3. *Ventriculopleural (VPL) shunts*: *They* divert CSF into the pleural cavity. The CSF gets added to the pleural fluid and is absorbed and excreted through normal urination.
4. *Lumboperitoneal (LP) shunts*: *They are unique in the sense that they* divert CSF from intrathecal space in spine and not from the brain. The CSF is diverted into the peritoneal cavity, reabsorbed into the bloodstream, and excreted through normal urination.

Fixed and Adjustable (Programmable) Valves

As ICP is higher in a hydrocephalus patient, a shunt is placed to divert the excess CSF and lower the ICP. The pressure control valves operate mostly on the principle of change in differential pressure (DP), i.e., difference between pressure at proximal catheter tip and that at the distal catheter tip.

Selection of a DP valve depends upon the following factors:

- Age of the patient
- Size of ventricles
- Amount of pressure that needs to be relieved
- Other clinical factors

Valves can be either set to a *fixed pressure* or they can be *adjustable/programmable* from outside the body.

Fixed pressure valves drain to a defined intracranial pressure. They regulate ICP using a one-way valve. Most of them have 3–5 possible settings: very low, low, medium, high, or very high pressures. Once implanted, the pressure setting cannot be changed without an additional surgery.

Adjustable/programmable valves: *They* regulate ICP based on a pressure setting which can be adjusted by the doctor using an external adjustment tool outside the body. This allows to non-invasively change or program the valve pressure during an office visit.

Most of these valves are adjusted by a strong magnetic field present in the external adjustment tool. Therefore, care must be taken to keep toys with magnets away from the implanted device.

Reservoirs

Many shunt systems have a reservoir located beneath the skin between the proximal catheter and the valve. The reservoir can be used to remove samples of CSF for testing, can be used by the doctor to inject fluid into the system for testing the flow and function of the shunt, and to measure pressure. It can also be pumped manually to help keep the proximal catheter open. If one pushes on the reservoir and it does not spring back, it might indicate an obstruction.

In case of clogging in the valve/distal catheter, the reservoir might feel stiff and more force is needed to depress it.

Overdrainage Control Devices
In some patients, standing or sitting can cause a siphoning effect, which "pulls" CSF out of the lateral ventricles or the lumbar region and results in overdrainage. To prevent this, a siphon control device is added, which can minimize excessive drainage, when the individual is upright.

Flushing Devices (CSF Flushing Devices)
Can also be added to an existing shunt system to provide a non-surgical means to alter the CSF flow in a non-flowing shunt.

Shunt Casings
The shunt system placement can leave bump under the skin. To overcome this, encase the shunt valve in a cranial implant/securing the valve, allowing it to be a flush to the cranium and avoiding stretching and thinning of the skin over time.

Complications The most common complications include failure, malfunction, and infection.

10.4 Implanted Intracerebral/Subcortical Stimulators

Deep brain stimulation (DBS) is a popular therapeutic approach for diverse neurological disorders including depression, Parkinson's disease (PD), obsessive compulsive disorder (OCD), and epilepsy. This involves implantation of an electrode in a specific brain structure in intracerebral/subcortical areas followed by electrical stimulation. The electricity stimulates the brain cells in that area. The current reaches the brain through one or more wires attached to a small device implanted underneath the skin (pulse generator). An MRI and CT are done before implantation to decide the best location where the implant can be placed.

Structure
The DBS apparatus consists of electrodes implanted adjacent to specific brain structures, and they are then connected to a pacemaker-like machine (pulse generator) that is implanted beneath the skin, via a subcutaneous wire.

Mechanism
The precise mechanism is unclear, but studies have shown that high-frequency stimulation (HFS) reduces the firing rate of neurons and is responsible for the suppression of symptoms seen in the various neurologic conditions.

Advantages
1. Used as a treatment option when other medications are not helpful
2. Relieves motor symptoms
3. Can be life-changing/life-saving treatment
4. Adjustable
5. Reversible

Disadvantages

1. Associated cognitive impairments
2. Complications during surgery like bleeding, coma, stroke, infection, and sepsis
3. Numbness and tingling
4. Seizures
5. Diplopia
6. Confusion and balance problems
7. Depression

10.5 Implanted Intracranial Pressure Monitors

Introduction

Intracranial pressure (ICP) is the pressure inside the skull that is exerted by the brain, blood, and cerebrospinal fluid. Monitoring ICP is crucial in the management of a variety of neurological conditions such as traumatic brain injury, intracranial hemorrhage, and hydrocephalus. Intracranial pressure monitors are medical devices that are used to measure the pressure inside the skull. These devices can be invasive, involving the insertion of a catheter into the brain or non-invasive, using external sensors to measure pressure. The accurate measurement of ICP is important in determining appropriate treatments and interventions to prevent further brain damage. In this age of rapidly advancing technology, intracranial pressure monitoring techniques are constantly evolving, with new devices and technologies being developed to improve accuracy and reduce complications. Understanding the different types of intracranial pressure monitors and their benefits and limitations is essential for clinicians who manage patients with neurological conditions.

Parts of Intracranial Pressure Monitor

The ICP monitor consists of several parts that work together to accurately measure ICP. Here are the main parts of an ICP monitor:

Sensor: The sensor is the component that directly measures the ICP. In invasive ICP monitoring, a small catheter is inserted through a hole drilled in the skull, and the sensor is located at the tip of the catheter. Non-invasive ICP monitors use sensors placed on the surface of the scalp or in the ear canal.

Transducer: The transducer converts the physical pressure signal detected by the sensor into an electrical signal that can be processed by the monitor.

Cable: The cable connects the sensor and the transducer to the monitor.

Monitor: The monitor displays the ICP measurement and alerts the clinician if the pressure exceeds a certain threshold. The monitor may also have additional features, such as the ability to store data for later analysis.

Power source: ICP monitors require a power source to operate. This may be a battery, an external power supply, or a combination of both.

In addition to these main components, invasive ICP monitors also require a burr hole or other surgical access to the brain in order to insert the catheter. This procedure must be performed by a trained neurosurgeon or neurologist.

Advantages

1. Improved patient outcomes: One of the primary benefits of ICP monitoring is the ability to improve patient outcomes by detecting and treating elevated ICP before it causes further brain damage. ICP monitoring has been shown to reduce mortality and improve functional outcomes in patients with traumatic brain injury.
2. Individualized treatment: ICP monitoring allows clinicians to tailor treatments to the individual needs of the patient. For example, if ICP is elevated, the clinician may adjust the patient's head position, administer medications to reduce swelling, or perform surgery to relieve pressure. Invasive ICP monitoring is especially useful for guiding surgical interventions.
3. Early detection of complications: ICP monitoring can help clinicians detect complications such as intracranial hemorrhage or hydrocephalus early, before they cause irreversible damage. This can lead to earlier intervention and improved outcomes. Non-invasive ICP monitoring has the advantage of being able to detect changes in ICP over time without requiring invasive procedures.
4. Real-time monitoring: ICP monitors provide real-time feedback on the patient's ICP, allowing clinicians to make immediate decisions about treatment. This can be especially important in critical care settings, where rapid intervention can be life-saving. The development of new technologies for ICP monitoring that provides even more rapid and accurate feedback.

Overall, ICP monitoring has many potential advantages in the management of neurological conditions, including improved outcomes, individualized treatment, early detection of complications, and real-time monitoring. However, as with any medical procedure, there are also risks and limitations that must be considered. Clinicians must carefully weigh the benefits and risks of ICP monitoring on a case-by-case basis to determine the most appropriate course of action for each patient.

Disadvantages

1. Invasiveness: Invasive ICP monitoring requires a surgical procedure to insert the catheter, which carries some risk of bleeding, infection, or other complications. In addition, the catheter may become dislodged or malfunction, requiring additional procedures to correct.
2. Cost: ICP monitoring can be expensive, in terms of both the equipment and personnel required to perform the procedure, as well as the ongoing costs of monitoring and treatment.
3. False readings: ICP monitors may produce inaccurate readings due to a variety of factors, including movement of the patient, air bubbles in the catheter, or other technical issues.
4. Clinicians must be trained to recognize and correct these issues to avoid unnecessary interventions or delays in treatment.

5. Limited availability: Invasive ICP monitoring requires specialized equipment and trained personnel, which may not be available in all hospitals or regions. Non-invasive ICP monitoring techniques are becoming more widely available, but may not be suitable for all patients or conditions.
6. Ethical considerations: Invasive ICP monitoring involves drilling a hole in the patient's skull, which carries some risk of harm and raises ethical questions about the balance between potential benefits and harms to the patient.

Overall, ICP monitoring is a complex medical procedure that requires careful consideration of the potential benefits and risks. Clinicians must carefully weigh the advantages and disadvantages of ICP monitoring on a case-by-case basis to determine the most appropriate course of action for each patient.

10.6 Implanted Spinal Cord and Nerve Stimulators

Introduction

Implanted spinal cord and nerve stimulators and electrodes are medical devices used to treat chronic pain and other neurological disorders. These devices work by delivering electrical impulses to specific nerves or areas of the spinal cord, which can block pain signals and provide relief for patients who have not responded well to other treatments.

Implanted spinal cord stimulators consist of a small device that is surgically implanted under the skin, typically in the lower back, and connected to electrodes placed near the spinal cord. The electrodes deliver electrical signals to the spinal cord, which can modify the pain signals and provide relief.

Implanted nerve stimulators, on the other hand, are typically used to treat conditions such as peripheral neuropathy or bladder dysfunction. They are similar to spinal cord stimulators but are placed closer to the nerves they are targeting.

Both types of devices have been shown to be effective in reducing pain and improving quality of life for many patients. However, they are not appropriate for everyone and should only be considered after other treatments have been tried and proven ineffective. As with any medical procedure, there are risks involved, and patients should carefully weigh the potential benefits and risks before deciding to undergo implantation.

Parts

The components of an implanted spinal cord or nerve stimulator system may vary depending on the specific device and manufacturer, but in general, it consists of the following parts (Figs. 10.2, 10.3, 10.4, 10.5 and 10.6):

- Pulse generator: This is a small battery-powered device that generates the electrical signals used to stimulate the nerves or spinal cord. The pulse generator is typically implanted under the skin, usually in the lower back or abdomen.

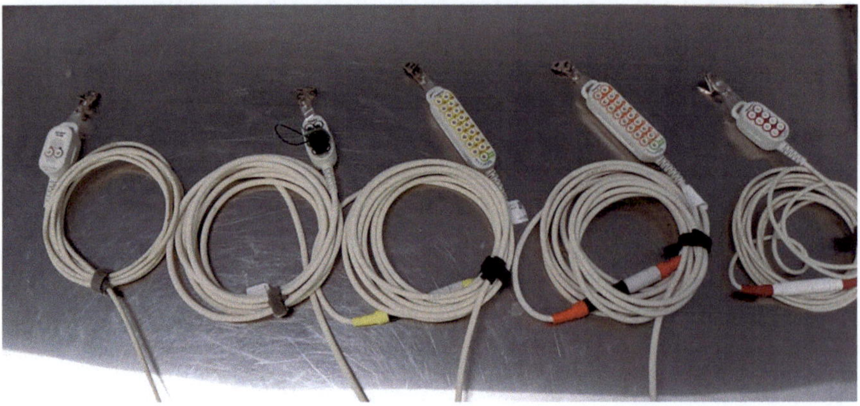

Fig. 10.2 Nerve stimulator electrodes

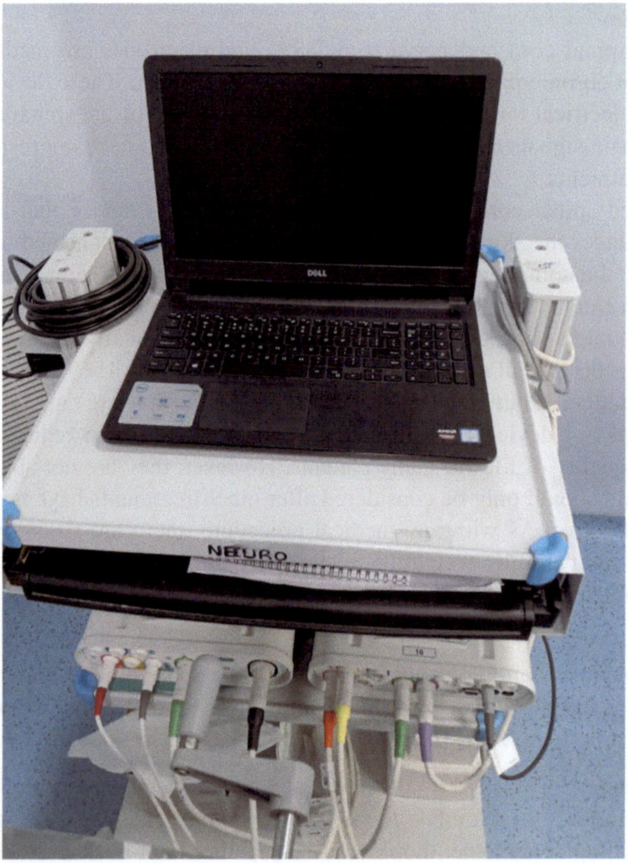

Fig. 10.3 Monitor

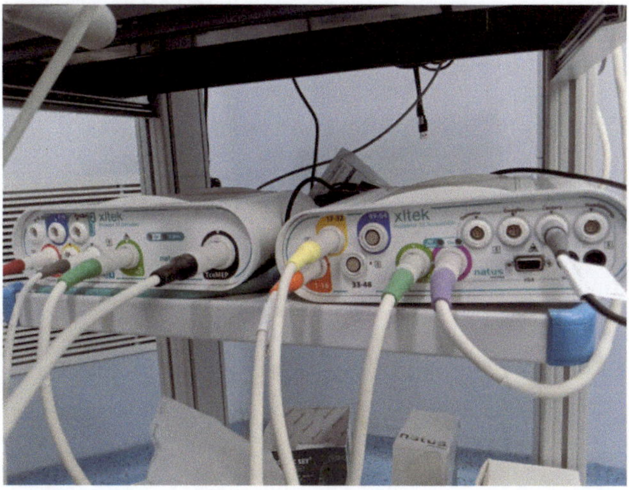

Fig. 10.4 Amplifier

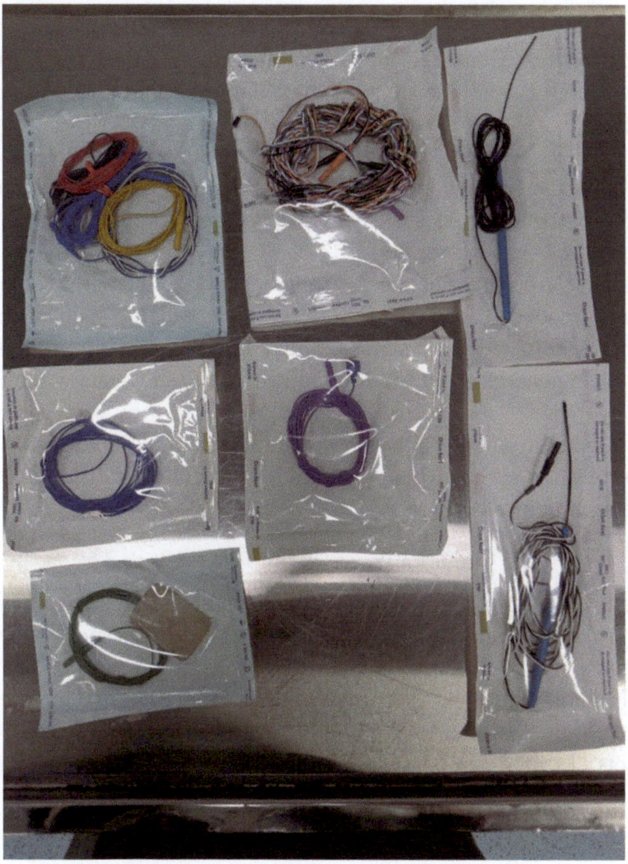

Fig. 10.5 Nerve stimulator electrodes

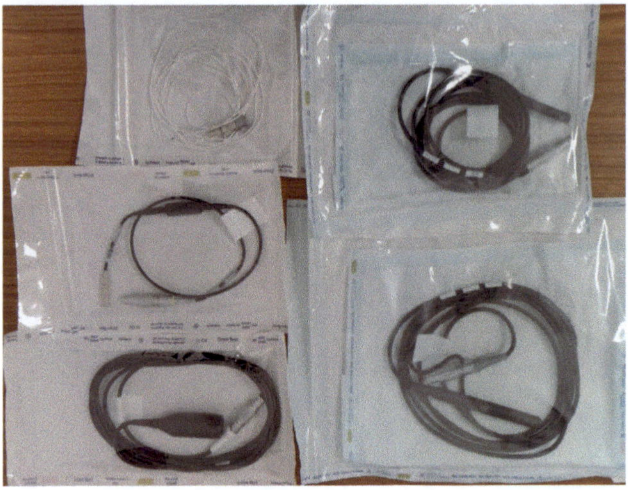

Fig. 10.6 Nerve stimulator electrodes

- Leads: Leads are thin wires that are connected to the pulse generator and carry the electrical signals to the targeted nerves or spinal cord. The leads are inserted through a small incision and guided to the desired location using X-ray guidance.
- Electrodes: Electrodes are small devices that are attached to the leads and placed near the nerves or spinal cord. The electrical signals generated by the pulse generator are delivered to the electrodes, which then stimulate the nerves or spinal cord.
- Programmer: A programmer is a handheld device that allows the patient or healthcare provider to adjust the settings of the stimulator, such as the frequency and intensity of the electrical signals. The programmer communicates wirelessly with the pulse generator, allowing for adjustments to be made without additional surgery.
- Remote control: Some systems may include a remote control that allows the patient to turn the stimulator on or off, adjust the intensity of the electrical signals, and activate pre-set programs.

Advantages

1. Implanted spinal cord and nerve stimulators and electrodes can provide significant relief for patients with chronic pain who have not responded well to other treatments.
2. Spinal cord stimulation has been shown to be more effective than repeated surgeries for chronic pain.
3. The effects of spinal cord stimulation can be sustained for at least 24 months.

4. Novel high-frequency spinal cord stimulation techniques have shown to be more effective than traditional low-frequency techniques.
5. Dorsal root ganglion stimulation has also shown to be effective for chronic neuropathic pain.

Disadvantages
1. As with any medical procedure, there are risks involved in the implantation of spinal cord and nerve stimulators and electrodes, including infection, bleeding, and nerve damage.
2. Some patients may not respond to the treatment or may experience a worsening of symptoms.
3. The cost of the procedure can be high, and insurance coverage may be limited.

Future Implications
There is ongoing research into new techniques and technologies for spinal cord and nerve stimulation that could potentially improve outcomes and reduce risks.

Researchers are also investigating the use of spinal cord and nerve stimulation for the treatment of other conditions, such as epilepsy, depression, and bladder dysfunction.

As the field continues to evolve, it will be important to carefully evaluate the long-term effectiveness and safety of these devices and ensure that they are used appropriately and ethically.

10.7 Neurological Catheters

Neurological catheters are medical devices that are used in the diagnosis and treatment of neurological conditions. These catheters are designed to access different parts of the nervous system and deliver drugs, fluids, or gases directly to the affected area. In this chapter, we will discuss the different types of neurological catheters, their applications, and considerations for their use.

Types of Neurological Catheters
1. Cerebrovascular Catheters: These catheters are used to access the blood vessels of the brain and are used in the diagnosis and treatment of cerebrovascular diseases such as stroke, aneurysms, and arteriovenous malformations (AVMs)
2. Occlusion Balloon Catheters: These catheters are used to temporarily block blood flow in specific blood vessels of the brain. They are used in the treatment of aneurysms and AVMs, as well as in the diagnosis of certain conditions such as vasospasm after subarachnoid hemorrhage.
3. Epidural Catheters: These catheters are placed in the epidural space, which is the space outside the spinal cord and within the spinal column. They are used to deliver local anesthetics, opioids, or other medications for pain management during surgery or labor and delivery.

4. Intrathecal Catheters: These catheters are placed within the cerebrospinal fluid (CSF) of the spinal cord and are used to deliver medications such as opioids, local anesthetics, or chemotherapy drugs directly to the spinal cord. They are used in the management of chronic pain, spasticity, and certain types of cancer.

Applications of Neurological Catheters

- Diagnosis: Neurological catheters can be used in the diagnosis of certain neurological conditions such as aneurysms and AVMs. They can also be used in the diagnosis of vasospasm after subarachnoid hemorrhage.
- Treatment: Neurological catheters are used in the treatment of cerebrovascular diseases such as stroke, aneurysms, and AVMs. They are also used in the treatment of chronic pain, spasticity, and certain types of cancer.

Considerations for Use

- Risks and Complications: Neurological catheterization carries a risk of complications such as infection, bleeding, and nerve damage. These risks should be carefully considered before the procedure is performed.
- Expertise and Training: The placement and management of neurological catheters require specialized training and expertise. Proper training and experience should be considered before performing these procedures.
- Patient Selection: Patient selection is an important consideration when using neurological catheters. Patients with certain medical conditions may not be candidates for these procedures due to the risks involved.

10.8 Transcutaneous Electric Nerve Stimulation (TENS) Devices

Transcutaneous electrical nerve stimulation (TENS) is a therapy used for pain relief that uses low voltage electrical current. It consists of a battery-powered device that delivers electrical impulses through electrodes placed on the surface of the skin. The electrodes are placed at trigger points or near the nerves where the pain is located.

Mechanism of Action

The exact mechanism of action of TENS is not completely understood, but it is thought to work through several different mechanisms, including the following:

- Gate Control Theory: TENS is thought to work by stimulating the sensory nerves in the area where pain is felt. This stimulation can activate the gate control system in the spinal cord, which can decrease the transmission of pain signals to the brain.
- Endorphin Release: TENS can also stimulate the release of endorphins, which are natural pain-relieving chemicals produced by the body. Endorphins can help to reduce the perception of pain and promote a sense of well-being.

- Nerve Stimulation: TENS can stimulate the nerves in the area where pain is felt, which can help to block the pain signals from reaching the brain. This can help to reduce the sensation of pain and provide temporary relief.

The main parts of a typical TENS unit include the following:

- Electrodes: These are the sticky pads that are placed on the skin over the area where the pain is felt. The electrical current from the TENS unit is delivered through the electrodes to the nerves.
- Lead Wires: These are the wires that connect the electrodes to the TENS unit.
- TENS Unit: This is the device that generates the electrical current. It typically has a display screen, buttons to adjust the settings, and a battery or power source.
- Settings: The TENS unit has various settings that can be adjusted to customize the treatment for each individual. These settings may include the frequency of the electrical current, the intensity of the current, and the duration of each treatment session.

Some TENS units may also have additional features, such as pre-set programs for specific conditions or wireless connectivity to a mobile app for remote control and tracking of treatment sessions.

Types of TENS

There are several types of TENS, including conventional TENS, acupuncture-like TENS, and burst TENS. Conventional TENS is the most common type of TENS and works by delivering a continuous electrical current at a low frequency (usually between 1 and 5 Hz). Acupuncture-like TENS uses a higher frequency (usually between 10 and 200 Hz) and is designed to mimic the effects of acupuncture. Burst TENS delivers a series of high-frequency bursts of electrical current followed by a period of no stimulation.

Clinical Applications

TENS can be used to treat a variety of conditions, including chronic pain, acute pain, and postoperative pain. TENS is often used as a non-pharmacological alternative to pain medication, particularly in patients who cannot tolerate or do not respond well to pain medication. TENS can also be used in combination with other therapies, such as physical therapy and massage, to enhance pain relief and promote healing.

Advantages

- Non-invasive: TENS therapy is non-invasive and does not require needles or surgery, which makes it a safe option for people who cannot tolerate invasive procedures.
- Easy to Use: TENS devices are generally easy to use and can be operated at home or in a clinical setting with minimal training.
- Cost-effective: TENS therapy is a relatively cost-effective option compared to other pain management treatments, such as surgery or medication.

- No side effects: Unlike medications, TENS therapy does not have any significant side effects, making it a safe option for people who cannot tolerate medication side effects.
- Portable: Many TENS devices are small and portable, making them convenient for use at home or while traveling.

Disadvantages

- Limited Effectiveness: TENS therapy may not be effective for everyone and may only provide temporary relief for some individuals.
- Skin Irritation: Some people may experience skin irritation or allergic reactions to the adhesive used to attach the electrodes to the skin.
- Dependence: Long-term use of TENS therapy may lead to dependence, and the effectiveness of the therapy may decrease over time.
- Not suitable for all types of pain: TENS therapy may not be effective for certain types of pain, such as pain caused by structural damage or inflammation.
- Interference: TENS therapy may interfere with other electronic medical devices, such as pacemakers, and should be used with caution in people with these devices.

10.9 Transcranial Doppler

Transcranial Doppler (TCD) is a non-invasive ultrasound technique that is used for analyzing the blood flow in the brain. TCD is used to diagnose and monitor a variety of neurological conditions, including stroke, vasospasm, and brain tumors. The use of ultrasound to measure blood flow in the brain was first described in the early 1960s. However, it was not until the 1980s that TCD became widely used in clinical practice. The development of portable and affordable Doppler ultrasound devices allowed for TCD to be performed at the bedside, making it a valuable tool in the diagnosis and management of neurological disorders.

Indications

TCD is used to diagnose and monitor a variety of neurological conditions, including the following:

- Stroke: TCD can be used to identify the presence and severity of cerebral arterial stenosis or occlusion, which can cause ischemic stroke.
- Vasospasm: TCD can detect changes in blood flow velocity in the cerebral arteries, which can indicate vasospasm after subarachnoid hemorrhage.
- Brain tumors: TCD can be used to identify changes in blood flow in and around brain tumors, which can help with diagnosis and treatment planning.
- Intracranial pressure: TCD can be used to estimate intracranial pressure indirectly by measuring blood flow velocity in the cerebral arteries.

Technique

TCD involves the use of a handheld ultrasound probe that is placed on the temporal bone just above the ear. The ultrasound waves are then transmitted through the skull to the cerebral arteries, where they are reflected back to the probe. The frequency of the reflected waves is then analyzed to determine the velocity of blood flow in the arteries (Figs. 10.7 and 10.8).

Interpretation

TCD provides information about blood flow velocity, which can be used to diagnose and monitor a variety of neurological conditions. Normal blood flow velocities vary depending on the location of the artery being examined and the age and sex of the patient. Abnormalities in blood flow velocities may indicate the presence of stenosis, occlusion, or vasospasm.

Types

There are two types of TCD equipment available: non-duplex and duplex. Non-duplex devices use audible Doppler shift and spectral display to identify arteries blindly based on standard criteria. Duplex devices, such as B-mode transcranial color-coded duplex (TCCD) and power motion-mode TCD (PMD/TCD), combine pulsed wave Doppler ultrasound with cross-sectional views to identify arteries more accurately. The angle of insonation is assumed to be <30° in TCD, but it can be measured and corrected in TCCD. PMD/TCD simplifies TCD handling by providing multi-gate flow information simultaneously in the power M-mode display. Although imaging TCD improves reliability, the clinical applications of these modalities are still under development, and non-duplex TCD remains the focus of this review. As duplex TCD technology advances, it will likely replace non-duplex TCD in clinical practice.

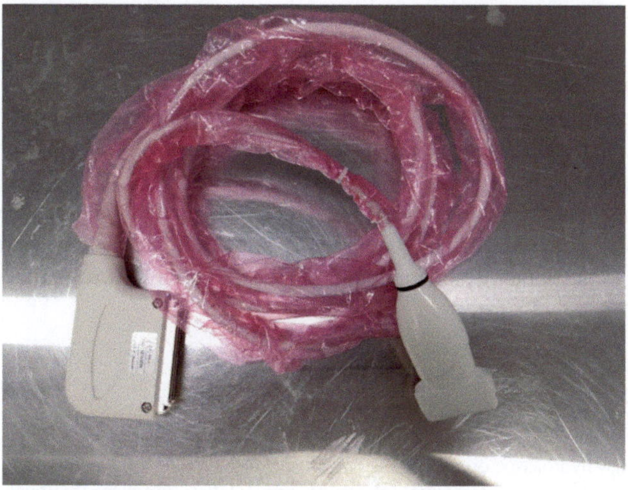

Fig. 10.7 Transcranial Doppler Probe

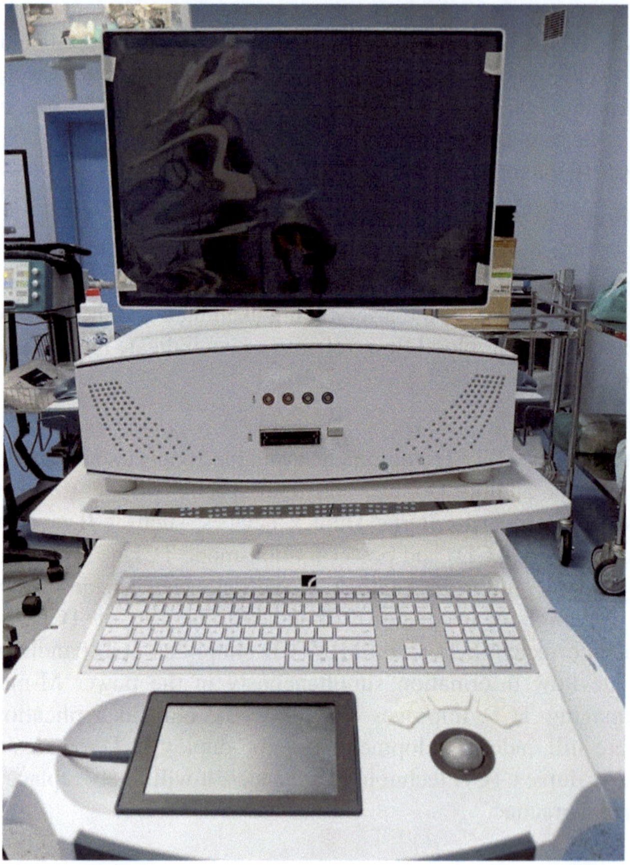

Fig. 10.8 Transcranial Doppler Image System

Limitations

TCD has several limitations that should be considered when interpreting the results: The technique is operator-dependent, and the results can vary depending on the skill and experience of the operator. The technique is limited to the arteries that can be accessed through the temporal window, which may not provide a complete picture of blood flow in the brain. The technique is not able to provide information about blood flow in the veins, which can be important in certain neurological conditions.

References

1. https://www.hydroassoc.org/shunt-systems/#:~:text=table%20of%20contents
2. Polat U, Biegon A. Intracranial electrode implantation produces regional neuroinflammation and memory deficits in rats. Exp Neurol. 2010;222(1) https://doi.org/10.1016/j.expneurol.2009.12.006.

3. https://www.physio-pedia.com/Deep_Brain_Stimulation
4. Marshall J, Geyer RG, Winn JL. Intracranial pressure monitoring: a review of invasive and non-invasive techniques.
5. Cooper JE, Fults DN. Current and future developments in intracranial pressure monitoring.
6. Robertson JS, Juul JM. Intracranial pressure monitoring in traumatic brain injury: a review of the evidence.
7. Gupta PN, Singh MK. Non-invasive methods for measuring intracranial pressure.
8. Korteweg KM, Vanwesenbeeck PJCM, Van Huffelen AC. Complications of intracranial pressure monitoring in children.
9. Deer TR, Mekhail N, Petersen E, Krames E, Staats P, Pope, Eisenach J. Neuromodulation: technology at the neural. Interface. 2014;17(6):515–50.
10. North RB, Kidd DH, Lee MS, Piantadosi S. Spinal cord stimulation versus repeated lumbosacral spine surgery for chronic pain: a randomized, controlled trial. Neurosurgery. 2005;56(1):98–107.
11. Kumar K, Taylor RS, Jacques L, Eldabe S, Meglio M, Molet J, North RB. Neurosurgery. 2015;77(5):822–31.
12. Kapural L, Yu C, Doust MW, Gliner BE, Vallejo R, Sitzman BT, Yearwood TL. Anesthesiology. 2020;132(2):272–85.
13. Li Y, Li H, Li H, Li X, Chen H. Long-term outcomes of dorsal root ganglion stimulation for chronic neuropathic pain: a systematic review and meta-analysis. Pain Pract. 2021;21(3):332–41.
14. Jha AN, Vitek JL. Neurological catheters and electrodes: current and future trends. Exp Rev Med Dev. 2015;12(6):677–89.
15. Levy EI, Turk AS. Aneurysm and arteriovenous malformation catheterization. Intervent Neurol. 2012;1(1):35–51.
16. Harden RN, Remble TA, Houle TT. Intrathecal drug delivery for treatment of chronic low back pain: report from the National Outcomes Registry for Low Back Pain. Pain Med. 2012;13(8):1090–101.
17. Johnson MI, Tabasam G. An optimal TENS machine for relieving pain in labour. Pract Midwife. 2019;22(4):20–4.
18. Hurlow A, Bennett MI, Robb KA, Johnson MI. Transcutaneous electric nerve stimulation (TENS) for cancer pain in adults. Cochrane Database Syst Rev. 2012;3:CD006276.
19. Proctor DN, Kowalske KJ, Mileski JP. TENS for postoperative pain after burn injury: a double-blind study. Pain. 1997;69(1-2):43–8.

Chapter 11
Significant Risk Medical Devices – Obstetrics and Gynecology

T. Y. Sree Sudha, K. G. Sruthi, Shruti Mutsaddi, K. S. B. S. Krishna Sasanka, Shikha Sahay, Adity Bansal, and Harminder Singh

Introduction to Obstetrics and Gynecology Instruments

Obs and Gynae, O&G, OB-GYN, and OB/GYN are the abbreviations for medical Obstetrics and Gynecology, which includes the two subspecialties of Obstetrics (covering pregnancy, delivery, and the postpartum period) and Gynecology (covering the health of the female reproductive system—vagina, uterus, ovaries, and breasts).

At Surgical Holdings, we provide a selection of equipment to obstetricians and gynecologists for various operations like colposcopy, loop electrical excision procedure (LEEP), endometrial biopsy, IUD insertion, dilation and curettage (D&C), tubal ligation, and ovarian cystectomy, and for investigational gynecological approaches, these can be utilized.

The present book tries to compile all the instruments used from older to newer in the field of OBG.

T. Y. Sree Sudha (✉) · H. Singh
Department of Pharmacology, AIIMS, Deoghar, India

K. G. Sruthi
KLE VK Institute of Dental Sciences, Belgaum, Karnataka, India

S. Mutsaddi
Subbaih Dental College, Shimoga, Karnataka, India

K. S. B. S. Krishna Sasanka
Department of ENT, AIIMS, Deoghar, India

S. Sahay
Department of Obstetrics and Gynaecology, AIIMS, Deoghar, India

A. Bansal
Department of Dentistry, AIIMS, Deoghar, India

P. S. Timiri Shanmugam et al. (eds.), *Significant and Nonsignificant Risk Medical Devices*, https://doi.org/10.1007/978-3-031-52838-5_11

11.1 Abdominal Decompression Chamber

11.1.1 Instrument Parts

The unit consists of four parts: (1) The Suit; (2) The Chair; (3) The Front Piece and Back Support; (4) The Pump.

The Suit
This is made of sturdy plastic and has an airtight zip in the center, with the upper part resembling a topless dress and the lower half an empty shopping bag. The suit must snugly fit around the chest in order to ensure an airtight seal during the surgery. Each patient undergoing "decompression" is given her own suit, which is designed to fit the underarm chest dimensions.

The Chair
It has a television-like aesthetic. The head, back, knees, and feet are supported by the chair and in comfort. Depending on the patient's desire, it may be either upright or completely horizontal. One can't help but see rows of chairs in maternity hospitals' admittance wards in the future, to be utilized in the early stages of labor instead of beds that are more anthropomorphic in design.

Front Piece and Back Support
The chair's concave support is positioned behind the patient's lumbar area and hips, and the front element is a sizable semicircular fiber-glass device that fits over the patient's abdomen. This support is only a structure with a hole cut out in the middle to permit lordosis during suit operation, which significantly reduces back pain.

Pump
It has been determined that the current Hoover Constellation vacuum cleaner pump is appropriate and effective. In only 6 seconds, the pressure within the suit may be dropped to 50 mm of mercury below atmospheric pressure, providing quick pain relief. The patient has full control over the device and may lower the pressure as needed, up to a maximum of a little more than 100 mm of mercury.

When the circumstances allow the use of the same device, or during surgery, abdominal decompression for peritonitis can be performed using the same tool [1].

11.1.2 Introduction

An abdominal decompression chamber is a hood-like device intended to relieve abdominal pain in pregnant patients during pregnancy or birth by lowering pressure on the abdomen. In November 1955, decompression was utilized during labor [1]. Earlier in May 1954, "certain observations were made on the form of an abdomen, the muscles of which had been studied with Scoline, in collaboration with Halliday" [1].

The variations in the abdomen's shape during uterine contraction were captured on a continuous cinematograph. Two points a rose:

1. It was essential to use anesthetics in the form of pentothal since diaphragm paralysis produces such severe agony. At this time, it was unknown how Pentothal might affect the uterus's normal function.
2. No other method was available other than the contraction of the uterus to capture electrical impulses in a dilated abdominal wall [1].

When scoline was applied to the first patient, the cervical was quickly dilated, which was a remarkable result, i.e., in 68 minutes, from having dubious labor to having three fingers dilated. The second case on the 3rd August, 1954, was equally dramatic, showing that the speed of dilation of cervix depended upon a relaxed abdominal wall, or, in other words, with a nonresistant abdominal wall, cervical dilation was more rapid than usual [1].

11.1.3 Mechanism of Instrument

The abdominal wall is forced forward by around four inches with each contraction when the suit's air is expelled and the pressure is decreased. Because of this, the uterus may rise up into a position directly above the pelvis and contract more effectively because it is a spherical organ. The cervix eventually disappears when the uterus' longitudinal fibers constrict.

The uterus appears spherical in shape, even though it is compressed by the lumber spine posteriorly and flattened by a tensed abdomen wall anteriorly. The cervix opens evenly and more rapidly. The cervix expands evenly and more rapidly if the uterus is compressed by a tensed abdominal wall anteriorly and lowered by the lumber spine posteriorly. It is believed that decompression before labor improves the fetus's oxygenation, which is especially important for essential tissues like the brain, and results in the delivery of a fetus in outstanding health [1].

11.1.4 Uses

1. To make contractions feel less painful and longer than they actually are. When the individual does not feel the beginning or the conclusion of the contraction, its length looks shorter and its peak is more tolerable [1].
2. The sacrospinalis muscle is significantly stretched at its attachments in order to produce a lordosis and a very deep inspiratory action with decompression to treat acute back pain. This reduces back spasm and relieves pain from dysmenorrhea, back pain from various causes, such as "slipped discs," and back pain that occurs during labor.

3. Cervical dilatation occurs more quickly to shorten the duration of the first stage of labor.
4. For the use of decompression in the second and third stages of labor, special equipment has been built. Although many patients have received care in this manner, this adjustment is still not widespread.

Single use or multiple The use of instrument in the patients can be of multiple times.

11.1.5 Benefits

For patients who want to decompress during the early stages of labor, the facility is employed. Prior to the start of labor, which usually occurs between weeks 36 and 38, the patients have six to eight one-hour runs in the suit. The physiological changes that take place right before the start of labor are hypothesized to be sped up by the regular intervals of decompression and relaxation (i.e., 40 seconds of decompression and 20 seconds of rest per minute) [1].

Adverse effect No major adverse effects noticed.

11.2 Antepartum Home Uterine Activity Monitors

Uterine Activity Monitor (HUAM)

11.2.1 Introduction

The home uterine activity monitor (HUAM) is a device that records uterine contractions in the privacy of the user's home during pregnancy, sends the information over the phone to a healthcare setting, and then receives and displays the information in a doctor's office [2].

11.2.2 Parts of the Instrument

The HUAM system consists of a tocotransducer, a home recorder, a modem, and a computer/monitor that can receive, analyze, and display data.

11.2.3 Use

The HUAM is a prescription-only device that is approved for use in conjunction with conventional high risk therapy for the daily at-home evaluation of uterine activity in pregnancies under 24 weeks in women with a history of prior preterm birth [2].

11.2.4 Benefits

The use of HUAM, or home uterine activity monitoring, is experimental and unproven. After considering each situation and any of the following factors, it may decide that HUAM is medically essential:

1. Uterine activity is visible from a distance to aid in the early detection of pre-term labor. Women with complicated pregnancies who cannot feel their contractions. HUAM may be deemed medically necessary in a particular circumstance if a pregnant woman has a gestational age greater than 18 weeks and in any of the following conditions: Cervical incompetence, as indicated by the necessity for cerclage or other symptoms (funneling on valsalva, silent shortening); or
2. Preterm delivery is more likely when certain risk factors are present, such as a placenta previa with hemorrhage, a history of a traditional caesarean section, or a deep myomectomy during the index pregnancy, which makes it impossible for the mother to bear contractions safely; or
3. Physiologic or anatomical conditions (such as paralysis or neuromuscular diseases like muscular dystrophy) that make it difficult to recognize one's own contractions; or
4. Larger multiple gestations, such as triplets, quadruplets, or more (however, HUAM is not considered medically necessary in twin gestations unless other extenuating circumstances exist); or
5. Women in premature labor whose attempts to stop the course of preterm labor using standard treatments have failed; or
6. Women who have a fetal fibronectin test result that is positive, who are 20 weeks or more pregnant but under 36 weeks, and who continue to experience progressive cervical alterations despite using repeated tocolytics [3].

11.2.5 Contraindications

There are no known risks associated with using this device.

11.2.6 Risks

The FDA has found five health hazards linked to this kind of gadget. These risks are:

1. Electrical shock and/or injury
2. Skin irritation and sensitization (from abdominal belt or to transducer)
3. Unnecessary evaluation and treatment (from overdiagnosis)
4. Potential harmful effects from treatment with tocolytics
5. Use on unproven patient subpopulations (with shifted risk-benefit)

Which circumstances permit the employment of the same instrument or during operations: Surgery cannot be performed with it.

11.2.7 Instructions for Use

Instructions for use should be given to both professionals and patients.

1. After a doctor's prescription for the device, the patient should get routine high risk care education on preterm labor signs and symptoms.
2. A certified medical professional should provide the patient with information on how to use the gadget properly. Tocotransducer or sensor implantation instructions should be included in these guidelines to enable the detection of uterine activity.
3. A licensed medical professional should provide the patient instructions on how to monitor her uterine activity as directed. The following factors were utilized in studies and device experience that showed early preterm labor detection success (which may be included in the labeling as guidelines for the practitioner):
4. One or two one-hour monitoring sessions per day
5. Use of the device in a reclining position
6. Transmission of data to the physician right away after each session If the patient notices any uterine activity at any other point during the day, tell her to call the doctor. She could then be told to (a) monitor immediately or (b) come in to the clinic [2, 3].

11.3 Cervical Dilatation Devices

A plastic shaft, a first inflatable part, and a second inflatable member are all included in the dilator assembly of the "Cervical Canal Dilator" invention. The shaft can be either stiff or extremely flexible. The second inflatable part has a maximum inflated diameter and is made of a non-elastic substance. The maximum inflated diameter of the second inflatable part is fixed to be between 4 and 20 mm. Moreover, a sheath may at least partially encase the dilating assembly [4].

A cervical canal dilator described in the patent "Fluid-filled Cervical Dilator" has an elongated tubular or cylindrical shaft with a distal end and a proximal end. The interior of the shaft is lined with internal cavities that communicate with anchor and dilation balloons in a way that allows the separate inflation of each. The anchor balloon is located on the distal end of the shaft and is capable of anchoring the dilator against the bottom of the cervical canal through the inflation of the anchor balloon [4].

A cervical canal dilation device is a patent that describes a device that causes fixed cervical dilatation while causing the patient the least amount of discomfort possible. The "dilating balloons" shut the inner and outer orifices of the cervical canal when the liquid or gel is administered into the cervix. The cervical canal dilation device is anchored to the desired position in the cervix canal for the dilation therapy by an inflating balloon at or near the device's tip [4].

11.3.1 Foley Catheter

This catheter was created in an effort to speed up the dilation process and improve dilator effectiveness. Two domains made up the project. First to develop and produce a novel kind of catheter based on an anchor bolt mechanism. And second, to create a cervical ripening device utilizing flexible silicone that helps speed up the dilation process [4].

11.3.1.1 Instrument Parts

Design and manufacture a device with four flaps:

1. Provide a balloon for anchoring at the top.
2. Put a covering over the flaps to prevent the propagation device from coming into contact with the cervix's linings.
3. Make the design economical.
4. Cervical Ripening Device, which uses flexible silicone to speed up the dilation process.

11.3.1.2 Uses

The Foley catheter can induce labor and deliver the baby more quickly than a twin balloon catheter when second trimester pregnancies are terminated.

In comparison to medical procedures, mechanically induced dilatation is more practical, beneficial, quick, and preferable. Osmotic dilation is a more time-consuming and unpleasant form of dilation. It is successful to use a foley catheter combined with other tools [4]. One of the complications with Foley catheter is urinary tract infection.

11.4 Contraceptive Devices

Spermicidal contraceptive techniques (creams, jellies, and suppositories) were commonly used prior to the development of contraceptive pills and intrauterine devices (IUDs) in the 1960s, frequently in conjunction with diaphragms, cervical caps, and condoms.

There have been more reports of negative side effects and hazards related to IUDs and oral contraceptives in recent years.

Consumers and doctors are thus thinking at mechanical contraception as alternatives.

3 Despite this, birth control methods that mechanically cover the cervix have received the least amount of research and evaluation. Many questions about the accessibility, efficacy, and safety of vaginal mechanical contraceptive devices have been sent to the Department of National Health and Welfare's bureau of medical devices [5].

11.4.1 Cervical Caps

11.4.1.1 Introduction

A cervical cap is a little cup (Fig. 11.1) with a sailor hat-like form made of soft silicone. In Europe as opposed to North America, the cervical cap is a more often used form of birth control. It is a thimble-shaped piece of rubber or plastic that is positioned over the cervix to act as a sperm barrier.

11.4.1.2 Mechanism

To completely cover the cervix, it must be positioned deep into the vagina. The edge of the cap's suction maintains it firmly on the cervix. A spermicidal substance is frequently used in conjunction with the cap. Chemicals in spermicide harm sperm.

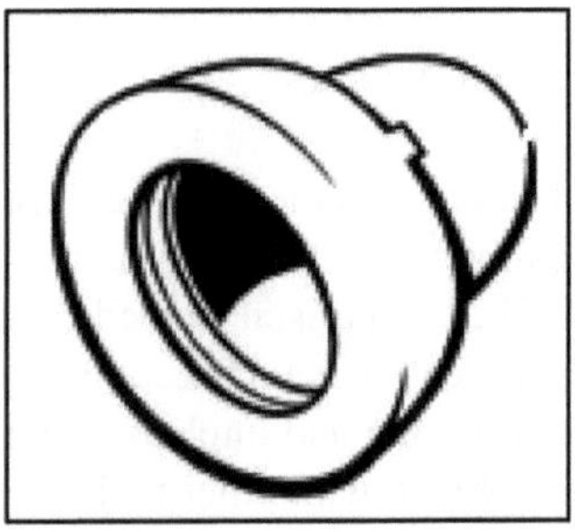

Fig. 11.1 Representing cervical cap [6]. (Courtesy: Ref. [6])

11.4.1.3 Parts

It features a spermicide-filled ring around the outside.

11.4.1.4 Uses

It helps to avoid pregnancy.

11.4.1.5 Contraindications

- Irregularly shaped cervix.
- History of surgery on your cervix or vagina.
- Uncomfortable putting your fingers in your vagina.
- Allergic to silicone or spermicide.
- You're on your menstrual period days.
- Pap test abnormalities.
- Cervicitis.
- Pelvic inflammatory disease (PID).
- Reproductive system infections.
- Toxic shock syndrome (TSS)

Single use or multiple By wiping it with clean hands and applying spermicide to the rim, it may be used repeatedly.

11.4.1.6 Benefits

- Cervical caps are lightweight, inexpensive, and reusable.
- Permitted to have multiple sex acts with a cervical cap within 48 hours.
- The majority of partners don't notice the cervical caps when they're in place.
- Cervical caps may be quickly reversed, so you can become pregnant after having it taken out.
- Cervical caps don't contain hormones, so they don't affect the menstrual cycle.
- Cervical caps can be applied 6 hours before intercourse to avoid interfering safe during lactation period.

11.4.1.7 Adverse or Toxic Effects

Before every intercourse, one should consistently and correctly wear cervical caps. Cervical caps may come loose during sexual activity, particularly when they are not placed correctly.

Some negative aspects include:

- Gets on prescription only.
- Spermicide can be unpleasant.
- Vagina may get inflamed by cervical caps.
- Must keep in mind to take it off after 2 days (48 hours).
- No protection is provided by cervical caps against STDs (STIs).
- Any pregnancy ending in a vaginal delivery.
- Abortion or miscarriage.
- Pelvic surgery.
- Significant weight loss or weight gain.

11.4.1.8 Duration

The cap can be put on for 48 hours or 2 days. The cap is kept on for at least 12 hours to avoid conception by any chance due to residual spermatozoa to stop any leftover viable spermatozoa from entering the uterus. Surgery does not utilize this device.

11.4.1.9 Risks

Using a cervical cap might raise certain illnesses, such as:

- Infections of the bladder and urinary tract.
- Inflammation of the cervix (cervicitis).
- Hazardous/Toxic shock syndrome (TSS).

To avoid TSS, one must remove cervical cap after 48 hours. TSS can also occur if cervical cap is used during menstruation. One must seek help from healthcare provider if there is difficulty in removing cervical cap [5, 6].

11.4.2 Condoms (for Men) [7]

11.4.2.1 Introduction

A polyurethane condom is an exterior condom made of a specific type of plastic. Polyurethane condoms are authorized by the Food and Drug Administration (FDA) for use in the prevention of pregnancy and STDs.

Polyurethane condoms are an alternative to the more widely used latex condom. However, they have some risks and might not be the best choice for everyone.

11.4.2.2 Mechanism

The male condom creates a physical barrier when it is rolled onto an erect penis before sexual contact, preventing semen from entering a sexual partner's body.

The FDA has categorized all condoms as Class II medical devices. The condoms must pass a number of tests, such as an airburst test, a water leak test, and a tensile test, in order to be approved (which measures the stretchability of a condom) [7].

This does not imply that all condoms are FDA-approved equally. The only condoms that have been authorized for use in preventing pregnancy and STDs including chlamydia, gonorrhea, and HIV are those made of polyurethane, latex, and polyisoprene.

Single use or multiple Single use.

11.4.2.3 Benefits

Due to a few characteristics, polyurethane is perfect for exterior condoms. Among them:

- More durable than latex is polyurethane.
- More resistant to heat exposure and more durable in storage is polyurethane.
- Latex condoms can be harmed by lubricants with an oil basis, whereas polyurethane is resistant to them.
- Those who don't like the scent of latex or lambskin prefer polyurethane since it has no fragrance.
- Since they are lighter than latex and polyisoprene condoms, polyurethane condoms may increase sensitivity.
- Because of its smaller profile and improved heat transmission compared to other external condoms, polyurethane may be more sensitive.
- Compared to other exterior condoms, polyurethane condoms are looser and more comfortable. These could be perfect for those who loathe how latex, lambskin, or polyisoprene condoms fit so tightly.
- Those who have latex allergy may benefit the most from polyurethane condoms. Considering that 4% of people have a latex allergy, it's critical to have a non-latex option [7].

11.4.2.4 Adverse Effects

- Less elastic and more prone to breaking
- May need more lubricant
- During intercourse, it is more prone to fall off.
- Costly
- Higher rate of contraceptive failure rate
- Maybe a little less successful in preventing STDs

11.4.3 Contraceptives In Vitro Diagnostics

11.4.3.1 Diaphragms [8]

Soft silicone was used to create a thin dome form. The United States has recently seen a surge in the use of diaphragm.

11.4.3.2 Parts

Diaphragms come in a variety of sizes, with their diameter ranging from 45 mm to 105 mm in 5 mm increments. This is because a precise fit is necessary to achieve optimal efficiency.

11.4.3.3 Types

Three distinct spring kinds allow for precise fitting for woman with various anatomical characteristics.

(a) A round, coiled metal-wire rim surrounds the coil-spring diaphragm.
(b) A flat metal-band spring is installed in the flat-spring diaphragm.
(c) Women with low vaginal muscle tone employ the arcing-spring diaphragm, which has a double metal-spring rim that applies significant pressure on the vaginal walls.

11.4.3.4 Mechanism

The diaphragm is a small rubber cup that is reinforced by a spring-loaded rim. It functions as a mechanical deterrent to sperm by fitting into the vagina and covering the cervix.

11.4.3.5 Uses

1. Spread approximately a teaspoon of spermicide over the rim and apply it to the dome.
2. Fold the diaphragm in half, ensuring that the rim's two sides contact and that the dome is pointing downward.
3. Use the other hand to hold the vagina open.
4. Aiming back toward the tailbone, push the folded diaphragm as deep into the vagina as you can.
5. Push the diaphragm's front rim up the behind pubic bone with one finger.

6. By placing finger on the dome, one can determine the diaphragm's location. It should allow the user to feel cervix. The area that feels solid but not bony is the cervix.
7. remove it and start over if it is not in the proper location. Just make sure of applying the spermicide again. The diaphragm should be in place for at least 6 hours after sexual activity and can be placed an hour before having sex. No more than 24 hours may be allowed for it to remain in the vagina.

Single use or multiple It is washable and has a long lifespan.

11.4.3.6 Benefits

The diaphragm has the advantages of being secure, efficient, transitory, natural, simple to use, practical, and affordable. There are no negative side effects, and it does not impair sexual experience. Women who use diaphragms had a lower incidence of dysplasia and in-situ cervical cancer than women who use oral contraceptives or IUDs, and they appear to be protected from sexually transmitted diseases [8].

11.4.3.7 Duration of Human Contact

A new application of spermicide is needed for each subsequent coital act, and the device should be implanted no later than 6 hours prior to the intended sexual activity. After the last coitus, the diaphragm shouldn't be removed for 6–8 hours [8].

11.4.3.8 Adverse Effect

- To some women, diaphragm cannot be fitted due to anatomical defects, uncomfortable, and it interferes with spontaneous sexual behavior.
- The issues with a diaphragm that are most frequently reported include allergic responses to the spermicide or the diaphragm, pain, bleeding, urine retention, bladder infections, constipation, offensive odor, and issues inserting or removing the device.
- Additional issues include the diaphragm being dislodged during sexual activity, developing a vaginal lesion as a result of the device's poor fit, aggravating recurrent cystitis, and excessive vaginal discharge. Yet, issues might arise. Diaphragm usage has been associated with vaginal discomfort and urinary tract infections (UTIs). A UTI can be avoided by urinating both before and after using the diaphragm.
- Both silicone sensitivity and a spermicide response can cause vaginal discomfort. It could be a good idea to use a different spermicide if vaginal irritability arises. Nonoxynol-9 is included in most spermicides. This might cause irritation if they use it frequently during the day or if they have HIV. It may make it more likely to get HIV and other STIs [8].

11.4.4 Female Condoms [9]

11.4.4.1 Introduction

One form of barrier device that prevents pregnancy and is helpful in preventing sexually transmitted diseases.

Woman used to insert a little tube composed of nitrile rubber or synthetic latex into vagina and remove it after each sexual intercourse.

11.4.4.2 Mechanism

The ends of female condoms are rimmed. The front of vagina is covered by the open end, which one should try to insert as far as within the vagina as possible. The condom tube allows partner's penis to enter vagina during intercourse. The condom prevents conception by keeping sperm from entering uterus.

Single use or multiple Only single time can be used female condom.

11.4.4.3 Benefits

The ability to purchase female condoms without a prescription

- This is advantageous for those who are allergic to latex.
- Has few to no negative consequences

11.4.4.4 Risks

The penis slides between the vagina and the condom's outer surface, the condom splits, or the condom slips out of the vagina.

The female condom may potentially cause pain during insertion, a burning sensation, itching, or a rash. The outer ring of the condom is inserted into the vagina during intercourse.

11.4.5 Intrauterine Devices [10]

11.4.5.1 Introduction

A tiny, T-shaped birth control device called an intrauterine device (IUD), sometimes known as an intrauterine contraceptive device (IUCD or ICD), or coil, is put into the uterus to prevent pregnancy.

11.4.5.2 Instrument

There are two varieties available in the UK;

- Nonhormonal: Copper-containing IUD (ParaGard and others)
- Hormonal: Progestogen-releasing IUD (Mirena and others)

Both copper and hormonal devices are classified as IUDs by the WHO ATC. The Anatomical Therapeutic Chemical (ATC) Classification System is a drug classification system that classifies the active ingredients of drugs according to the organ or system on which they act and their therapeutic, pharmacological and chemical properties. Its purpose is an aid to monitor drug use and for research to improve quality medication use [11].

More than ten distinct kinds of copper IUDs are offered in the United Kingdom.

11.4.6 New Electrosurgical Instrument for Tubal Coagulation

11.4.6.1 Introduction

To sterilize females, a procedure known as a tubal ligation, sometimes referred to as having one's "tubes tied," involves surgically plugging/cutting, or removing the fallopian tubes. This prevents sperm from fertilizing eggs, preventing a fertilized egg from being implanted as a result. Tubal ligation is a permanent method of sterilization and fertility control [12].

11.4.6.2 Mechanism

It is possible to do tubal ligation by blocking through an open abdominal procedure or a laparoscopic method, or a hysteroscopic approach. The patient will require local, general, or spinal (regional) anesthetic depending on the method chosen. A "postpartum" or "postabortion tubal ligation" is when the operation is done right away after a pregnancy ends. An "interval tubal ligation" is when it is done more than 6 weeks later. Depending on the kind of technique being employed, several processes will be included in the sterilizing process. If the patient decides to have a postpartum tubal ligation, the mode of birth will also affect the surgery. The surgeon will often remove some or all of the fallopian tubes 1 or 2 days after the delivery, while the patient is still in the hospital, if the patient delivers vaginally and requests a postpartum tubal ligation. If the patient opts for an interval tubal ligation, the surgery is normally carried out in a hospital environment while the patient is under general anesthesia. With an incision in the umbilicus and zero, one, or two smaller incisions in the lower sides of the abdomen, laparoscopic procedures are used to perform the majority of tubal ligations. Without a laparoscope, the procedure can also be carried out with bigger abdominal incisions. A hysteroscopic interval tubal ligation can also be done, and it can be done while you're under local anesthetic, mild sedation, or complete general anesthesia. Essure and Adiana systems were formerly utilized for hysteroscopic sterilization, and research studies are looking at novel hysteroscopic techniques, despite the fact that no hysteroscopic sterilization techniques are currently on the market in the United States as of 2019 [12].

11.4.6.3 Benefits and Advantages for Use as Contraception

- High effectiveness

The majority of female sterilization techniques are at least 99% efficient in preventing conception. These rates are somewhat less successful than vasectomy-based permanent male sterilization, and about similar to the effectiveness of long-acting reversible contraceptives such intrauterine devices and contraceptive implants. These rates are much greater than those of other contemporary methods of contraception that need the user's ongoing active participation, including oral contraceptives or male condoms [12].

- Avoidance of hormonal medications

Using progesterone and/or estrogen to suppress the menstrual cycle is a common component of female-controlled contraception. Tubal ligation offers very efficient birth control without the use of hormones for people who desire to forego hormonal drugs due to personal medical contraindications such as breast cancer, intolerable side effects, or personal preference.

- Reduction of pelvic inflammatory disease risk

It is less likely that a sexually transmitted infection may pass from the vagina to the abdominal cavity and cause pelvic inflammatory disease (PID) or a tubo-ovarian abscess if both fallopian tubes are blocked or removed. The risk of PID is not completely eliminated by tubal ligation, nor does it provide protection from STDs.

- Reduction of ovarian and fallopian tube cancer risk

The lifelong risk of getting ovarian or fallopian tube cancer later in life is decreased by partial tubal ligation or full salpingectomy (a tubal ligation procedure that requires the actual removal of the fallopian tube). This holds true for both patients who have already been identified as having a high risk of developing ovarian or fallopian tube cancer as a result of genetic abnormalities and women who are at normal population risk.

- Risks associated with surgery and anesthesia

The majority of tubal ligation procedures entail making incisions in the abdominal wall to reach the abdominal cavity and call for some kind of anesthetic. Significant risks associated with laparoscopic surgery include the requirement for blood transfusions, infections, conversion to open surgery, and the unexpected need for extra major surgery. Anesthesia-related risks include hypoventilation and cardiac arrest. With mortality rates in the United States estimated at 1–2 patient fatalities per 100,000 operations, major problems during female sterilization are rare, occurring in an estimated 0.1–3.5% of laparoscopic surgeries. Patients having a history of prior abdominal or pelvic surgery, obesity, and/or diabetes are more likely to experience these problems [12].

11.4.6.4 Adverse Effect [13]

- Menstrual changes

The menstrual cycles of patients who have had female sterilization operations have changed very little or not at all. They had a higher likelihood of feeling as though their menstrual cycle had improved, with minimal bleeding and less discomfort.

- Ovarian reserve

Research on hormone levels and ovarian reserve have shown inconsistent or no significant changes following female sterilization. There isn't much proof that women who become sterilized start going through menopause early.

- Sexual function

Compared to non-sterilized women, the sexual function of females following sterilization seems to have remained the same. Hysterectomy.

It has been discovered that patients who had tubal occlusion procedures are four to five times more likely to have hysterectomies later in life than patients whose spouses had vasectomy surgery. There is a connection across all tubal ligation techniques but no known biochemical mechanism to indicate a causal relationship between tubal ligation and eventual hysterectomy.

11.4.7 New Devices for Occlusion of the Vas Deferens

11.4.7.1 Introduction

Vas-occlusive contraception is a sperm transit in the vas deferens, the tubes that carry sperm from the epididymis to the ejaculatory ducts, is blocked by a method of male contraception.

In order to identify an acceptable alternative to vasectomy and potential hormonal contraception therapies that are now being explored, a variety of vas-occlusive contraceptive techniques have been researched for human use, with an interest in both reversible and irreversible procedures.

Clips, plugs, valves, polymers, hydrogels, and other devices are examples of potential techniques [14].

11.4.7.2 Types

- Intravasal control valve (ICV)

An intravasal control valve is a reversible valve that is placed in the vas deferens. Depending on the position of the device, it may either prevent or allow sperm flow. A T-shaped intravasal control valve made of gold and stainless steel for human usage has been created by Bionyx. A perforated ball within the T may be rotated to either allow or prevent sperm passage. The ICV implant requires expert microsurgeon.

- Chemosterilization

Due to scarring on the vas deferens wall, injection of non-toxic, sclerotic substances into the vas deferens may hinder the transfer of sperm. To accomplish sterilization, at least 26 distinct chemical fusions have been tried. By adherence to the luminal surface, a mixture of carbolic and n-butyl alpha cyanoacrylate completely blocked the vas deferens. This chemical mixture caused 96% azoospermia and 99% conception prevention in people 8 years after injection. The 4% failure rate of azoospermia is caused by the scar tissue's inconsistency and failure to completely occlude the vas deferens. Injections of the chemical mixture ethanol and formaldehyde also rendered persons sterile. These chemical reactions, however, cannot be reversed.

- Vas-occlusive plugs

There are two types of *vas-occlusive plugs*: injectable plugs and non-injectable plugs. 96% of males who had a medicinal polyurethane (MPU) injection to create a plug in the vas deferens experienced azoospermia; however, this effect was not apparent until 24 months following the injection. Research on more than 130 men who had the plug removed within 5 years revealed good fertility restoration. The Shug, an acronym for "silicon plug," is an example of a non-injectable plug, although research on the Shug has been scarce or nonexistent since 2008. The Shug is made up of two silicon plugs that adhere to the vas deferens wall with the help of nylon tails. It must be surgically implanted into the vas deferens and surgically removed. A 97% reduction in sperm motility via the vas deferens was seen in Shug human studies [14]. In contrast to conventional vasectomy, plugs have been demonstrated to be less effective overall [14].

- Intravasal thread (IVT)

It has been demonstrated that placing an Intravasal thread (IVT), often referred to as an intravasal device (IVD), in the vas deferens mechanically obstructs sperm movement, and that removing the IVT restores vas deferens patency. A plastic IVT had no significant impact on sperm count in pigs. A plastic and polyethylene IVT consistently blocked sperm transport in dogs during the experimental period, and removal of the device restored sperm transport after a period of 2 weeks. A urethane-coated nylon IVT had a significant impact on sperm count in human subjects, but it could not ensure absolute sterility at the same rate as a metal IVT [6, 15].

11.4.8 Sponges

11.4.8.1 Introduction

The spermicidal and barrier approaches are combined in the contraceptive sponge to prevent conception. Sponge use has two variations. To cover the cervix and keep sperm from accessing the uterus, the sponge is first placed into the vagina. Second, spermicide is used in the sponge.

11.4.8.2 Mechanism

Before to sexual activity, the sponges are inserted vaginally and must cover the cervix in order to be successful. Infections that are spread sexually cannot be prevented by sponges. Sponges can serve as contraception for several sexual acts [16].

11.4.8.3 Uses

To use the sponge, moisten it, squeeze it, fold it, and place it over the cervix inside the vagina. After inserted, a sponge functions for 24 hours, during which the female may engage in act of intercourse. After having intercourse, the sponge needs to be left in place for 6 hours.

Single use or multiple multiple times. After once it is used, it is washed and left for some time to dry and it can be used.

11.4.8.4 Duration of Body Contact

More than 30 hours should not be spent with a sponge in the vagina. This means 24 hrs and 6 hrs after the intercourse hence 30 hrs of body contact is present with the sponge.

Fig. 11.2 It is representing sponges used by women as barrier for birth control [6]. (Courtesy: Ref. [6])

11.4.8.5 Adverse Effect

Individuals who are sensitive to Nonoxynol-9, a component of the spermicide used in the sponge, may have unpleasant irritability and may have a higher chance of contracting STDs. Users of sponges may be somewhat more likely to get toxic shock syndrome (Fig. 11.2) [6].

11.5 Intrapartum Fetal Monitors Using New Physiological Markers [17]

11.5.1 Continuous Electronic Fetal Monitoring (CEFM)

11.5.1.1 Introduction

Electronic foetal monitoring is a technique that continuously logs your heartbeat and that of your unborn child. During labor and delivery, it may be an indication of fetal discomfort. Although physicians only advise it for pregnancies with a high risk of problems, providers frequently utilize EFM.

11.5.1.2 Mechanism

The blood arteries that feed baby with oxygen-rich blood is constricted during labor contractions. In most cases, newborns' oxygen levels are adequate throughout delivery. Baby's heart rate will alter if blood oxygen levels fall, though. Service provider can spot problems and safeguard infant by keeping an eye on baby's heart rate. Rarely, decreases in oxygen levels might result in fetal discomfort [17]. It operates by connecting the gadgets to an external monitor, which records the action on a paper or electronic readout. The healthcare professional monitor the infant's baseline heart rate and reassess the reading as needed. It's typical for the fetal heart rate to fluctuate during contractions. Modifications might be problematic include:

- A prolonged length of time with a heart rate that is either abnormally tachycardia or bradycardia.
- Irregular heartbeat rhythms during contractions.

The EFM equipment may also be equipped with alarms that alert labor and delivery team when baby's heart rate changes [17].

11.5.1.3 Use

It is utilized just for pregnancies with a high risk of difficulties, as advised by professionals.

11.5.1.4 Benefit

It contributes to preventing low heart rate fetal stillbirths.

11.5.1.5 Adverse Effect

Mobility will be restricted to bed and chair by external EFM can walk about more freely if a hospital offers wireless telemetry monitoring for external EFM. need to stay in bed with internal EFM [17].

11.5.1.6 Risk

Electronic fetal monitoring can:

- Lead to erroneous warnings that might be stressful.
- Limiting mobility will be helpful while giving birth.
- Lead to a surgical birth utilizing a vacuum or forceps or a caesarean section due to a misplaced worry for fetal distress [17].

EFM conducted internally has further dangers. The internal EFM wire might result in:

- A scalp injury to your child.
- Maternal infection.
- HIV or genital herpes from a pregnant parent to the child [17].

The catheter used to assess the pressure inside your uterus might:

- Entangle oneself with the umbilical cord (uncommon).
- Piercing the uterus (rare).
- Make the placenta rip (rare).

11.6 Devices to Prevent Postoperative Pelvic Adhesions

11.6.1 Introduction

After abdominal surgery, postoperative adhesions are a common occurrence that may result in chronic pelvic discomfort, intestinal blockage, and infertility. Several products, in the form of film or fluid, are frequently used to inhibit the formation of postoperative adhesions. These items often operate as barriers to prevent the surfaces of injured tissue from coming into touch. Polylactic acid film, Seprafilm 1, medical sodium hyaluronate, medical chitosan, and other items are often used in clinics. They were discovered to be able to lessen adhesion development in a variety of animal models and several therapeutic procedures.

11.6.2 Materials Available in the Market

11.6.3 Mechanism

There have been evaluations of several anti-adhesion barriers. In order to prevent the development of fibrous bands, anti-adhesion barriers often prevent contact between the surgical site and adjacent locations. Gels, solutions, and films are the three categories into which these materials fall. In order to treat surgical regions and prevent adhesion, functional anti-adhesion barriers have recently been developed by including medications that inhibit adhesion.

11.6.4 Uses

To reduce tissue ingrowth into the scaffold materials, which may result in the effective blockage of the tissue link, avoid post-operative adhesion (Table 11.1).

Single use or multiple Single use.

Table 11.1 materials used for adhesion

Biomaterial	Product name	form
Oxidesed regenerated cellulose (ORC)	Surgicel Interceed	Film
Carboxymethyl cellulose (CMC)	Suprafilm	Film
Hyluronic acid(HA)/CMC	Supragel	Solution
HA/CMC	Guardix-sol	Hydrogel
HA/CMC	Sepraspray	Powder
HA derivate	Incert	Film
HA	Hyalobarrier	Hydrogel
HA	Supracoat	Solution
HA derivate	ACP gel	Solution
Ferric HA	Lubricoat	Solution
Ferric HA	Intergel	Solution
HA derivate	Carbylan-SX	Film/spray
Icodextrin	Adept	Solution
Dextran	Hyscon	Hydrogel
Collagen	COVA +	Hydrogel
Polyethylene Glycol(PEG)	Spray shield	Spray
PEG	Spray gel	Spray
PEG	Coseal	Spray

11.6.5 Benefits

Prevent physically unnecessary adhesion.

11.6.6 Risk

- On the operative site, barriers are firmly fixed.
- It might be difficult to simultaneously accomplish minimal tissue ingrowth and stable fixation at the surgical site.

11.7 Embryoscope and Devices Intended for Fetal Surgery

Embryoscope:

11.7.1 Introduction

The incubators, which take care of the embryos as they develop while simulating the environment of a mother's womb, are the heart of in vitro fertilization (IVF) facility. Usually, incubators can only show the culture conditions, and the embryos must be removed every day for inspection, which disrupts the well-regulated culture environment [18].

11.7.2 Mechanism

The embryoscope TM is an innovative IVF incubator with a built-in microscope that blends the newest incubator technology with time-lapse imaging of the embryos during culture. This keeps the optimal conditions in this cutting-edge incubator while allowing high-resolution analysis of embryo growth. The best embryos are chosen for transfer using the detailed and accurate information about the embryo developing patterns that is obtained through analysis of the time-lapse photographs.

11.7.3 Use

Although the Embryoscope may be utilized with any kind of patient receiving assisted reproduction treatment, patients who produce more embryos have the best chance of improving their outcomes since there is a larger potential for selection [18].

Single use or multiple use It can be used multiple times.

11.7.4 Benefits [19]

The 24-hour observation reduces the chance of accidents by preventing the embryos from being disturbed during their evaluation, which improves the culture conditions.

Embryologists can choose the embryo with the best likelihood of implanting in the uterus by analyzing the time-lapse pictures and movies, which improves your odds of becoming pregnant.

This gives pregnant parents access to the first images and videos of their unborn child.

11.8 Devices Intended for Fetal Surgery

Fetoscopes:

11.8.1 Introduction

Contrarily, the current range of fetoscopes has diameters of 1.0–3.8 mm and working lengths of 20–30 cm to permit operating anywhere in a (polyhydramniotic) uterine cavity (Fig. 11.3) [20].

11.8.2 Parts and Mechanism

Sheaths, Cannulas and Uterine Access

(a) *Fetoscopic sheaths*

Endoscopes are used within a sheath with various purposes. It serves as both a guide and a safeguard for the scope. It may occasionally be bent to drive the fiber endoscope into the proper goal angle.

Operational canals expand in diameter in order to retain instruments in a fixed location to the scope. Sheaths made of two parallel tube parts might be irregular, circular, oval, or any other shape.

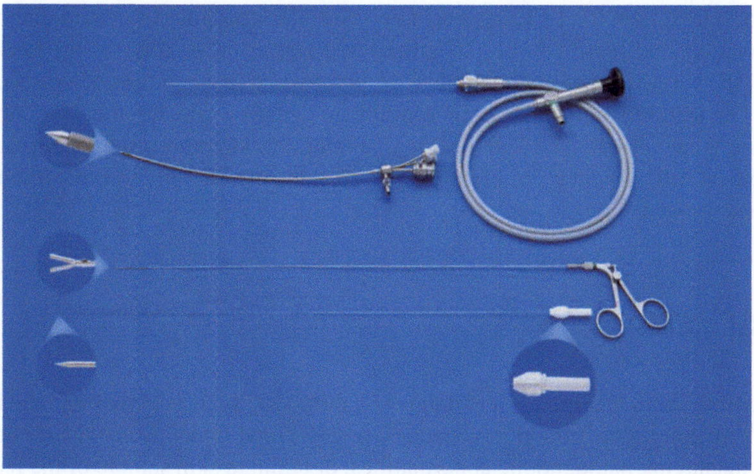

Fig. 11.3 Fetoscope cannula to access uterus [20]. (Courtesy: Van der Veeken et al. [20]))

(b) *Cannulas.*

This makes it possible to replace instruments more frequently and, in theory, might lessen friction and membrane dehiscence. Moreover, the cannula can be utilized for drainage, irrigation, or fluid exchange. The added usage of a cannula, however, inevitably results in a larger overall outer diameter. Reusable cannulas are available in a variety of lengths and diameters. Metal cannulas have a generally thick wall and are not flexible. Thin-walled plastic cannulas that were initially created for vascular access were used which are flexible, featured with a leak-proof port housing, measure 13 cm in length, and are available in diameters ranging from 4 to 15 Fr (1.33–5.0 mm), allowing the diameter to be selected in accordance with the intended surgery [20].

(c) *Uterine access*

It is possible to insert fetal sheaths directly, without the need of a cannula. They are equipped with sharp obturators that enable uterine puncture under the direction of ultrasonography for this reason.

(d) *Camera system and double image display*

A tiny diameter light cable that has been specially designed to fit the light transmission bundle and has a small diameter of 2.5 mm is used to link the fetoscope to a high-quality cold light source, such as Xenon light. It is desirable to use longer cables (230 cm, for example) so that the patient's equipment may be placed in various places [21].

(e) *Distension medium*

The amniotic fluid environment can be employed for fetoscopy; distension media may be used to expand the working area or enhance visibility.

Laser coagulation for TTTS (Twin to twin transfusion syndrome)

The best treatment for TTTS is chorionic vessel laser coagulation. According to the supposition that these can be recognized and run at the surface of the chorionic plate, the surgical purpose is to ablate all intertwined anastomoses of vessels. This calls for suitable visualization in addition to a safe angle for the laser fiber to approach the vessels. The fetal position, amniotic fluid composition, and placental placement, all of which are often outside of the surgeon's control, determine the aforementioned.

11.8.3 Use

Utilized for both therapeutic and diagnostic purposes, such as obtaining fetal blood for the diagnosis of hemoglobinopathies, giving blood under direct visual supervision, displaying pathognomic abnormalities, or taking a skin sample of the fetus.

Other use It has also been suggested to examine the lower urinary tract using fetoscopy. Nonetheless, this is still in its early phases, and better instrumentation is unquestionably needed.

It is still difficult to steer the endoscope percutaneously to the bladder neck since the device is still too big (which is the area of interest). This makes it difficult for even early researchers to distinguish between urethral valves and atresia.

11.8.4 Contraindications

Because it requires specialized knowledge, is intrusive, and lacks the necessary tools, it is not employed [20].

11.9 Endometrial Ablation System

11.9.1 Introduction

Endometrial ablation is a treatment modality in which a thin layer of endometrium inside uterine cavity is removed. This procedure helps improve symptoms and decrease excessive menstrual bleeding. To access the uterus, the healthcare practitioner inserts tiny instruments through the vagina [21].

11.9.2 Mechanism

- *Electricity (electrical or electrocautery).* routinely suppliers employ this technique, which involves passing an electric current through a wire loop or roller ball. To damage the uterine lining, the current is applied to it.
- *Fluids (hydrothermal).* This process uses hot fluid in order to remove the lining, it is pushed into the uterus.
- *Balloon therapy.* A tiny tube (catheter) is inserted into the uterus by healthcare professional. The catheter's tip is capped with a balloon. The balloon is heated and filled with fluid. The lining is destroyed by the hot fluid.
- *High-energy radio waves (radiofrequency ablation).* doctor inserts an electrical mesh into uterus using this technique. He or she makes it larger. The lining is then destroyed by radio waves used.
- *Cold (cryoablation).* doctor uses a probe that is extremely cold to freeze the lining.
- *Microwaves (microwave ablation).* To remove the lining, provider uses microwave radiation to pass through a tiny probe [21].

With a device known as a hysteroscope, certain endometrial ablations are performed.

11.9.3 Benefit

Menstrual bleeding is reduced or totally stopped after endometrial ablation.

11.9.4 Disadvantage

After endometrial ablation, might not be able to become pregnant. This is due to the removal of the endometrial lining, which is where the fertilized egg implants.

11.9.5 Contraindicated

In women who wants to get pregnant.

11.9.6 Risks

- Bleeding
- Infection
- Tearing of the uterine wall or bowel
- Overloading of fluid into the bloodstream

11.10 Falloposcopes and Falloposcopic Delivery System

11.10.1 Introduction

A transcervical, transvaginal procedure called a falloposcopy (Fig. 11.4) allows to see the fallopian tube's lumen from the uterotubal junction to the fimbria. Kerin and colleagues published the first accounts of the effective use of this microendoscopic technology in 1990 [22].

11.10.2 Uses

The evaluation of an infertile woman with probable proximal tubal illness following hysterosalpingography is the main reason to undergo a falloposcopy [23].

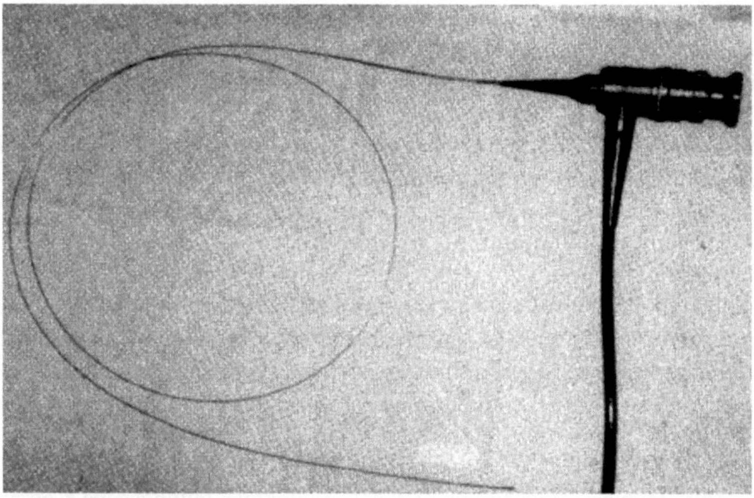

Fig. 11.4 This falloposcope is specifically designed with features that ensure safe navigation through the human fallopian tube. Its optical fiber is 1 meter long, with a diameter of 0.5 millimeters. The polished tip reduces the risk of trauma, while its flexibility allows it to maneuver through the curves of the fallopian tube with ease. Additionally, it includes an eyepiece for direct viewing or connection to a camera video monitoring system, as well as an illuminating bundle connection for clear visibility [22]. (Courtesy: Kerin et al. [22])

11.10.3 Contraindication

Patients with an active pelvic infection, uterine haemorrhage, or those who are unable to tolerate local anesthetics when conscious sedation is used during the operation should not undergo falloposcopy.

11.10.4 Mechanism

When menstruation has stopped, the midfollicular phase of the menstrual cycle is the best time to undergo a falloposcopy. Antibiotics are given to patients as a preventive measure. There have been two methods used.

1. a coaxial approach based on conventional cardiovascular access techniques and
2. a really creative method using a catheter that "evates linearly" [22, 23].

By using the coaxial technique, the UTO (Urinary tract obstruction) is initially seen by inserting a flexible hysteroscope into the uterine cavity while being seen on camera. This allows for a long axis view that is between 1 and 2 mm from the tubal ostia. Cervical dilatation is seldom necessary and is to be avoided if feasible to avoid leaking. Several hysteroscopes with ODs (Optical density) ranging from 1.5

to 2.0 mm and just one operational channel have been used (Olympus Corp., Lake Success, NY; Intramed Laboratories, San Diego, CA; Mitsubishi Cable Industries, Itami, Japan). Via extension tubing attached to one arm of an attached Tuohy-Borst-type Y-connector, lactate Ringer's solution or gamete culture medium is injected as a distention medium (Cook Ob-Gyn, Spencer, IN). The UTO is then entered through the second arm of the Y-connector with a flexible platinum-tipped tapered guidewire of 0.3- to 0.8-mm OD (Target Therapeutics, San Jose, CA; Conceptus, San Carlos, CA; Cook Ob-Gyn; Glidewire Medi-Tech, Watertown, NH) and advanced until either a point of resistance or a distance of 15 cm is reached. It is important to take precautions to prevent passing the wire through the UTO when the ostial muscles are in a spasm. While being relatively straight across its 1.5–2.5 cm length, the intramural part of the fallopian tube may form an acute angle with the cavity. The guidewire may be able to pass through this area more easily with a light torque action.

The catheter's proximal end is then connected to a second Tuohy-Borst Y-connector. The second Y-straight connector's arm is used to insert the falloposcope. This enables safeguarding of the very flexible falloposcope's atraumatic leading end, which is 120–130 cm in length and 0.3–0.5 mm in OD (Olympus Corp; Mitsubishi; Intramed Laboratories, San Diego, CA; Medical Dynamics, Englewood, CO). Via the angled arm of this second Y-connector, lactated Ringer's solution or culture media is injected. It needs a xenon light source, a camera chip, and a high-definition video display. In order to maintain the falloposcope flush with the catheter's distal entrance, seeing the tubal lumen is often done retrogradely. If the lens comes into contact with the tubal endothelium lining, a white-out happens. It is beneficial to use dual video monitoring for falloposcopy and hysteroscopy. In order to perform this procedure in an office environment, very little volumes of distending medium must be infused and catheters and guidewires must be moved incrementally in an exceedingly delicate manner. For effective atraumatic cannulation in cases with kinked fallopian tubes, an aid conducting concurrent laparoscopy may be needed to gently straighten the tubes [22].

Steps

1. Uterotubal Ostium (UTO) visualization using flexible hysteroscope
2. UTO guidewire cannulation
3. Installation of an over-the-wire catheter
4. Taking off the guidewire
5. Falloposcope insertion through catheter
6. Backwards visualization [23]

11.10.5 Instrument Parts

0.3–0.8 mm in outside diameter, flexible platinum-tipped tapered guidewire (Target Therapeutics, San Jose, CA; Conceptus, San Carlos, CA; Cook Ob-Gyn; Glidewire Medi-Tech, Watertown, NH).

1.2–1.3 mm OD Teflon-coated catheter (Target Therapeutics, Conceptus, Cook Ob/Gyn).

The catheter's proximal end is then connected to a second Tuohy-Borst Y-connector. Camera for visualization.

Linear everting catheter system:

- Outer catheter
- Inner catheter
- Balloon

An innovative alternate method for transcervical cannulation and fallopian tube imaging is the linear everting catheter system (Imagyn, Laguna Niguel, CA). 1~men. This tool has been used the bulk of the time for falloposcopy in a professional environment. An outer and an inner catheter body are connected distally by a balloon to form the linear everting catheter. By pressurizing a connecting membrane, which gently unrolls the catheter tip as the inner body is advanced, this balloon everts.

By inserting the falloposcope through the catheter's distal lumen, the UTO may be seen without the necessity of a separate hysteroscope. The catheter is progressively everted once the balloon is inflated. With the help of this catheter technology, the balloon may more precisely follow the curved course of the fallopian tube while being freed from lateral shear stresses. The 0.5-mm OD falloposcope is then inserted via the everted catheter once the tube has been entirely catheterized to the point at which resistance is met or until the catheter has been progressed for a distance of 15 cm.

Single use or multiple It can be used multiple times.

11.10.6 Risk

Tubal perforation is the main danger of falloposocopy, and it happens very seldom (10w). Exaggerated sharp angles created by the tube's junction with the UTO, peritubal adhesions that limit tubal flexibility, and fibrotic blockage that causes the lumen to shrink all increase the risk of perforation [22, 23].

11.11 Fundal Pressure Belt: (for Vaginal-Assisted Delivery)

11.11.1 Introduction

A novel medical gadget called the Baby-guard system (Cabel, Pistoia, Italy) was developed using biomechanics and biophysics research after obstetric semiotics. The Baby-guard device provides pressure to the uterine fundus toward the pelvic outlet during the second stage of labor via its ergonomic three-chamber inflatable abdominal belt (Fig. 11.5) [24].

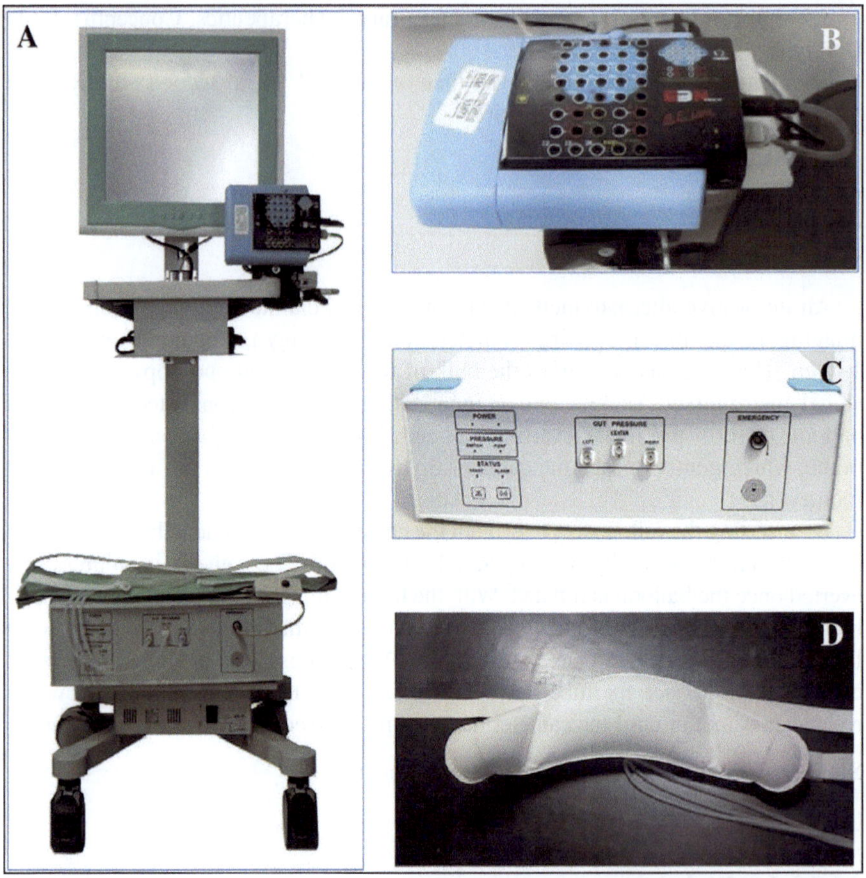

Fig. 11.5 Baby-guard composition. (**a**) The Baby-guard device with (**b**) the equipment for fetal heart-rate monitoring, (**c**) the pump for the belt, and (**d**) fundal the belt [24]. (Courtesy: Acanfora et al. [24])

11.11.2 Parts with Mechanism

A disposable ergonomic three-chamber inflatable belt and a detector of electrophysiologic signs of uterine activity from the mother's belly make up the Baby-guard system (i.e., fetal and maternal heart signals). To move the fetus, the three chambers of the belt each be separately inflated. The fetus can be gently positioned in the proper position toward the pelvis using these chambers, which are filled in accordance with the pressures established by the operator (midwife or doctor). When the uterus contracts, all three chambers are expanded simultaneously after the proper fetal position has been achieved. The maternal and fetal heart monitoring system consists of a medical touch-screen computer that monitors electrophysiologic data from the mother (uterine contractions and maternal heart rate) and the fetus (fetal heart rate). The fetal heart's Doppler parameters can also be recorded using the cardiotocograph. The Baby-guard system may keep all characteristics and signals it detects on the computer's hard disc [24].

11.11.3 Use

Utilized to enhance intrauterine pressure and uterine forces during the second stage of labor, but there may be clinical consequences from this method.

11.11.4 Benefit

It facilitates natural vaginal birth.

The favorable results shown during the second stage of labor may be attributed to the three-chambered inflatable belt's surface area for applying uterine fundal pressure.

Consequently, the Baby-guard system may reduce maternal tiredness and exhaustion by facilitating maternal pushing and reducing the length of the second stage of labor.

11.11.5 Risk

Manual maneuvers may be harmful to the fetus or baby, as well as pregnant mothers. Contrarily, the Baby-Guard system in the current research had the overall effect of eliminating this practice of manual pushing by substituting regulated fundal pressure applied in the direction of the pelvic outlet during spontaneous uterine contractions in the second stage of labor [24].

11.12 Gametes and Embryo Surgical System

11.12.1 Introduction

A potential technology for assisted reproduction includes electrical stimulation of gametes and embryos as well as on-chip modification of microdroplets of culture media (ARTs).

Electrowetting on dielectrics and dielectrophoresis (DEP) and associated electrorotation (ER) phenomena are used in microdevices that handle and analyze mammalian gametes and early embryos in vitro using an applied electric field (EWOD) [25].

11.12.2 Mechanism

The term "assisted reproductive technology" (ART) refers to a group of medical techniques mostly used to treat infertility. In vitro fertilization (IVF), intracytoplasmic sperm injection (ICSI), cryopreservation of gametes or embryos, and/or the use of fertility drugs are some of the methods covered by this topic [25].

- In vitro fertilization methods often utilized include: A thin needle is introduced into the back of the vagina and directed into the ovarian follicles using ultrasonography to collect the fluid that contains the eggs. This procedure is known as transvaginal ovum retrieval (OVR). The procedure stage known as "embryo transfer" involves inserting one or more embryos into the female's uterus in an effort to start a pregnancy.
- Less popular in vitro fertilization methods include:

The embryo is put through assisted zona hatching (AZH) just before being placed in the uterus. In order to facilitate the embryo's emergence from the egg and implantation of the developing embryo, a tiny hole is formed in the layer of tissue surrounding the egg. Intracytoplasmic sperm injection (ICSI) is beneficial in the case of male factor infertility, where sperm counts are very low or failed fertilization occurred with previous IVF attempt(s). The ICSI procedure involves a single sperm carefully injected into the center of an egg using a microneedle. With ICSI, only one sperm per egg is needed. Without ICSI, you need between 50,000 and 100,000. This method is also sometimes employed when donor sperm is used [25].

- Patients who have had unsuccessful IVF treatments in the past or who have low embryo quality may benefit from autologous endometrial coculture. A layer of cells from the patient's own uterine lining is put on top of the patient's fertilized eggs to provide a more natural environment for embryo growth.
- Zygote intrafallopian transfer (ZIFT) involves removing the woman's ovaries' egg cells, fertilizing them in a lab, and then inserting the resultant zygote into the fallopian tube.
- The procedure known as "cytoplasmic transfer" involves injecting sperm and the contents of a donor's fertilized egg into the patient's infertile egg.
- Egg donors are available for women who lack eggs because of illness, treatment, genetics, low egg quality, failed IVF rounds, or advanced maternal age. In the egg donation procedure, the recipient's partner's sperm is used to fertilize the donor's eggs in a lab, and the healthy embryos that result are then placed back into the recipient's uterus.
- If the male partner cannot generate sperm, suffers from an inherited illness, or if the woman undergoing treatment has no male partner, sperm donation may serve as a source for the sperm utilized in IVF treatments.
- Preimplantation genetic diagnosis (PGD) uses genetic screening technologies, including comparative genomic hybridization (CGH) and fluorescence in-situ hybridization (FISH), to detect genetically defective embryos and enhance healthy results.
- Twinning can be done with embryo splitting to enhance the number of available embryos.

11.12.3 Uses

Helps to Overcome with Infertility [25].

11.13 New Devices to Facilitate Assisted Vaginal Delivery

11.13.1 Odon Device

11.13.1.1 Introduction

Two million stillbirths and neonatal deaths each year, as well as around half of all maternal fatalities, are caused by intrapartum problems. A prolonged second stage of labor is a key contributor to stillbirths and the morbidity and death of newborns. It is also connected with potentially deadly maternal problems such as hemorrhage and infection. It also raises the danger of perineal and vaginal infections [26].

11.13.1.2 Instruments

Forceps, vacuum extractor, and caesarean section.

Sometimes, following an unsuccessful vacuum extraction, forceps are utilized. Moreover, vacuum extraction is not advised for women who are HIV-positive or who are unsure of their status.

The Odon gadget is envisioned as a potential replacement for current technology. It could be a safe alternative to some caesarean procedures and perhaps safer and simpler to use than forceps or vacuum extractors for assisted births. It might be widely used in low-resource environments, even by mid-level providers. The Odon device will be the first advancement in surgical vaginal delivery since the separate creation of forceps and the vacuum extractor centuries ago, if it is shown to be safe and successful [27].

11.13.1.3 Benefit

By enabling successful management of intrapartum problems in settings lacking surgical capability and/or people appropriately trained in the use of forceps and the vacuum extractor, the Odon device might significantly improve intrapartum obstetric care in low resource settings [28].

11.13.1.4 Mechanism

The baby's head is where the inserter is placed. Perfect fit to the fetal head is ensured by a soft plastic bell, which also avoids damage. The baby's head is gradually encircled by the Odon device as it is inserted. Placement happens as the baby's head is softly wrapped around the folded sleeve's two surfaces as they slide through the birth canal. A marking on the insertion handle becomes plainly visible in the reading window when the Odon device is properly positioned. An air chamber in the inner surface is pumped with a small and self-limited volume of air. This results in a firm grip being formed around the baby's head before it is removed. The two sides of the sleeve's folded surface. Surface lubrication makes the extraction procedure much easier. Up to 19 kg of traction can be used if necessary (which is equivalent to the force applied with the metal vacuum extractor) (Fig. 11.6) [27, 28].

11.13.1.5 Benefits

Safe and successful delivery

11.13.1.6 Contraindication

1. Maternal infection, suspected or confirmed: If any of the following conditions exist, a suspected infection will be identified:
   (a) Fever of at least 38 degrees Celsius
   (b) Amniotic fluid with a foul odor
2. Extended membrane rupture (greater than 24 hours)
3. Meconium stained
4. History of coagulation disorder

a

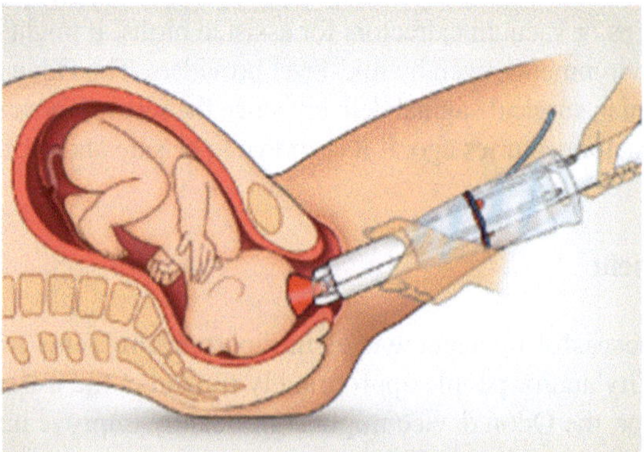

Fig. 11.6 Representing baby delivery using Odon device [29]. (**a**) The inserter is placed on the head of the baby. The small plastic ball is assumed to attach perfect adaptation to the head of the baby. (**b**) The inserter progressively takes position and Odon's device is around the head of the baby, and it is positioned as the inserter gently positioned by the sliding of the two surfaces of the folded sleeves along the birth canal and the head of the baby. (**c**) A marking on the insertion window is clearly visible when Odon's device is properly placed. (**d**) Firm grip being formed around the baby's head before it is removed. (Courtesy: Schvartzman et al. [29])

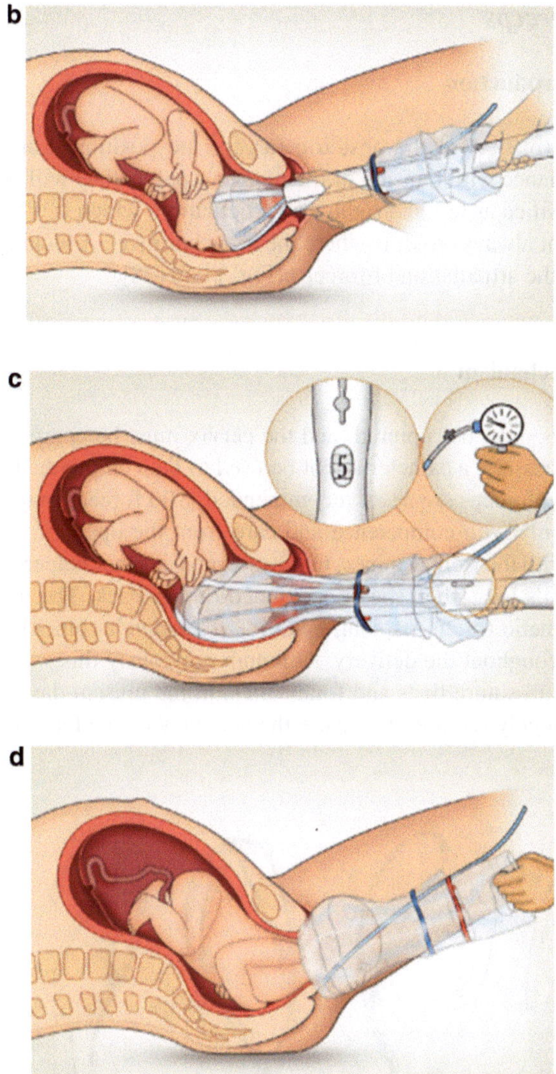

Fig. 11.6 (continued)

5. Previous uterine surgery
6. Intrapartum hemorrhage
7. Suspected cephalopelvic disproportion or need of operative delivery
8. Recognized recto-vaginal positive culture of *Streptococcus B Hemolyticus*
9. Known HIV positive status
10. Any situation that could endanger the mother or the fetus' safety in the investigator's assessment [28]

11.13.2 Forceps

11.13.2.1 Introduction

Two branches (blades) of the obstetric forceps are positioned around the fetus's head. In accordance with the side of the mother's pelvis they will be applied, these blades are classified as left and right. The articulation, which is where the branches typically but not always cross, is where the branches meet. A locking mechanism is often found at the articulation of forceps (Fig. 11.7) [28].

11.13.2.2 Mechanism

The membranes must be ruptured, and the cervix must be completely dilated and retracted. If necessary, a catheter might be used to help empty the bladder of urine. Nowadays, high forceps are never recommended. Mid forceps may occasionally be indicated but only with competence and care from the operator. The head station must be level with the ischial spines. Often, stirrups or other helpers are used to support the woman's legs when she is positioned on her back. The mother is given a localized anesthetic (often a spinal, epidural, or pudendal block) to help her feel comfortable throughout the delivery. Although previously this was done by touching the fetal skull suture lines and fontanelles, in the present day, confirmation by ultrasound is largely required to ensure the accuracy of the fetal head position. At

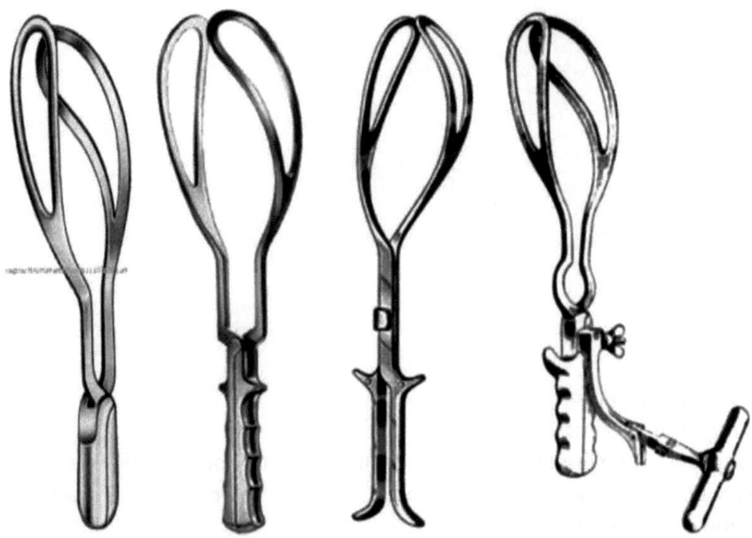

(a)Wrigley's (b) Simpson's (c) Kielland's (d) Neville-Barnes-Simpson's

Fig. 11.7 The common types of obstetric forceps [30]. (Courtesy: Adaji and Ameh [30])

this stage, each of the forceps' two blades is independently inserted, with the left blade going in first for the most typical occipito-anterior position and the posterior blade going in first in a transverse position. On the infant's head, the posture is examined. If it isn't already there, the fetal head is then rotated to the anterior position of the occiput. If required, an episiotomy can be done. The infant is then delivered with light traction in the pelvic axis [28].

11.13.2.3 Parts

Blades, handle.

11.13.2.4 Uses

1. Maternal fatigue.
2. Labor's second stage taking a long time.
3. Maternal illness such as heart disease, hypertension, glaucoma, aneurysm, or other conditions that make pushing difficult or dangerous.
4. Hemorrhaging.

Single use or multiple Multiple.

11.13.2.5 Risks

- Cuts and bruises.
- Increased risk of facial nerve injury (usually temporary).
- Increased risk of clavicle fracture (rare).
- Increased risk of intracranial hemorrhage—sometimes leading to death: 4/10,000.
- Increased risk of damage to cranial nerve VI, resulting in strabismus.

To Mother,

- Increased risk of perineal lacerations, pelvic organ prolapse, and incontinence.
- Increased risk of injury to vagina and cervix.
- Increased postnatal recovery time and pain.
- Increased difficulty evacuating during recovery time.

11.13.3 Vacuume Extractor

Vacuum extraction (*VE*), also known as *ventouse*, is a method to assist delivery of a baby using a vacuum device [31].

11.13.3.1 Use

If labor has not advanced sufficiently, it is employed in the second stage. It could serve as an alternative to a caesarean birth and forceps delivery. It cannot be utilized for preterm deliveries or when the infant is in the breech position. Although VE is normally safe to use, sometimes it might have unfavorable consequences on either the mother or the kid. The name is derived from the French meaning "suction cup."

11.13.3.2 Mechanism

The woman is put in the lithotomy position and pushes to help out all the way through. The baby's head is covered with a suction cup, and the cup's suction pulls the skin from the scalp into it. For a vacuum extraction to be successful, the cup must be positioned correctly above the flexion point, approximately 3 cm anterior to the occipital (posterior) fontanelle. Handles are on ventouse devices to provide traction. The contraption detaches after the baby's head is born, allowing the woman and the birthing assistant to finish delivering the child.

The head must be positioned in the delivery canal, the maternal cervix must be fully dilated, and the ventouse must be used correctly. To complete the process safely, the vacuum extractor's operator preferably needs expertise. The infant shouldn't be premature, have had their scalp sampled before, or have had a forceps delivery attempt fail. Forceps or a caesarean surgery may be required to deliver the baby if the ventouse attempt fails.

11.13.3.3 Benefit

- Perhaps an episiotomy is not necessary.
- Still actively involved in the birth process is the mother.
- No particular anesthetic is necessary.
- Compared to forceps and caesarean section, there is decreased chance of mother trauma.

11.13.3.4 Risks

- The infant will have a temporary chignon, or bulge, on its head.
- Cephalohematoma development and potentially fatal subgaleal bleeding are also possible outcomes.
- Compared to using forceps, there is a larger chance of the baby not being delivered, as well as a higher chance of perineal damage [31].

11.14 Operative Hysteroscopy and Laparoscopy

11.14.1 Hysteroscopy Introduction

The term "hysteroscopy" refers to the endoscopic examination of the uterine cavity through the cervix. It facilitates surgical intervention and permits the detection of intrauterine disease (operative hysteroscopy) [32].

11.14.2 Parts of Hysteroscope

- An endoscope with optical and light tubes or fibers is called a hysteroscope.
- A sheath that offers a conduit for air to enter and leave the uterus during insufflation.
- For the introduction of scissors, graspers, or biopsy tools, an operational channel may be present.
- Similar to a transurethral resectoscope, a hysteroscopic resectoscope permits entrance of an electric loop to remove tissue, such as to remove a fibroid.
- Hysteroscopes without distention medium are known as touch hysteroscopes.

11.14.3 Mechanism

Hysteroscopy has been performed in medical offices, surgical facilities, and hospitals. It is best performed right after a menstrual cycle, when the endometrium is still somewhat thin. For carefully chosen patients, diagnostic and straightforward surgical hysteroscopy can be performed in a clinic or office environment. You can utilize local anesthetic. Painkillers are not always required. A lidocaine injection in the upper region of the cervix can produce a paracervical block. Hysteroscopic intervention can also be performed while the patient is under general anesthesia using an endotracheal or laryngeal mask (MAC). Antibiotics for prevention are not required. Throughout the operation, the patient is in a lithotomy posture [33].

11.14.4 Uses

- Syndrome Asherman (i.e., intrauterine adhesions). The procedure known as hysteroscopic adhesiolysis involves lysing uterine adhesions with either microscissors (preferred) or thermal energy modalities. Laparoscopy or other techniques can be used with hysteroscopy to lessen the chance of perforation during the surgery.

- To remove endometrial polyps in the uterus (Polypectomy)
- Abnormal bleeding in the uterus
- Adenomyosis
- Endometrial ablation (several more recent endometrial ablation devices, such the Novasure, were created expressly for this purpose and do not require hysteroscopy)
- For uterine fibroids, a myomectomy
- Mullerian abnormalities, which are congenital uterine malformations such as septate uterus
- Evacuation of retained fetal products in case of abortion
- Removal of embedded IUDs

Due to embolism or fluid overload with electrolyte imbalances, the use of insufflation (also known as distending) media can result in significant and even deadly consequences. The danger of fluid overload with electrolyte imbalances, notably hyponatremia and heart failure, as well as pulmonary and cerebral edema, is increased, particularly by electrolyte-free insufflation medium. The primary causes of fluid excess during hysteroscopy include:

- The insufflation media's hydrostatic pressure
- Quantity of exposed blood vessels, which is enhanced during myomectomy and endometrial ablation
- How long does the hysteroscopy process take

11.14.5 Laparoscopy

11.14.5.1 Introduction

Modern surgical procedures such as laparoscopic surgery, often known as minimally invasive surgery, bandage surgery, or keyhole surgery [33].

11.14.5.2 Parts

It comprises a lengthy fiber-optic cable system that, by snaking the cable from a further but more convenient site, enables observation of the impacted region.

11.14.5.3 Used in Other Surgeries

Whereas thoracoscopic surgery is conducted through a small incision in the chest or thoracic cavity, laparoscopic surgery involves procedures in the abdominal or pelvic cavities. Obstetrical forceps, scissors, probes, dissectors, hooks, and retractors are a few of the specific surgical tools used in laparoscopic surgery. Laparoscopic and thoracoscopic surgery is a subset of endoscopy, a more general term.

11.14.5.4 Mechanism

There are two types of laparoscopes:

1. A three- or single-chip CCD video camera with a telescopic rod lens mechanism attached.
2. A digital laparoscope without a rod lens system instead attaches a tiny digital video camera at the end of the instrument.

The second type of mechanism is mostly employed to increase the picture quality of flexible endoscopes, which take the place of traditional fiberscopes. Laparoscopes are rigid endoscopes, despite this. Clinical practice necessitates rigidity. Due to their fine optical resolution (50 m generally, depending on the aperture size employed in the objective lens), rod-lens-based laparoscopes predominate significantly in practice. If necessary, the image quality can even surpass that of a digital camera. In the laparoscope market and in hospitals, the second type of laparoscope is quite uncommon.

A fiber optic cable system that illuminates the surgical field using a "cool" light source (halogen or xenon) that is introduced through a 5 mm or 10 mm cannula or trocar is also attached. Carbon dioxide gas is often inhaled into the belly. In order to provide a space for working and seeing, the abdominal wall is raised over the internal organs. Since it occurs naturally in the human body, tissue can absorb it, and the respiratory system can expel it, CO_2 is utilized. Also, it is non-flammable, which is crucial because laparoscopic surgeries frequently utilize electrosurgical instruments.

11.14.5.5 Uses

- Decreased bleeding, which lowers the likelihood of having a blood transfusion.
- A smaller incision causes less post-operative scarring and less post-operative discomfort, which speeds up healing.
- Reduced discomfort means less need for painkillers.
- In contrast to the general anesthetic used for many nonlaparoscopic operations, regional anesthesia can result in fewer difficulties and a speedier recovery when used for laparoscopic surgery (with the advice of employing a combination spinal and epidural anesthetic).
- Although hospital stays are shorter and procedure periods are often longer, same-day discharges frequently result, allowing for a quicker return to normal activities.
- Lower risk of infection due to decreased access of internal organs to potential external pollutants.

11.14.5.6 Risk [34]

- The major problems during laparoscopic surgery are related to the cardiopulmonary effects of pneumoperitoneum, systemic carbon dioxide absorption, venous gas embolism, unintentional injuries to intra-abdominal structures, and patient positioning.
- The trocar is normally put blindly; therefore, the greatest danger is from injuries sustained during insertion into the abdominal cavity. A hematoma in the abdominal wall, umbilical hernias, an infection of the umbilical wound, and bowel or blood vessel penetration are among the injuries. Patients with low body mass indices or a history of abdominal surgery are more likely to sustain such injuries. Although these injuries are uncommon, serious consequences might happen, and they are mostly connected to the location where the umbilical cord was inserted. Vascular injuries can result in bleeding that may be life-threatening. Bowel injuries may result in a delayed peritonitis. It is crucial to identify these injuries as soon as possible.
- There is a chance of port site metastases during oncologic laparoscopic operations, especially in individuals with peritoneal carcinomatosis. With particular precautions like trocar site protection and midline trocar implantation, this risk of iatrogenic cancer spread may be decreased.
- By using electrodes that leak electricity into the surrounding tissue, surgeons working on some patients failed to notice electrical burns they were causing. The ensuing wounds may cause punctured organs as well as peritonitis.
- Due to increased exposure to chilly, dry gases during insufflation, 20% or more patients have hypothermia following surgery and peritoneal trauma. It has been demonstrated that using surgical humidification treatment, which involves inhaling heated and humidified CO2, lowers this risk.
- During surgery, not all of the CO_2 that is injected into the abdominal cavity is taken out through the incisions. Gas has a tendency to ascend, and when a pocket of CO_2 rises in the abdominal cavity, it presses on the diaphragm (the muscle that divides the abdominal from the thoracic chambers and aids in breathing) and may put pressure on the phrenic nerve. This generates a sense of discomfort that may extend to the patient's shoulders in roughly 80% of women, for example. The discomfort is always momentary since the body's tissues will take up the CO_2 and expel it by breathing.
- Coagulation issues and thick adhesions (scar tissue) from prior abdominal surgery are relative contraindications to laparoscopic surgery and may increase the risk.
- Both laparoscopic and open surgery carry the potential of intra-abdominal adhesion development, which is a serious unsolved issue. Adhesions are fibrous deposits that, after surgery, join tissue to organ. These often happen in 50–100% of abdominal operations, with both procedures carrying the same risk of adhe-

sion development. Consequences of adhesions include persistent pelvic discomfort, intestinal blockage, and female infertility. The most serious issue is small bowel blockage in particular. The frequency of adhesion development during laparoscopic surgery may be reduced by the use of surgical humidification treatment. The use of physical barriers like films or gels, or broad-coverage fluid agents to separate tissues during healing after surgery are other methods to decrease the formation of adhesions.

- During surgical operations, both the smoke and the gas used to create space might seep into the operating room through or around equipment as well as access devices. The gas plume may introduce particles and perhaps diseases, such as virus particles, into the patient's and surgical team's shared airspace [33, 35].

11.15 Uterine Artery Embolization

11.15.1 Introduction

A catheter is used by an interventional radiologist during uterine artery embolization to inject tiny particles that obstruct the uterine body's blood flow. The surgery is used to treat adenomyosis and uterine fibroids. This minimally invasive method, also known as uterine fibroid embolization, is frequently used to treat uterine fibroids.

11.15.2 Uses

Utilized to treat adenomyosis, uterine fibroids, or uncomfortable bulk-related symptoms as well as irregular or severe uterine bleeding. Three factors that might determine whether a treatment is successful are fibroids' size, quantity, and placement. The long-term results of patient satisfaction are comparable to those of surgery. There is some preliminary evidence that conventional surgery may improve fertility. Also, it looks that uterine artery embolization necessitates more follow-up treatments than original surgery would. Its recuperation times are quicker. It is believed that uterine artery embolization works because uterine fibroids have atypical vasculature and inappropriate reactions to hypoxia (inadequate oxygenation to tissues). In addition to treating fibroids, uterine artery embolization helps manage severe uterine bleeding brought on by conditions including adenomyosis and postpartum obstetrical hemorrhage.

11.15.3 Mechanism

Under conscious anesthesia, an interventional radiologist performs the operation. The radial or femoral arteries can be accessed most frequently by the wrist or groin, respectively. The artery is accessible using the Seldinger method after anesthetizing the skin above the desired artery. Afterward, a guidewire and access sheath are inserted into the artery. A guiding catheter is frequently utilized and inserted into the uterine artery under X-ray fluoroscopy guidance to identify the uterine arteries for future embolization. Once the catheter has been positioned at the level of the uterine artery, a contrast angiography is done to verify the location of the catheter before the embolizing agents (spheres or beads) are released. The fibroid will shrink as a result of a considerable reduction or cessation of blood supply to it. As many arteries that supply the fibroid may undergo this procedure many times, unilateral uterine artery embolizations carry a significant chance of failure; this is done bilaterally from the point of first puncture. When the fibroids absorb the majority of the embolization material, they shrink in size and vascularity and avoid uterine necrosis when both uterine arteries are blocked.

The treatment typically takes little more than an hour to complete and can be carried out in a hospital, surgical center, or office environment. However, many patients are discharged the same day, with some staying in the hospital for a single-day admission for pain management and observation. If access was gained through a femoral artery puncture, an occlusion device can be used post-procedure to speed healing of the puncture site, and the patient is asked to remain with the leg extended for several hours. If the radial artery was used to achieve access, the patient will be able to get up from the operating table and leave right away. Because it does not need surgery, the uterus may be retained in situ without experiencing many of the difficulties that come with it [35].

11.15.4 Adverse Effect

- Misembolization occurs when microspheres or polyvinyl alcohol particles inadvertently travel into unintended tissues or organs, potentially causing harm. This can affect various areas of the body, including the ovaries, bladder, rectum, uterus, vagina, and, in rare cases, the labia.
- Infertility, decreased ovarian function, and organ loss
- Failure—continued fibroid growth, regrowth within 4 months
- Menopause—iatrogenic, abnormal, cessation of menstruation, and follicle-stimulating hormones elevated to menopausal levels
- A hysterectomy may be necessary as a result of infection, discomfort, or failure of embolization in post-embolization syndrome, which is characterized by acute and/or persistent pain, fevers, malaise, nausea, vomiting, and intense night sweats.
- Severe, ongoing discomfort necessitating the use of morphine or other synthetic opioids

11.15.5 Risks

Hematoma, a blood clot at the site of the incision, bloody vaginal discharge, bleeding from the site of the incision, bleeding from the vagina, fibroid expulsion (fibroid pushing out through the vagina), unsuccessful fibroid expulsion (fibroid trapped in the cervix, causing infection and requiring surgical removal), a potentially fatal allergic reaction to the contrast material, and uterine adhesions are all possible complications. Death caused by an embolism or sepsis (multiple organ failure brought on by the presence of pus-forming or other pathogenic organisms, or their toxins, in the blood or tissues). Endometritis (uterine infection) brought on by infection from the tissue death of fibroids necessitates a protracted hospital stay for the administration of intravenous antibiotics [36].

References

1. Heyns OS. Use of abdominal decompression in pregnancy and labour to improvefoetal oxygenation. Dev Med Child Neurol. 1958–1962;4:473–82.
2. Hill WC, Fleming AD, Martin RW, Hamer C, Knuppel RA, Lake MF, Watson DL, Welch RA, Bentley DL, Gookin KS, et al. Home uterine activity monitoring is associated with a reduction in preterm birth. Obstet Gynecol. 1990;76(1 Suppl):13S–8S. PMID: 2359574.
3. Urquhart C, Currell R, Harlow F, Callow L. Home uterine monitoring for detecting preterm labour. Cochrane Database Syst Rev. 2017;2:CD006172.
4. A cervix dilation device useful in presence of minimal resource settings. Eur J Mol Clin Med. ISSN 2515-8260. 2020;7(07):1646.
5. Cervical cap. In: Wikipedia [Internet]. 2022 [cited 2023 Mar 31]. Available from: https://en.wikipedia.org/w/index.php?title=Cervical_cap&oldid=1128252573.
6. https://www.fda.gov/consumers/free-publications-women/birth-control.
7. Condom. In: Wikipedia [Internet]. 2023 [cited 2023 Mar 31]. Available from: https://en.wikipedia.org/w/index.php?title=Condom&oldid=1145408294.
8. Diaphragm (mechanical device). In: Wikipedia [Internet]. 2022 [cited 2023 Mar 31]. Available from: https://en.wikipedia.org/w/index.php?title=Diaphragm_(mechanical_device)&oldid=1111008569
9. Peters A, Jansen W, van Driel F. The female condom: the international denial of a strong potential. Reprod Health Matters. 2010;18(35):119–28.
10. Lanzola EL, Ketvertis K. Intrauterine device. [Updated 2022 Jul 4]. In: StatPearls [Internet]. Treasure Island: StatPearls Publishing; 2023. Available from: https://www.ncbi.nlm.nih.gov/books/NBK557403/.
11. Creinin MD, Zite N. Female tubal sterilization: the time has come to routinely consider removal. Obstet Gynecol. 2014;124(3):596–9.
12. Lohiya NK, Manivannan B, Mishra PK, Pathak N. Vas deferens, a site of male contraception: an overview. Asian J Androl. 2001;3(2):87–95. PMID: 11404791
13. Clausen S, Lindenberg S, Nielsen ML, Gerstenberg TC, Praetorius B. A randomized trial of vas occlusion versus vasectomy for male contraception. Scand J Urol Nephrol. 1983;17(1):45–6. https://doi.org/10.3109/00365598309179779. PMID: 6346477.
14. Kuyoh MA, Toroitich-Ruto C, Grimes DA, Schulz KF, Gallo MG. Sponge versus diaphragm for contraception. Cochrane Database Syst Rev. 2002;(3):CD003172.
15. Hatcher RA, Kowal D. Birth control. In: Walker HK, Hall WD, Hurst JW, editors. Clinical methods: the history, physical, and laboratory examinations. 3rd ed. Boston: Butterworths; 1990. Chapter 174. Available from: https://www.ncbi.nlm.nih.gov/books/NBK283/.

16. Park H, Baek S, Kang H, Lee D. Biomaterials to prevent post-operative adhesion. Materials. 2020;13(14):3056.
17. Rubio I, Galán A, Larreategui Z, Ayerdi F, Bellver J, Herrero J, Meseguer M. Clinical validation of embryo culture and selection by morphogenetic analysis: a randomized, controlled trial of the EmbryoScope. Fertil Steril. 2014 Nov;102(5):1287–1294.e5.
18. Winkler SM, Harrison MR, Messersmith PB. Biomaterials in fetal surgery. Biomater Sci. 2019;7(8):3092–109.
19. Cullen MT, Albert Reece E, Whetham J, Hobbins JC. Embryoscopy: description and utility of a new technique. Am J Obstet Gynecol. 1990;162(1):82–6. https://doi.org/10.1016/0002-9378(90)90826-s.
20. Van der Veeken L, Russo FM, De Catte L, et al. Fetoscopic endoluminal tracheal occlusion and reestablishment of fetal airways for congenital diaphragmatic hernia. Gynecol Surg. 2018;15:9. https://doi.org/10.1186/s10397-018-1041-9.
21. Kerin J, Surrey E, Daykhovsky L, Grundfest WS. Development and application of a falloposcope for transvaginal endoscopy of the fallopian tube. J Laparoendosc Surg. 1990;1(1):47–56.
22. Kerin J, Daykhovsky L, Segalowitz J, Surrey E, Anderson R, Stein A, Wade M, Grundfest W. Falloposcopy: a microendoscopic technique for visual exploration of the human fallopian tube from the uterotubal ostium to the fimbria using a transvaginal approach. Fertil Steril. 1990;54(3):390–400. https://doi.org/10.1016/s0015-0282(16)53750-9.
23. Ezzati M, Djahanbakhch O, Arian S, Carr BR. Tubal transport of gametes and embryos: a review of physiology and pathophysiology. J Assist Reprod Genet. 2014;31:1337–47.
24. Acanfora L, Rampon M, Filippeschi M, Marchi M, Montisci M, Viel G, Cosmi E. An inflatable ergonomic 3-chamber fundal pressure belt to assist vaginal delivery. Int J Gynecol Obstet. 2013;120:78–81. https://doi.org/10.1016/j.ijgo.2012.07.025.
25. Hotton EJ, Alvarez M, Lenguerrand E, et al. The Odon Device™ for assisted vaginal birth: a feasibility study to investigate safety and efficacy—the ASSIST II study. Pilot Feasibil Stud. 2021;7:72.
26. Hotton EJ, Alvarez M, Lenguerrand E, Wade J, Blencowe NS, Draycott TJ, Crofts JF. The Odon Device™ for assisted vaginal birth: a feasibility study to investigate safety and efficacy—the ASSIST II study. Pilot Feasibil Stud. 2021;7(1):1–0.
27. Vannevel V, Swanepoel C, Pattinson RC. Global perspectives on operative vaginal deliveries. Best Pract Res Clin Obstet Gynaecol. 2019;56:107–13.
28. Evanson SM, Riggs J. Forceps delivery. [Updated 2022 Jul 17]. In: StatPearls [Internet]. Treasure Island: StatPearls Publishing; 2023. Available from: https://www.ncbi.nlm.nih.gov/books/NBK538220/.
29. Schvartzman JA, Krupitzki H, Merialdi M, et al. Odon device for instrumental vaginal deliveries: results of a medical device pilot clinical study. Reprod Health. 2018;15:45. https://doi.org/10.1186/s12978-018-0485-8.
30. Adaji SE, Ameh CA. 16 Operative vaginal deliveries in contemporary obstetric practice. In: From preconception to postpartum; 2012.
31. Chan SCS, Fraser IS. The role of diagnostic hysteroscopy in modern gynecological practice. HKMJ. 1995;1:161–6.
32. Santos-Paulo A. Psychosocial and clinical characteristics predicting women's acceptance of office hysteroscopy: an observational study, 2016. https://doi.org/10.13140/RG.2.2.34669.79840.
33. Papastefanou I, Nowacka U, Buerger O, Akolekar R, Wright D, Nicolaides KH. Evaluation of the RCOG guideline for the prediction of neonates that are small for gestational age and comparison with the competing risks model. BJOG. 2021;128(13):2110–5.
34. Daniilidis A, Hatzis P, Pratilas G, Dinas K, Loufopoulos A. Laparoscopy in gynecology-how why when. In: Advanced gynecologic endoscopy. IntechOpen; 2011.
35. Young M, Coffey W, Mikhail LN. Uterine fibroid embolization. [Updated 2022 Nov 28]. In: StatPearls [Internet]. Treasure Island: StatPearls Publishing; 2023.
36. Taylor M, Jenkins SM, Pillarisetty LS. Endometritis. [Updated 2023 Feb 3]. In: StatPearls [Internet]. Treasure Island: StatPearls Publishing; 2023. Available from: https://www.ncbi.nlm.nih.gov/books/NBK553124/.

Chapter 12
Significant Risk Medical Devices – Ophthalmics

K. P. G. Uma Anitha, T. S. Subashini, K. S. Sridevi Sangeetha, Saranya Sankar, and Ranjitha Dhevi V. Sundar

12.1 Introduction

According to a survey conducted by the World Health Organization (WHO) in 2015, approximately 217 million people in the world who are aged 18 years or older experience an ophthalmic disease that could cause visual impairment, which can eventually lead to irreversible blindness. Some of the ophthalmic diseases and disorders that were associated with the eye are glaucoma, cataracts, diabetic retinopathy, age-related macular degeneration, keratoconus, retinal detachment, uveitis, retinal pigmentation, etc. The widespread prevalence of visual impairment has been increasing over the past decades because of changes in the lifestyles, environments, and genetic and hereditary conditions and because of aging. Uncorrected refractive errors like myopia, hyperopia, and astigmatism affect 49% of the world population. Vision acuity in these conditions is restored with the help of glasses or contact lenses. The most commonly affected ocular disorder is cataract, followed by age-related macular degeneration and glaucoma. With innovations in technology and advances in research, many novel devices have been designed by the scientific community these visual errors and disorders, and some of the devices that are used as such in ophthalmology appear later in this chapter.

K. P. G. Uma Anitha (✉) · S. Sankar · R. D. V. Sundar
School of Biosciences and Technology, Vellore Institute of Technology, Vellore, India
e-mail: aanitha@vit.ac.in

T. S. Subashini
Department of Pharmacology, SRM Dental College, SRM Institute of Science & Technology, Ramapuram, Chennai, India

K. S. Sridevi Sangeetha
Faculty of Allied Health Sciences, Meenakshi Academy of Higher Education and Research, Chennai, India

P. S. Timiri Shanmugam et al. (eds.), *Significant and Nonsignificant Risk Medical Devices*, https://doi.org/10.1007/978-3-031-52838-5_12

12.2 Medical Devices: Ophthalmic Devices

An ophthalmic medical device can be defined as a device that satisfies the medical purposes of ophthalmology and optometry. These devices include noninvasive devices and instruments often used for diagnoses, invasive devices such as contact lenses (and their associated cleaning and care products), and implantable devices such as intraocular lenses and glaucoma stents. Surgical systems, including lasers, phacoemulsification machines, and surgical instruments are further examples of these devices [1].

12.2.1 Contact Lenses

Ocular surface disease could be difficult to manage and cause discomfort and vision loss. Using therapeutic contact lenses is important treatment option that often ends up neglected because they cause dryness or irritation in the eyes and are therefore not good candidates [2].

Contact lenses are artificial devices applied on the front surface of the eye, acting as replacements for the anterior corneal surface. These lenses correct the refractive error of the cornea, irregularities, and surface abnormalities. The design and the idea of contact lenses started in 1508 with Leonardo da Vinci, who thought of neutralizing the cornea by introducing a new refracting surface. In 1946, contact lenses were made of polymethylmethacrylate (PMMA) and started to become more popular than before. Contact lenses are classified according to their anatomical position, their mode of wear, and the material used in and the water content of the lens. Contact lenses can be single cut or lenticular cut. How well contact lens perform depends on the inherent properties of lenses, like wettability, refractive index, water content, oxygen permeability, oxygen transmission, light transmission, resistance to heat, and mechanical properties like lens flexure. Contact lenses come in several types: optical, therapeutic, preventive, diagnostic, operative, cosmetic, and occupational [3].

Contact lenses are artificial prosthetic devices worn on the eye's front surface to act as replacements for the anterior corneal surface. Contact lenses help correct the refractive error and irregularities of the corneal surface [4]. Contact lenses differ on the basis of various specifications, like overall diameter; optic zone diameter; base curve; central, peripheral, and intermediate curves; edge; power; thickness; and tint. Contact lenses come in other types too: soft, hard, and rigid gas permeable [5].

On the basis of the material used for their manufacture, contact lenses are divided into two other types: focons and filcons. An ideal contact lens should be biocompatible, gas permeable, moldable, sterile, and stable and should have good optical properties and tolerance and good surface chemistry. Each type of contact lens has various advantages and disadvantages [6].

The medical use of contact lenses is the only remedy for many complex ocular conditions, like high refractive error, irregular astigmatism, primary and secondary corneal ectasia, disfiguring disease, and ocular surface disease. The development of high oxygen permeability and soft and rigid materials has extended the suitability of contact lenses for such applications. Evidence has shown that bandage soft contact lenses, especially silicone hydrogel lenses, improve epithelial healing and reduce pain after trauma or surgery and in conditions like persistent epithelial defects and corneal dystrophies. Contact lenses hold promise in improving topical therapy in drug-delivery applications. Modern scleral lenses have achieved great success in both visual rehabilitation and therapeutic applications, including those requiring the retention of a tear reservoir or protection against an adverse environment [7].

Therapeutic or Bandage Contact Lenses

Therapeutic or bandage contact lenses are used to treat conditions like ocular discomfort, to support the cornea after surgery for healing, to help treat the cornea, and to protect the cornea from the environmental or any mechanical interaction with the eyelids.

Rehabilitative Contact Lenses

Rehabilitative contact lenses are advised for conditions that prevent a patient from having adequate visual function with spectacles because of high, irregular, or asymmetric refractive errors. Partially or completely occlusive lenses that improve the function or cosmesis after a trauma, surgery, or stroke also fall into this category [8].

12.2.2 Intraocular Lenses

Intraocular lenses were first developed in 1949 by British ophthalmologist Sir Harold Ridley in consultation with John Pike, an optical expert for the Rayner Optical Company. Cataract surgery already existed, but at the time, no replacements were available for the lenses that this surgery removed.

Currently, the surgery is most commonly performed as a surgical procedure in ophthalmology. Removing the opaque cataract lens and replacing it with an artificial lens to achieve near to normal postoperative visual acuity is now an expectation. The lenses used for this purpose are called intraocular lenses. Cataract surgery, the most common surgery in the world today, has a memorable history. The earliest example, cataract "couching," was initially reported around 3000 years ago in India in an ancient text, the Mahabharata. Couching is a process that should be consigned to history: It requires pressing cloudy lenses down into the vitreous by using a thorn or a needle. It leaves the patient aphakic but with some visual function and requires a high hyperopic prescription lens. Reducing the size of the incision and using phacoemulsification revolutionized cataract surgery. This was the second major development in the implantation of intraocular lenses [9].

Cataract is a disorder in which the crystalline lens becomes opacified. It is one of the leading causes of visual impairment in all but the most developed countries, where cataract surgery is often the most common surgical procedure performed. The etiology of the vast majority of cataracts is age. Genetic factors probably account for around 50% of age-related cataracts, along with systemic and environmental factors. Diabetes mellitus, corticosteroid use, heavy alcohol use, smoking, and lifetime ultraviolet light exposure are among the many risk factors linked to cataract formation. The major types of cataracts include nuclear, cortical, and posterior subcapsular, though mixtures of these types of opacities are common in any given cataract. Although all types of cataracts can be simply age related, posterior subcapsular cataracts are commonly caused by toxicities, specifically uveitis- or corticosteroid-induced cataracts.

In a single-piece design, the entire intraocular lens (IOL) is manufactured from the same material. In multipiece lenses (usually described as three-piece lenses, with an optic and two loops), the material of the haptic components is different from that of the optic. More recently, many special designs have appeared, including open-bag lenses, fluid-filled lenses, and modular lenses.

Intraocular lenses have been developed for the implantation into the anterior or posterior chambers, according to the site of fixation. Each chamber offers various fixation possibilities, and the IOL design has to be customized for each of them [10].

The central zone used for viewing is called an optic. This is a clear spherical disc measuring 5.5–6.5 mm in diameter (about 1/4 inch). The optic features the optical power of the lens, and about 40 powers are available to choose from. The unit for optical power is the diopter.

Hook-like structures appear on both the sides of the optic: two flexible struts, called haptics. These haptics act like tension-loaded springs that automatically center the lens within the eye compartment where they are implanted. Haptics come in various shapes, where the shape depends on the type of IOL. The most common lenses used are the three-piece posterior-chamber IOLs. They are called three-piece IOLs because their round optic is fused with two plastic haptics that are shaped like curved wires [11].

Types of Intraocular Lens (IOL) Implants for Cataract Surgery

Monofocal Intraocular Lens Implants

A monofocal lens is a standard type of IOL that has a fixed focal distance. Usually, this lens is used for distance vision where no glasses are generally required. However, reading glasses are required for close-up work such as reading. Glasses are also needed for intermediate work such as using a computer. The spherical design of monofocal lenses means that they are capable of providing vision correction only for nearsightedness or farsightedness. Although attempting to correct one eye for distance vision and one for near vision (monovision) is possible, monofocal lens recipients generally require reading glasses or bifocals for close reading vision after surgery.

Multifocal Intraocular Lens Implants

As the name implies, a multifocal IOL offers more than one lens power. The technology functions similarly to that of progressive eyeglasses and multifocal contact lenses. The difference is that the multifocal IOL is surgically implanted into the eye and offers a permanent, maintenance-free solution to presbyopia. Multifocal IOLs contain multiple zones that can focus light at various distances, providing a continuous range of vision without glasses. Many patients achieve excellent results, including the ability to see distance and up close without glasses. Some patients also complain of seeing rings or halos around lights with these lenses, which may impede some tasks performed in low light levels, such as when driving at night. But most patients can adapt to this side effect.

Both eyes require a multifocal IOL implant for glasses-free vision. Multifocal IOLs address this issue by offering a lens-replacement solution that boasts an aspherical design capable of restoring vision across varying distances [12].

Toric Intraocular Lenses

Toric IOLs are used to treat astigmatism. Astigmatism occurs when the shape of the cornea becomes irregular. In astigmatism, this irregularity can be corrected with a toric IOL implant. This correction imparts good distance vision without glasses but still requires that patients wear glasses for reading, after the operation. This type of IOL can correct mild to moderate astigmatism. In the case of high astigmatism, it may also require that tiny relaxing cuts be made in the cornea called limbal relaxing incisions.

Aspheric IOLs

Traditional monofocal intraocular lenses (IOLs) provide good vision by correcting either nearsightedness or farsightedness, but they do not correct other optical refractive errors, such as astigmatism, presbyopia, or spherical aberration. Spherical abberation is a type of optical imperfection that can increase glare and reduce the overall quality of vision in low light and darkness. Lenses that are used to correct spherical aberration are called aspheric lenses. Reduction in night vision clarity before IOL care might be only partially corrected by using standard monofocal IOLs, because most IOLs these days incorporate aspheric optics.

Aspheric IOLs are monofocal lenses that are used to correct spherical aberration. The result is a lens that provides better overall vision than traditional IOLs, especially at night [13].

12.2.3 Autorefractor or Retinoscope

An autorefractor or optometer is an instrument that is used in the automated assessment of refraction. Using such an instrument is an alternative method for observing the refractive errors, in contrast to using conventional refractive techniques. This procedure is called refractometry or optometry. The autorefractors come in various types early autorefractors (Badal refractors) and modern refractors. The various indications used in determining whether to perform optometry are myopia, hyperopia, astigmatism, presbyopia, spectacle prescription, and contact lens prescriptions, which serve as a starting point for ophthalmologists and optometrists when assessing subjective refraction in children and people who require glasses. Various commercially available subjective and objective autorefractors are available, and they provide information to healthcare providers who work with ocular disorders that the autorefractors can address. As with any subjective refraction, the initial objective assessment of refractive status uses an autorefractor or retinoscope as a starting point. Retinoscopy can also provide invaluable information when qualitatively viewing the red reflex over the lenses. Features such as optical zone edges and bifocal segments within the entrance pupil can be observed and correlated with patient complaints. Distortions, such as small visual obstructions induced by lens deposits, scratches, warped lenses, or lens lift-off from the corneal surface, can also be detected via retinoscopy. Lastly, ocular pathology, such as a posterior subcapsular cataract, can be easily detected by viewing the red reflex [14].

Indications of Autorefractors

Autorefractors are used to treat many refractive errors, like myopia, hypermetropia, astigmatism, and presbyopia and can replace the need for glasses or contact lenses. They mark the starting point for ophthalmologists and optometrists when assessing subjective refraction, pediatric refraction, and refraction in people with a disability requiring glasses.

Contraindications of Autorefractors

Using autorefractors also comes with several disadvantages for many patients, such as some mentally ill patients; patients with postural problems; patients with gross vision loss; patients with acute traumatic injury to the eye, conjunctivitis, keratitis, uveitis, episcleritis, corneal edema, an anophthalmic socket, an artificial prosthesis, phthisis bulbi, or atrophic bulbi; small children; and patients with accommodation anomalies [14].

Optical Principles

Schiener's Principle

In 1619, Schiener became the first to describe the refractive error of eyes that can be determined by employing a double pinhole aperture in front of the pupil. He observed that when parallel rays of light were passed through the eye from a distant object, the rays were limited to two small bundles when a double pinhole was placed in front of the eye. In hypermetropic eyes, the bundle of rays is intersected by the retina before the rays join, and two small light spots can be observed [15].

In myopic eyes, the two bundles of light rays cross each other before falling on the retina, and thus, two small spots of light can be seen. The two points of light can be merged to form a single point by taking the double pinhole at the eye's far point. Hence, the ocular refractive error can be determined from the far point of the eye [16].

Development of Optometers

For a long time, Schiener's principle and their modifications have been used in automating refraction. Today, autorefraction is a proven and well-established technique, and computerized autorefractors are now widely used, whereas the older ones are rarely used.

The development of optometers can be divided into two types of refractometers: early and modern [17].

Early Subjective Optometers

Early subjective optometers were first developed between 1895 and 1920. The subjective optometers required the patient to adjust to the instrument to yield the best focus and alignment with the target. They became unpopular because of instrument accommodation. Examples include Badal and Young optometers [18].

Early Objective Optometers

Early objective optometers were designed as alternative methods for assessing the optical correction needed in the eye. However, these were less accurate than retinoscopy. These objective optometers will depend on the examiner's decision when the image is transparent or needs a coincident setting. They are more common in Europe and employ both optometers and Schiener's principle [19].

12.2.4 Optical Coherence Tomography (OCT)

Optical coherence tomography (OCT) is an optical analog of ultrasound imaging that uses a low coherence interferometry to produce cross-sectional images of the retina. It captures the optical scattering from the tissue to decode the spatial details of the tissue microstructures. It uses infrared light from a super-luminescent diode that has two parts. One of the lights is reflected from a reference mirror, and the other light is scattered from the biological tissue. The two reflected beams of light are made to produce interference patterns, which obtain the echo time delay and the amplitude information that make up an A scan. A scans that are captured at adjacent retinal locations via a transverse scanning mechanism are combined to produce a two-dimensional (2D) image.

OCT imaging plays a vital role in the field of diagnosing retinal diseases. This light-based and noninvasive imaging modality provides a high-quality cross-sectional analysis of the retina and has revolutionized the diagnosis and management of retinal and choroidal diseases. Since its introduction in the early 1990s, OCT technology has continued to advance to provide quicker acquisition times and higher resolutions. The emerging innovations include wide-field OCT, adaptive optics OCT, polarization-sensitive OCT, full-field OCT, handheld OCT, intraoperative OCT, at-home OCT, and more. The applications of these rising OCT systems and techniques allow for closely monitoring of chorioretinal diseases and treatment response, more robust analysis in basic science research, and further insights into surgical management. In addition, these innovations to optimize the visualization of the choroid and the retina offer a promising future for advancing our understanding of the pathophysiology of chorioretinal diseases.

OCT is an imaging modality that has revolutionized the field of ophthalmology. As a noninvasive imaging technique, OCT utilizes light and light interference to capture high-resolution cross-sectional tomographic information on biological tissues, such as the retina and choroid, at the micron level. This technology was first introduced in 1991 [20], and it has been adopted into clinical practice in diagnosing retinal diseases. Diagnostic evaluation in retinal and choroidal diseases is often conducted with OCT, including neovascular age-related macular degeneration (AMD), central serous chorioretinopathy (CSCR), vascular retinal disorders [21], and other vitreoretinal disorders [22]. OCT biomarkers have also been instrumental in further understanding and monitoring chorioretinal disease status, and these biomarkers include central macular thickness, subretinal/intraretinal fluid, neurosensory detachment height, subfoveal choroidal thickness, choroidal vessel diameter, and choroidal vascularity index [23, 24]. Recently, to optimize the design of early interventional clinical trials for nonneovascular AMD, several structural OCT biomarkers, such as intraretinal hyperreflective foci, subretinal drusenoid deposits, drusen with a hyporeflective core, and high central drusen volume, have been described as high-risk for AMD progression to late stages [25, 26]. The emerging innovations made in this imaging modality will help advance several critical aspects in retinal care, including image-acquisition times, field of view, portability/accessibility, and intra-operative management [27].

Visible Light OCT (Vis-OCT)

Visible light OCT (vis-OCT) utilizes visible light, rather than near-infrared (NIR) light, for OCT illumination to capture images. This technique improves the resolution of the biological features of the retina thanks to its capturing shorter illumination wavelengths [28].

Adaptive Optics (AO) in OCT (AO-OCT)

Adaptive optics was initially developed to reduce the dynamic wavefront errors in astronomical imaging. It was able to quantify and eliminate the high-order monochromatic aberrations from the light passing through ocular tissues such as the cornea and the lens. These aberrations could cause poor lateral resolution in ophthalmic imaging and previously limited the clinical applications of various ophthalmic imaging systems. AO systems are composed of a wavefront sensor (generally a Shack-Hartmann wavefront sensor) that is used for measuring distortions, a wavefront corrector (typically a deformable mirror) that alters its shape to cancel out the aberrations, and a controller that connects all these elements [29, 30].

Polarization-Sensitive (PS) OCT

Polarization-sensitive OCT was introduced in 1992, and it functions by analyzing the polarization state of backscattered light and measures the birefringence in tissue samples. Different tissues can change the polarization state of the OCT light source [31]. Initial PS-OCT schemes were based on TD-OCT. However, PS is now employed in both Swept-source OCT (SS-OCT) and SD-OCT to capture images of various ocular structures, such as the macula and peripheral retina [32, 33]. PS-OCT carries many promising applications in both basic and clinical ophthalmic research, particularly in the automated segmentation of retinal structures, such as Retinal pigment epithelium (RPE). Fibrotic tissues, which contain collagen, are particularly birefringent and are thus imaged well by PS-OCT. PS-OCT has proven useful for the evaluation of RPE lesions in choroidal neovascularization in eyes with neovascular AMD [34]. Retinal fibrosis growth in the setting of neovascular AMD can be tracked by using PS-OCT and segmentation algorithms [35].

High-Resolution OCT (High-Res OCT)

Heidelberg engineering is one of the recent developments is the introduction of high-resolution OCT (high-res OCT). High-res OCT increases the bandwidth of the OCT light source, which increases optical axial resolution [20]. High-res OCT is capable of 3 μm axial resolution, allowing it to capture clearer images of the small vasculature, including the choriocapillaris [20, 21]. The choriocapillaris plays an important role in many retinal diseases, so a more detailed visualization of this microvasculature will likely advance the understanding of its dysfunction in these diseases. The utilization of this advancement in OCT imaging may help to provide an additional insight into the microstructures and microvasculature of the retina in chorioretinal diseases [36, 37].

Full-Field OCT (FFOCT) and Dynamic FFOCT (D-FFOCT)

Full-field OCT (FFOCT) captures 2D enface scans of ocular tissue at different depths. These can be used to reconstruct three-dimensional (3D) volumetric images with resolutions of up to 1 micron. The setup most commonly relies on incoherent illumination and a Linnik interferometer, with two microscope objectives in the reference and sample arms. FFOCT has clinical value as an optical microscopy tool because it can capture images of subcellular structures for tissue examination in a noninvasive, efficient manner. Its current utilization has bolstered basic science research in cellular-resolution analysis. D-FFOCT was recently used in the 3D imaging of retinal organoids (ROs). Derived from human-induced pluripotent stem cells, ROs are tissues that form 3D structures such as the developmental optical vesicle and optic cups and, ultimately, the retina. The design and implementation of ROs have been groundbreaking in ophthalmologic research because they closely mimic the structure and functionality of the human retina. Areas of study that benefit from using ROs include retinal transplantation [38, 39], drug delivery, and treating optic nerve diseases [40].

Wide-Field and Ultrawide-Field OCT (WF-OCT and UWF-OCT)

The anatomy of peripheral retinal changes such as the ischemic areas, retinal vein occlusions, the site of retinal breaks, peripheral retinal detachment, retinoschisis, and choroidal lesions can be easily obtained. Pupillary and ciliary shadowing-related artifacts reduce image quality. Increased peripheral retinal curvature, inter-individual variation in retinal curvature, and the need for a very high A-scan rate (> 1 million A scans per second) are the other variables that need to be addressed to obtain analyzably dense wide-field OCT scans. WF-OCT and UWF-OCT provide clinicians with high-quality, noninvasive, in vivo tomographic details of the chorio-retinal layers and aid in the management of these chorioretinal disorders [41].

Handheld and Intraoperative OCT (iOCT)

The most common use for handheld OCT is to help remove barriers to care for OCT imaging in infants and young children [42, 43]. Because commercial OCT systems are typically not designed for infants, the use of a portable OCT can help to address this limitation to identifying vision-threatening diseases in this patient population. In addition, handheld OCT can address limitations to imaging bedridden patients. Handheld OCT usually have two components: a lighter, handheld piece and a bulkier, base unit that contains the light source, reference arm, spectrometer, computer, and computer display. An overall reduction in the size of the base unit and its constituents and transferring the interferometer to a handheld device have led to significant reductions in cost. Low-cost OCT can therefore be available at market prices of approximately USD 5000–7000—i.e., a reduction of > 70% compared to the commercially available OCT devices [44, 45]. Several challenges for a handheld OCT system include operator variability, hand movement, and manual alignment [46].

OCT technology continues to progress in addressing certain limitations observed in the current standards of care for choroidal and retinal diseases. In addition, these advances will help to improve basic science research and our understanding of the pathophysiology of chorioretinal diseases. As evidenced by technologies such as ultrawide-field OCT, these applications can help in detecting and monitoring retinal diseases with OCT capabilities at the periphery. As observed with full-field OCT, this innovation allows for analyzing the individual human retinal ganglion cell axon. OCT advances in currently available technologies, such as intraoperative OCT, allow for gaining further insight into the surgical management of chorioretinal disorders.

12.2.5 Tonometers

Tonometry is a common method for measuring intraocular pressure in patients. According to the World Health Organization, glaucoma is the leading cause of irreversible blindness worldwide. Although intraocular pressure (IOP) is no longer considered to be a defining feature of the disease, its reduction remains the only treatment option for glaucoma. Therefore, the accurate and precise measurement of IOP is the cornerstone of glaucoma.

Intraocular pressure is a highly dynamic physiological parameter with individual circadian rhythms. The main limitation of current tonometry methods remains the static and mostly office-based nature of their measurements. This review provides a brief historical overview of tonometry and discusses current tonometry instruments. In recent years, approaches to 24-hour IOP monitoring have been introduced, and they may become part of routine clinical management in the future [47]. Intraocular pressure (IOP) is an important measurement that needs to be taken during ophthalmic examinations, especially for patients with ocular hypertension, patients with glaucoma, and patients with risk factors for developing glaucoma. Tonometry is a common procedure employed by healthcare professionals to measure intraocular pressure (IOP) by using a calibrated instrument. IOP is important in the diagnosis, screening, and management of ocular hypertension and glaucoma. Instruments measuring intraocular pressure assume that the eye is a closed globe with uniform pressure distributed throughout the anterior chamber and vitreous cavity. The normal range of intraocular pressure is 10–21 mm of mercury (mm HG). This activity shows the multiple old and new methods of tonometry currently available to assess intraocular pressure and identify potential diseases causing abnormalities in eye pressure [48].

Numerous instruments, called tonometers, have been proposed since the nineteenth century to obtain IOP measurements. These instruments can be divided into two main groups in accordance with their respective operating principles: indentation tonometers and applanation tonometers [49–51].

Indentation Tonometers

The principle of indentation tonometry is that a force or a weight will indent or sink into a soft eye. The prototype of the indentation tonometers was the Schiotz tonometer, which was introduced many years ago and is no longer currently used. By using this instrument, the cornea can be indented by a plunger that loads with different weights. The IOP is based on the depth of indentation. The values are shown on a scale ranging from 0 to 20 units, in which the protrusion of the plunger of 0.05 mm represents each unit of measurement. The value indicated on the handle needs to be converted into mm Hg by using a conversion scale. The coefficient of ocular rigidity, which can differ among eyes, would be taken into consideration to obtain corrected measurements of IOP. The Schiotz tonometer is a simple and relatively inexpensive instrument. It is still sometimes used in developing countries [52] and in children under general anesthesia. This tonometer, however, is subject to several sources of error, which include improper positioning on the eye, defective or dirty instruments, and high variability in comparison with other devices and measurements influenced by individual ocular rigidity. Moreover, for this tonometer to work properly, patients must be in a supine position while measurements are taken [53, 54].

Schiotz Tonometer

A Schiotz tonometer consists of a curved footplate, which is placed on the cornea of a supine subject. A weighted plunger attached to the footplate sinks into the cornea to an extent that is indirectly proportional to the pressure in the eye. The plunger sinks into the cornea of a soft eye further than it does into a harder eye. A scale at the top of the plunger gives a reading that depends on how much the plunger sinks into the cornea, and a conversion table converts the scale reading into IOP measured in mm Hg [55].

Pneumotonometers

A pneumotonometer is an applanation tonometer with some aspects of indentation tonometry. It is a 5 mm in diameter and slightly convex and has a silicone tip on the end of a piston that rides on a stream of air. The cornea is indented by the silicone tip. When the cornea and the tip are flat, the pressure pushing forward on the tip is equal to the IOP. The device at this point measures the pressure within the system, and the pressure is displayed in mm Hg. The readings correlate well with Goldmann applanation tonometry within normal IOP ranges [56].

Tono-Pens

The Tono-Pen involves both the applanation process and the indentation process. It is a small handheld, battery-powered portable device. The tonometer has an applanating footplate with a tiny plunger minimally protruding from the center. As the tonometer makes contact with the eye, the plunger encounters resistance from the cornea and the IOP, where a strain gauge generates a record of rising force. At the moment of applanation, the force is shared by the foot plate and the plunger, resulting in a small momentary decrease from the steadily increasing force. This is the point of applanation that is recorded electronically. Multiple readings are averaged. Because the area of applanation is known, the IOP can then be calculated. The readings correlate well with Goldmann tonometry within normal IOP ranges [57].

Pascal Dynamic Contour Tonometers (DCTs)

A Pascal Dynamic Tonometer (Zeimer Ophthalmic Systems AG, Port, Switzerland) utilizes a piezoelectric sensor embedded in the tip of a tonometer to measure the dynamic pulsatile fluctuations in IOP. In contrast to the Goldmann tonometer, measurements with the DCT are reported to be influenced less by corneal thickness and perhaps also by corneal curvature and rigidity. These claims are supported by in vitro and in vivo manometric studies. DCT can also be used to measure the ocular pulse amplitude. Disposable covers are used for each measurement, and the digital display provides a Q value that denotes an assessment of the quality of the measurements [58].

Rebound Tonometers

The newest version of the rebound tonometer, which has a plastic ball on a stainless-steel wire, is held in place by an electromagnetic field in a handheld, battery-powered device. When a button is pushed, a spring rapidly drives the wire and ball forward. When the ball hits the cornea, the ball and wire decelerate; the deceleration is more rapid if the IOP is high and slower if the IOP is low. The speed of deceleration are measured and are converted by the device into IOP. No anesthetic is necessary. This device shows good agreement with Goldmann and Tono-Pen readings. IOP measurements obtained with this tonometer have also shown to be influenced by central corneal thickness, where higher IOP readings are yielded from thicker corneas [59]. This tonometer has been shown to be affected by other biomechanical properties of the cornea, including corneal hysteresis [31, 60].

Applanation Tonometers

Applanation tonometry is based on the Imbert–Fick principle, which states that the pressure inside an ideal sphere—a dry, thin-walled sphere—equals the force necessary to flatten its surface divided by the area of flattening ($P = F/A$, where P = pressure, F = force, and A = area). In applanation tonometry, the cornea is flattened, and the IOP is determined by varying the applanating force or the area flattened [61].

Applanation tonometers are currently considered the most reliable instruments for accurately measuring IOP. Such tonometers use the Imbert–Fick law: $P = F/S$, where P represents pressure, S represents the surface of the flattened area, and F represents the force needed to flatten a fixed corneal area. Apart from the tonometer made by Maklakoff and several other instruments that are no longer currently in use, in which the force is provided by the weight of the tonometer itself, applanation tonometry is based on the area of a flattened cornea, which is calculated and converted into mm Hg. In all instruments of this type, the F value is varied to obtain the proper corneal applanation for a predetermined area. The Goldmann applanation tonometer (GAT) was first invented in 1948 by Hans Goldmann and is still considered the gold standard to date. The tonometer needs to be positioned on a slit lamp [62].

A truncated cone, with a 7.35 mm^2 surface area and a dimeter of 3.06 mm, illuminated by a blue light, is pushed on the center of the anesthetized cornea. A doubling prism embedded in the cone divides the circular meniscus on the surface of the flattened cornea, e, into two arcs, which need to be aligned to obtain a precise and standardized applanation. The force needed to flatten the corresponding surface of the cornea is directly proportional to the IOP, expressed in mm Hg, which can be directly read in the scale of the measuring drum or in the posterior window for the digital version.

Goldmann and Perkins Applanation Tonometers

A Goldmann applanation tonometer measures the force necessary to flatten a corneal area that is 3.06 mm in diameter. At this diameter, the resistance of the cornea to flattening is counterbalanced by the capillary attraction of the tear film meniscus for the tonometer head. The IOP (in mm Hg) equals the flattening force (in grams) multiplied by 10. Fluorescein dye is placed on the patient's eye to highlight the tear film. A split-image prism is used to divide the image of the tear meniscus into a superior arc and an inferior arc. The intraocular pressure is taken when these arcs are aligned such that their inner margins just touch.

Applanation tonometry measurements are affected by central corneal thickness (CCT). When Goldmann designed his tonometer, he estimated an average corneal thickness of 520 microns to cancel the opposing forces of surface tension and corneal rigidity, which would allow for indentation. Instead, wide variation exists among individuals' respective corneal thicknesses. A thicker CCT may give an artificially high IOP reading, whereas a thinner CCT may give an artificially low reading.

Other errors that may affect the accuracy of readings from a Goldmann tonometer include excessive or insufficient fluorescein in the tear film that affects the thickness of the overlapping arcs, high astigmatism, an irregular or scarred cornea, pressure from a finger on the eyelid while taking the measurement, and breath holding or a Valsalva maneuver by the patient during measurement.

The Perkins tonometer is a portable Goldmann applanation tonometer that can be used with the patient in either the upright or the supine position [63].

Noncontact Tonometry

Air Puff Tonometers

In air puff tonometry, the applanating force is a column of air that is emitted with gradually increasing intensity. At the point of corneal flattening, the air column is shut off, and the force at that moment is recorded and converted into mm Hg. Readings from these machines may underestimate IOP at high ranges and overestimate IOP at low ranges, in contrast to those from Goldmann applanation tonometers. A minimum of three readings should be averaged to estimate the mean IOP because IOP varies during the cardiac cycle.

Ocular Response Analyzers

An ocular response analyzer is a newer type of noncontact tonometer. This device also uses a column of air of increasing intensity as the applanating force. An ocular response analyzer notes the moment of applanation, but the air column continues to emit with increasing intensity until the cornea is indented. The force of the air column then decreases until the cornea has returned to a point of applanation. The difference in the pressures at the two applanation points is a measure of the corneal elasticity (e.g., hysteresis). Mathematical equations can be used to "correct" the applanation point for high or low elasticity. This "corrected" IOP is thought to be less dependent on corneal thickness than other forms of applanated pressures [64].

12.2.6 Aberrometers

An aberrometer is a diagnostic device that measures refractive aberrations of the eye. It is used in laser eye surgery to ensure high accuracy. This devices passes light through the eye and then measures that light as it exits the eye. This enables an ophthalmologist to measure wavefront—the change in the shape of the front of light waves as they exit the cornea. A wavefront aberrometer measures every variation of the eye, from the cornea to the retina, helping eyecare practitioners to diagnose and cure eye ailments. An aberrometer helps to quantify glare, halos, and night vision disturbances and helps doctors screen patients whose vision may be compromised by these conditions. It also allows physicians to measure highly distorted eyes. A wavefront aberrometer records data from several spots on the surface of the cornea, producing a map of imperfections there. Information from the aberrometer is used to provide customized laser eye surgery. Wavefront aberrometry is used not only to determine causes of visual disturbances but also to program ablations in wavefront-guided procedures. Using wavefront aberrometry to conduct such procedures is becoming increasingly common because they are much more effective than conventional methods. Wavefront aberrometers are also sometimes used therapeutically, to correct spherical aberration in patients with night vision disturbances [65]. One of the most powerful clinical applications of aberrometry is wavefront-guided refractive surgery. This concept led to a paradigm shift in refractive error correction, and the same ideas were applied to design the power and shape of intraocular and contact lenses. Other applications include the diagnosis of irregular astigmatism and the assessment of the optical quality of the eye. Because the higher-order aberrations of the eye are expressed as total root-mean-square errors, a set of coefficients for the Zernike terms, Strehl ratio, point spread functions, modulation transfer functions, and other types of metrics can be determined; hence, the deterioration in the quality of vision can be easily estimated. Simulations of the retinal images are also useful to understand some of the symptoms in patients with irregular astigmatism. With corneal topographic analyses, the origin of irregular astigmatism from the cornea or inside the eye, or both, can be specified via aberrometry. Regular astigmatism was once considered to be a refractive error that could not be corrected with conventional glasses. Corneal topography was used to diagnose corneal irregular astigmatism, and rigid gas permeable (RGP) contact lenses or corneal transplantations were used to treat these eyes. On the other hand, irregular astigmatism that was due to having a crystalline lens or an intraocular lens (IOL) was not studied in detail because of the difficulties, until recently, in measuring irregular astigmatism caused by the internal optics [66].

Types of Abberometers

Hartmann-Shack Wavefront Sensor (Outgoing Reflection Aberrometery)

A Hartmann-Shack wavefront sensor is used to measure the wave aberrations of the human eye by sensing the wavefront emerging from the eye produced by the retinal reflection of a focused light spot on the fovea. It comprises a microlenslet array subdividing the reflected wave of light into multiple focused beams. The incident plane wave results in a square grid of spots captured by an image detector in the focal plane of the lenslet array. The distorted wavefront causes lateral displacements of the spots on the Charge coupled device (CCD) array. Thus, from the spot pattern, the shape of the incident wavefront can be reconstructed [67, 68].

Tscherning Aberrometers

Tscherning aberrometers, on the other hand, measure light as it enters the eye. A Tscherning aberrometer shines a fine bundle of red laser beams into the eye. These beams form a pattern on the retina. As the beams traverse through the eye's optical system, they are bent or distorted by the optical aberrations in the lens and cornea. An image of the laser grid pattern on the retina is captured by a highly sensitive video camera and then analyzed. Because the aberrometer is based on Tscherning's principle, measurements of the wavefront aberrations of human eyes are available with high accuracy and reproducibility for standard diagnostic investigations. Measurements are presented in terms of Zernike coefficients and as height maps that can be directly converted into ablation profiles for wavefront-guided laser treatments. The Tscherning aberrometer is a simple optical device with high accuracy appropriate for routine clinical investigations into optical aberrations of the human eye [69].

In 1894, Tscherning published his investigations into optical aberrations of the human eye. He concluded that such optical aberrations may deteriorate vision, but unfortunately, there seemed to be no way to correct them. This situation has since changed thanks to the introduction of Argon fluride laser (ArF) excimer lasers for corneal laser surgery, so now these aberrations can be corrected via wavefront-guided laser in situ keratomileusis (LASIK) or photorefractive keratectomy (PRK) [70].

Ray Tracing (Retinal Imaging Aberrometry)

Ray tracing is a technique that measures the position of a thin laser beam projected onto the retina. The beam is directed into the eye parallel to the visual axis. Each entrance point provides its own projection on the retina. A set of entrance points forms a set of projections. From these data, a refraction map and the point spread function of the eye are reconstructed. The total time of scanning over the whole aperture of the eye takes between 10 and 20 ms and depends on the number of test points at the eye entrance and on the number of independent measurements for each point. Configuration options for the scanning pattern can be chosen by the operator. It may contain 60–400 points, each checked one to five times [71].

A ray-tracing aberrometer combines both the wavefront aberrometry and Placido disc–based corneal topography. Hence, it has an advantage over other aberrometers in providing individual results for corneal and internal aberrations, in addition to total aberrations. Recently, iTrace has gained immense popularity for preoperative evaluation and planning for cataract and refractive surgery patients. Ray-tracing aberrometry is a flexible technology for eye analysis. It can be adapted to any laser technique for vision correction. Its further development should be oriented toward laser-linked applications in refractive surgery [72].

Aberrometry based on the principle of ray tracing is a two-step, serial technique that uses forward projection and that can be used either subjectively or objectively. The ray-tracing method uses a laser beam parallel to the line of sight through the pupil. It measures the exact location where the laser beam reaches the retina via the retroreflected light captured by reference lineal sensors X and Y. Local aberrations in the path of the laser beam through the cornea and the internal structures cause a shift in the location on the retina. Once the first position has been determined, the laser beam is shifted to another position, which is then located in the retina. This process continues until several separated points are projected into the entrance pupil. This way, a connection is obtained between the direction that the light beams have taken while entering and leaving, allowing for a reconstruction of the real wavefront error. This principle measures "forward" aberrations of the light that goes through the eye. It is more physiological to measure these anterior aberrations as the natural trajectory of the light in the eye is analyzed [73].

12.2.7 Phaco

The phacoemulsification of the lens using a "phaco" machine can be approached via many methods. Pump systems between phaco machines vary, from peristaltic to Venturi to hybrid systems. Additional technologies that vary the delivery of phaco energy, such as torsional ultrasound, have also been adapted.

Typical phaco instrumentation consists of a phaco handpiece and a phaco tip. The phaco tip serves to both deliver ultrasound energy and aspirate (i.e., vacuum in material) from its open end. Surrounding the phaco tip is a sleeve, which is often of a rubbery consistency. The sleeve allows for irrigation around the tip, lowering the resulting temperature to avoid wound burns. Irrigation typically exits the sleeve from side ports. The sleeve can be adjusted to expose the amount of phaco tip that the surgeon desires. Some surgeons prefer to expose a lot of phaco tip; others minimize tip exposure. The sleeve can act as a barrier to tip penetration into the lens material. Beginning surgeons are often advised to minimize tip exposure, to increase safety. The irrigating ports of the tip can also be aligned according to the surgeon's preferences. Holding the phaco handpiece such that the phaco tip is bevel up, the irrigation ports on the sleeve are frequently oriented horizontally, such that irrigation is aimed sideways and not directly at the corneal endothelium. The irrigating ports are placed at 45-degree angles, such that the maneuvers require rotating the bevel, which can still be accomplished with minimal direct irrigation aimed at the endothelium.

Also, instrument designs are quite varied. Traditional divide-and-conquer techniques often use a Drysdale manipulator. This instrument has a paddle-like tip to facilitate surface-area contact within a groove. Other instruments include choppers, such as the Seibel chopper, which we often use. The Seibel has a rounded tip, which avoids bringing any sharp edges near the capsular bag. The Seibel can also serve as a divide-and-conquer second instrument.

Settings for phacoemulsification have become somewhat more complex with the advent of torsional ultrasound systems. The roles of torsional energy, traditional coaxial energy, and pulse/burst modes are often intertwined in the more efficient modern phaco systems.

Irrigation

When no viscoelastic is in the anterior chamber, maintaining irrigation is crucial. Failure to do so will allow fluid egress through the open wound and collapse the chamber. Continuous irrigation is sometimes needed and maintained by the continuous inflow of opthalmic viscosurgical device (OVD). This ensures chamber maintenance and reduces the level of fluctuation in intraocular pressure.

Irrigation and Aspiration

In addition to irrigation, aspiration is also applied at the phaco tip. This vacuum force allows for lens material to be brought to the tip and held there upon the occlusion of the tip with material. The flow of material moves toward the phaco tip.

Irrigation, Aspiration, and Phacoemulsification

The phaco settings will determine in which manner the energy is delivered at the phaco tip. In the case of emulsifying the lens material, it thus should not be entered unless there is lens material at the tip. A phaco without lens material can deliver phaco energy to nearby structures, including a capsule or an iris, resulting in unnecessary damage. Phaco power can be applied as "fixed" such that a consistent level of energy is maintained throughout the process. Any additional modulation of this energy can be applied by using pulse and burst modes or newer torsional technologies.

Technique

The tip of phaco hand piece is sharp, and thus, any contact with fragile structures such as the iris, lens capsule, or corneal endothelium can be damaging. Utmost care must be taken during tip position in the eye. When phaco energy is applied, this energy can be transmitted beyond the tip. Therefore, even when the tip is in proximity to, not necessarily contact with, delicate structures, it can have adverse effects.

The phaco tip can be held too far away, such as an instrument's handle. This can result in less control, by translating small hand movements into larger displacements at the instrument tip. The grasping of the instruments closer to the tip, or "choking up," can help achieve better control.

A divide-and-conquer technique consists of exactly those tasks: *Divide* the lens (grooving and cracking), then *conquer* through quadrant removal. During the grooving and cracking portion of phaco, aspiration and vacuum settings are low so that phaco energy can be modulated as needed on the basis of lens density. These settings are often called "sculpt" settings. During quadrant removal, higher aspiration and vacuum settings are typically desirable to facilitate lens removal with less energy usage [74].

Lasers

Ophthalmology was the first medical specialty to utilize lasers, where the first report described using a ruby laser to treat ocular lesions almost a year after the invention of the laser, and it still has the most laser procedures of any specialty with the use of lasers, permeating all subspecialties both diagnostically and therapeutically [2]. Therefore, understanding the principles of lasers is integral to the foundational knowledge of ophthalmologists. Although lasers are used in various therapeutic techniques, they are also integral parts of diagnostic imaging in ophthalmology. For instance, scanning laser ophthalmoscopy (SLO) uses confocal laser scanning microscopy to diagnostically image the retina or cornea. It uses lasers to rapidly scan a sample whose reflected light is imaged through a pinhole that suppresses reflections outside of those in the focal plane [75].

LASIK

Laser-assisted in situ keratomileusis (LASIK) eye surgery is the best-known and most frequently performed laser refractive surgery to correct vision problems. LASIK can be an alternative to glasses or contact lenses.

During LASIK surgery, a special type of cutting laser is used to precisely change the shape of the dome-shaped clear tissue at the front of an eye (cornea) to improve vision. In eyes with normal vision, the cornea precisely bends (refracts) light onto the retina at the back of the eye. But with nearsightedness (myopia), farsightedness (hyperopia), or astigmatism, the light is bent incorrectly, resulting in blurred vision. Glasses or contact lenses can correct vision, but reshaping the cornea itself will also provide the necessary refraction [76].

The refractive errors of the eyes are as follows:

- In nearsightedness (myopia), an eyeball is slightly longer than normal or the cornea curves too sharply: Light rays focus in front of the retina and blur distant vision. It can see objects that are close fairly clearly, but not those that are far away.

- In farsightedness (hyperopia), an eyeball is a shorter than average or a cornea is too flat: Light focuses behind the retina instead of on it. This makes near vision, and sometimes distant vision, blurry.
- In astigmatism, the cornea curves or flattens unevenly, resulting in astigmatism, which disrupts focus in near and distant vision [77].

LASIK and other forms of laser refractive surgery, such as PRK and LASEK, all use a highly specialized excimer laser to reshape the cornea and correct refractive errors, including myopia (nearsightedness), hyperopia (farsightedness), and astigmatism.

Excimer lasers have revolutionized the field of laser eye surgery. And over several decades, advances in excimer laser technology have increased the safety, efficacy, and predictability of corneal refractive surgery.

Excimer lasers can remove, or "ablate," microscopic amounts of tissue from the cornea with a very high degree of accuracy and without damaging the surrounding corneal tissue. The following excimer lasers have been approved by the US Food and Drug Administration (FDA) for use in vision correction surgery performed in the United States:

- The STAR S4 IR Excimer Laser System and the iDesign Advanced WaveScan Studio System (Johnson & Johnson)
- The Allegretto WAVE Eye-Q Excimer Laser System (Alcon)
- The TECHNOLAS 217Z Zyoptix System for personalized vision correction (Bausch + Lomb)
- The Nidek EC-5000 with Navex Quest customization (Nidek)
- The MEDITEC MEL 80 Excimer Laser System (Zeiss) [76]

Femtosecond Laser

The emergence of femtosecond laser technology has revolutionized lamellar flap creation. The refractive surgery community continues to debate whether to use either a blade or a laser to create corneal flaps. The femtosecond laser system uses a neodymium-doped aluminium garnet (YAG) laser operating in the infrared wavelength to produce ultrashort pulses of energy to create adjacent areas of microcavitation (the separation of tissue at the molecular level) at a specified depth in the cornea. The created bubble leaves a cavitation volume of 2–3 cubic μm. Thousands of these tiny bubbles, created in a raster pattern across the cornea, define the interface plane between the flap and the stromal bed. Bubbles are then stacked, starting around the edge of the interface and proceeding up through the epithelium to the corneal surface, creating the side cut and completing the flap creation [77].

The benefits of femtosecond laser–assisted cataract surgery include shorter cumulated phacoemulsification time and less endothelial cell loss, the perfect centration of the capsulotomy, and an opportunity to perform precise femtosecond laser–assisted accurate keratotomy incisions. The major disadvantages of femtosecond laser–assisted cataract surgery are the high cost of the laser and the disposables for surgery, intraoperative capsular complications specific to femtosecond laser–assisted cataract surgery, (corneal burn) the risk of intraoperative miosis, and its deep learning curve [78].

References

1. https://www.bsigroup.com/en-IN/medical-devices/technologies/ophthalmic-devices/
2. Wang X, Jacobs DS. Contact lenses for ocular surface disease. Eye Contact Lens: Sci Clin Pract. 2022;48(3):115–8. https://doi.org/10.1097/ICL.0000000000000879.
3. Gurnani B, Kaur K. Contact lenses. [Updated 2022 Dec 6]. In: StatPearls [Internet]. Treasure Island (FL): StatPearls Publishing; 2023. Available from: https://www.ncbi.nlm.nih.gov/books/NBK580554/
4. Kumar P, Mohamed A, Bhombal F, Dumpati S, Vaddavalli PK. Prosthetic replacement of the ocular surface ecosystem for corneal irregularity: visual improvement and optical device characteristics. Cont Lens Anterior Eye. 2019;42(5):526–32. [PubMed].
5. Alipour F, Khaheshi S, Soleimanzadeh M, Heidarzadeh S, Heydarzadeh S. Contact lens-related complications: a review. J Ophthalmic Vis Res. 2017;12(2):193–204. [PMC free article] [PubMed].
6. Musgrave CSA, Fang F. Contact lens materials: a materials science perspective. Materials (Basel). 2019;12(2). [PMC free article] [PubMed].
7. Jacobs DS, Carrasquillo KG, Cottrell PD, Fernández-Velázquez FJ, Gil-Cazorla R, Jalbert I, Pucker AD, Riccobono K, Robertson DM, Szczotka-Flynn L, Speedwell L, Stapleton F. CLEAR – medical use of contact lenses. Cont Lens Anterior Eye. 2021;44(2):289–329. https://doi.org/10.1016/j.clae.2021.02.002. Epub 2021 Mar 25. PMID: 33775381.
8. https://www.contactlensjournal.com/article/S1367-0484(21)00016-3/fulltext#secsect0185
9. Kumari R, Srivastava MR, Garg P, Janardhanan R. Intraocular lens technology – a review of journey from its inception. Ophthalmol Res: Int J. 2019;11(3):1–9. Article no.OR.53726 ISSN: 2321-7227.
10. Arthur SN, et al. Effect of heparin surface modification in reducing silicone oil adherence to various intraocular lenses. J Cataract Refract Surg. 2001;27(10):1662–9.
11. https://www.changcataract.com/cataract-information-center-los-altos/selecting-your-lens-implant/intraocular-lenses-for-cataract-surgery/
12. https://www.bettervisionguide.com/multifocal-iols/
13. https://www.la-sight.com/services/cataracts/lens-implant-choices/aspheric-iols/
14. Gurnani B, Kaur K. Autorefractors. [Updated 2022 Dec 6]. In: StatPearls [Internet]. Treasure Island (FL): StatPearls Publishing; 2023. Available from: https://www.ncbi.nlm.nih.gov/books/NBK580520/
15. Fitzke FW, Hayes BP, Hodos W, Holden AL. Electrophysiological optometry using Scheiner's principle in the pigeon eye. J Physiol. 1985. [PubMed PMID: 4093879].
16. Chen J, Lu F, Qu J, Li LP. [The development of a polarized vernier optometer for tonic accommodation measurement]. Zhongguo Yi Liao Qi Xie Za Zhi = Chin J Med Instrum. 2002. [PubMed PMID: 16104153].
17. Otero C, Aldaba M, Pujol J. Clinical evaluation of an automated subjective refraction method implemented in a computer-controlled motorized phoropter. J Optom. 2019. [PubMed PMID: 30389250].
18. Hervella L, Villegas EA, Prieto PM, Artal P. Assessment of subjective refraction with a clinical adaptive optics visual simulator. J Cataract Refract Surg. 2019. [PubMed PMID: 30309774].

19. Polse DA, Kerr KE. An automatic objective optometer. Description and clinical evaluation. Arch Ophthalmol. 1975. [PubMed PMID: 1094996].
20. Huang D, Swanson EA, Lin CP, Schuman JS, Stinson WG, Chang W, Hee MR, Flotte T, Gregory K, Puliafito CA, et al. Optical coherence tomography. Science. 1991;254:1178–81.
21. Boned-Murillo A, Albertos-Arranz H, Diaz-Barreda MD, Orduna-Hospital E, Sánchez-Cano A, Ferreras A, Cuenca N, Pinilla I. Optical coherence tomography angiography in diabetic patients: a systematic review. Biomedicine. 2021;10:88.25.
22. Majumdar S, Tripathy K. Macular hole. In: StatPearls. Treasure Island (FL): StatPearls; 2022.
23. Metrangolo C, Donati S, Mazzola M, Fontanel L, Messina W, D'Alterio G, Rubino M, Radice P, Premi E, Azzolini C. OCT biomarkers in neovascular age-related macular degeneration: a narrative review. J Ophthalmol. 2021;2021:9994098.
24. Dhurandhar DS, Singh SR, Sahoo NK, Goud A, Lupidi M, Chhablani J. Identifying central serous chorioretinopathy biomarkers in coexisting diabetic retinopathy: a multimodal imaging study. Br J Ophthalmol. 2020;104:904–9.
25. Nassisi M, Fan W, Shi Y, Lei J, Borrelli E, Ip M, Sadda SR. Quantity of intraretinal hyperreflective foci in patients with intermediate age-related macular degeneration correlates with 1-year progression. Investig Ophthalmol Vis Sci. 2018;59:3431–9.
26. Lei J, Balasubramanian S, Abdelfattah NS, Nittala MG, Sadda SR. Proposal of a simple optical coherence tomography-based scoring system for progression of age-related macular degeneration. Graefes Arch Clin Exp Ophthalmol. 2017;255:1551–8.
27. Nassisi M, Lei J, Abdelfattah NS, Karamat A, Balasubramanian S, Fan W, Uji A, Marion KM, Baker K, Huang X, et al. OCT risk factors for development of late age-related macular degeneration in the fellow eyes of patients enrolled in the HARBOR study. Ophthalmology. 2019;126:1667–74.
28. Shu X, Beckmann L, Zhang H. Visible-light optical coherence tomography: a review. J Biomed Opt. 2017;22:1–14.
29. Liang J, Williams DR, Miller DT. Supernormal vision and high-resolution retinal imaging through adaptive optics. J Opt Soc Am A Opt Image Sci Vis. 1997;14:2884–92.
30. Roorda A, Romero-Borja F, Donnelly Iii W, Queener H, Hebert T, Campbell M. Adaptive optics scanning laser ophthalmoscopy. Opt Express. 2002;10:405–12.
31. Chi WS, Lam A, Chen D, et al. The influence of corneal properties on rebound tonometry. Ophthalmology. 2008;115:80–4.
32. Pircher M, Hitzenberger CK, Schmidt-Erfurth U. Polarization sensitive optical coherence tomography in the human eye. Prog Retin Eye Res. 2011;30:431–51.
33. Ueno Y, Mori H, Kikuchi K, Yamanari M, Oshika T. Visualization of anterior chamber angle structures with scattering- and polarization-sensitive anterior segment optical coherence tomography. Transl Vis Sci Technol. 2021;10:29.
34. Schutze C, Teleky K, Baumann B, Pircher M, Gotzinger E, Hitzenberger CK, Schmidt-Erfurth U. Polarisation-sensitive OCT is useful for evaluating retinal pigment epithelial lesions in patients with neovascular AMD. Br J Ophthalmol. 2016;100:371–7.
35. Schranz M, Roberts PK, Motschi AR, Hollaus M, Mylonas G, Sacu S, Pircher M, Hitzenberger CK, Schmidt-Erfurth U. Tracking of fibrosis growth in neovascular age related macular degeneration. Investig Ophthalmol Vis Sci. 2022;63:1025–F0272.
36. Imaging that enlightens. Deeper insights into retinal structures with High-Resolution OCT. Ophthalmologist. 2020. Available online: https://theophthalmologist.com/subspecialties/imaging-that-enlightens. Accessed on 1 Aug 2022.
37. Spaide RF, Lally DR. High resolution spectral domain optical coherence tomography of multiple evanescent white dot syndrome. Retin Cases Brief Rep. 2021. 1;17(3):227–230.
38. Singh R, Cuzzani O, Binette F, Sternberg H, West MD, Nasonkin IO. Pluripotent stem cells for retinal tissue engineering: current status and future prospects. Stem Cell Rev Rep. 2018;14:463–83.
39. Ahmad I, Teotia P, Erickson H, Xia X. Recapitulating developmental mechanisms for retinal regeneration. Prog Retin Eye Res. 2020;76:100824.

40. Wright LS, Pinilla I, Saha J, Clermont JM, Lien JS, Borys KD, Capowski EE, Phillips MJ, Gamm DM. VSX2 and ASCL1 are indicators of neurogenic competence in human retinal progenitor cultures. PLoS One. 2015;10:e0135830.
41. Ong J, Zarnegar A, Corradetti G, Singh SR, Chhablani J. Advances in optical coherence tomography imaging technology and techniques for choroidal and retinal disorders. J Clin Med. 2022;11(17):5139. https://doi.org/10.3390/jcm11175139.
42. Nicholson R, Osborne D, Fairhead L, Beed L, Hill CM, Lee H. Segmentation of the foveal and parafoveal retinal architecture using handheld spectral-domain optical coherence tomography in children with Down syndrome. Eye. 2022;36:963–8.
43. Maldonado RS, Izatt JA, Sarin N, Wallace DK, Freedman S, Cotten CM, Toth CA. Optimizing hand-held spectral domain optical coherence tomography imaging for neonates, infants, and children. Investig Ophthalmol Vis Sci. 2010;51:2678–85.
44. Malone JD, El-Haddad MT, Yerramreddy SS, Oguz I, Tao YK. Handheld spectrally encoded coherence tomography and reflectometry for motion-corrected ophthalmic optical coherence tomography and optical coherence tomography angiography. Neurophotonics. 2019;6:041102.
45. Chopra R, Wagner SK, Keane PA. Optical coherence tomography in the 2020s-outside the eye clinic. Eye. 2021;35:236–43.
46. Wang KL, Chen X, Stinnett S, Tai V, Winter KP, Tran-Viet D, Toth CA. Understanding the variability of handheld spectral-domain optical coherence tomography measurements in supine infants. PLoS One. 2019;14:e0225960.
47. Brusini P, Salvetat ML, Zeppieri M. How to measure intraocular pressure: an updated review of various tonometers. J Clin Med. 2021;10(17):3860. https://doi.org/10.3390/jcm10173860. PMID: 34501306; PMCID: PMC8456330.
48. Bader J, Zeppieri M, Havens SJ. Tonometry. In: StatPearls [Internet]. Treasure Island (FL): StatPearls Publishing; 2022. 2022 Jul 14. PMID: 29630277 Bookshelf ID: NBK493225, 65.TonometryStatPearlsJuly2022.pdf.
49. Kniestedt C, Punjabi O, Lin S, Stamper RL. Tonometry through the ages. Surv Ophthalmol. 2008;53:568–91. https://doi.org/10.1016/j.survophthal.2008.08.024.
50. Stamper RL. A history of intraocular pressure and its measurement. Optom Vis Sci. 2011;88:E16–28. https://doi.org/10.1097/OPX.0b013e318205a4e7.
51. Brusini P. Intraocular pressure and its measurement. In: Choplin NT, Traverso CE, editors. Atlas of glaucoma. 3rd ed. Boca Raton: CRC Press; 2014. p. 29–36.
52. Nagarajan S, Velayutham V, Ezhumalai G. Comparative evaluation of applanation and indentation tonometers in a community ophthalmology setting in Southern India. Saudi J Ophthalmol. 2016;30:83–7. https://doi.org/10.1016/j.sjopt.2015.11.002.
53. Lasseck J, Jehle T, Feltgen N, Lagrèze WA. Comparison of intraocular tonometry using three different non-invasive tonometers in children. Graefes Arch Clin Exp Ophthalmol. 2008;246:1463–6. https://doi.org/10.1007/s00417-008-0863-y.
54. Ohana O, Varssano D, Shemesh G. Comparison of intraocular pressure measurements using Goldmann tonometer, I-care pro, Tonopen XL, and Schiotz tonometer in patients after Descemet stripping endothelial keratoplasty. Indian J Ophthalmol. 2017;65:579–83. https://doi.org/10.4103/ijo.IJO_31_17.
55. IOP and Tonometry – EyeWiki (aao.org).
56. Bhan A, Browning AC, Shah S, et al. Effect of corneal thickness on intraocular pressure measurements with the pneumotonometer, Goldmann applanation tonometer, and Tono-Pen. Invest Ophthalmol Vis Sci. 2002;43(5):1389–92.
57. Stamper R. A history of intraocular pressure and its measurement. Optom Vis Sci. 2011;88(1):E16–28.
58. Kniestedt C, Lin S, Choe J, et al. Clinical comparison of contour and applanaion tonometry and their relationship to pachymetry. Arch Ophthalmol. 2005;123:1532–7.
59. Pakrou N, Gray T, Mills R, et al. Clinical comparison of the Icare tonometer and Goldmann applanation tonometry. J Glaucoma. 2008;17(1):43–7.

60. Poostchi A, Mitchell R, Nicholas S, et al. The Icare rebound tonometer: comparisons with Goldmann tonometry, and influence of central corneal thickness. Clin Exp Ophthalmol. 2009;37:687–91.
61. American Academy of Ophthalmology. Basic and clinical science course section 10: glaucoma. Singapore: American Academy of Ophthalmology; 2008.
62. Goldmann H, Schmidt T. Über Applanationstonometrie. Acta Ophthalmol. 1957;134:221–42.
63. Herndon LW, Choudhri SA, Cox T, Damji KF, Shields MB, Allingham RR. Central corneal thickness in normal, glaucomatous, and ocular hypertensive eyes. Arch Ophthalmol. 1997;115:1137–41.
64. Wolfs RC, Klaver C, Vingerling JR, Grobbee DE, Hofman A, de Jong PT. Distribution of central corneal thickness and its association with intraocular pressure: the Rotterdam study. Am J Ophthalmol. 1997;123:767–72.
65. Aberrometer diagnostic device | Refractive aberrations | Aberrometer laser eye surgery (accuvision.co.uk).
66. Maeda N. Clinical applications of wavefront aberrometry – a review. Clin Exp Ophthalmol. 2009;37(1):118–29.
67. Bille JF. Preoperative simulation of outcomes using adaptive optics. J Refract Surg. 2000;16(5):S608–10.
68. Bille JF, Harner CFH, Loesel FF. Aberration-free refractive surgery: new frontiers in vision. 2nd ed. Berlin: Springer; 2004.
69. Kaemmerer M, Mrochen M, Mierdel P, Krinke H-E, Seiler T. Clinical experience with the Tscherning aberrometer. J Refract Surg. 2013;16(5):S584–7.
70. Molebny VV, Panagopoulou SI, Molebny SV, Wakil YS, Pallikaris IG. Principles of ray tracing aberrometry. J Refract Surg. 2000;16(5):S572–5. https://doi.org/10.3928/1081-597X-20000901-17. PMID: 11019876.75.
71. Sinha A, Goel S, Gupta V, Kumawat D, Sahay P. iTrace – a ray tracing aberrometer. Delhi J Ophthalmol. 2019;30(1):72–5. https://doi.org/10.7869/djo.489.
72. Rozema JJ, Dirk EM, Van Dyck PD, Tassignon MJ. Clinical comparison of 6 aberrometers. Part I: technical specifications. J Cataract Refract Surg. 2005;31:1114–27.
73. Navarro R, Moreno-Barriuso E. Laser ray-tracing method for optical testing. Opt Lett. 1999;24:951–3.
74. Phacoemulsification. Department of Ophthalmology Academic Resources | Boston University (bu.edu).
75. LASIK eye surgery – Mayo Clinic.
76. LASIK lasers: which laser is best? – All about vision.
77. Creating LASIK flaps: femtosecond laser vs. mechanical microkeratome – American Academy of Ophthalmology (aao.org).
78. Kanclerz P, Alio JL. The benefits and drawbacks of femtosecond laser–assisted cataract surgery. Eur J Ophthalmol. 2021;31(3):1021–30. https://doi.org/10.1177/1120672120922448. Epub 2020 Jun 7. PMID: 32508179.

Chapter 13
Significant Risk Medical Devices – Orthopedics and Restorative

T. S. Subashini and K. S. Sridevi Sangeetha

13.1 Introduction

The medical specialty of orthopaedics covers a wide range of issues involving the musculoskeletal system [1]. Despite being referred to be a type of surgery in medical circles, the majority of orthopaedic operations don't entail intrusive techniques [2]. Numerous orthopaedic operations only involve muscle manipulations and other methods of treating, preventing, and expediting recovery from an accident or illness [3]. Additionally, orthopaedic professionals can offer crucial support and care to patients in need of elbow, knee, and hip replacements following an injury or sickness [1].

13.2 Devices Used for Orthopaedics Fixation

Accident victims and those with joint conditions like osteoarthritis, rheumatoid arthritis, and post-traumatic arthritis may require surgery involving implants like total hip and knee replacements [4]. Additionally, temporary fracture-fixing tools

T. S. Subashini (✉)
Department of Pharmacology, SRM Dental College, SRM Institute of Science & Technology, Ramapuram, Chennai, India

Department of Pharmacology, SRM Dental College, Ramapuram, Bharathi Salai, Chennai, India
e-mail: subashis@srmist.edu.in

K. S. Sridevi Sangeetha
Faculty of Allied Health Sciences, Meenakshi Academy of Higher Education and Research, Chennai, India

P. S. Timiri Shanmugam et al. (eds.), *Significant and Nonsignificant Risk Medical Devices*, https://doi.org/10.1007/978-3-031-52838-5_13

and parts such as plates, screws, pins, wires, and nails are included in orthopaedic implants. Orthopaedic implants are intended to rebuild the strength and functionality of broken joints and bones [5]. The biomaterials should have the desired mechanical qualities, wear resistance, corrosion resistance, biocompatibility, and occasionally osseointegration in order to create safe implants with a long lifespan and without eliciting rejection [6]. The US Food and Drug Administration defines biocompatibility as the property that the materials have no detectable negative effects on the host [7]. According to Hallab et al. (2005), metal ions released from metallic implants to surrounding tissues may trigger the immune system's hypersensitive response and implant failure [8]. Therefore, when designing and making new metallic implants, harmless elements should be chosen as alloying components. By improving the corrosion resistance, it is possible to keep the released amounts of trace elements that are naturally present in the human body to relatively low levels during the lifespan of the implants [9].

Large numbers of foreign bodies are ingested by the local immune system from polymer, metal, or ceramic implants' wear, corrosion, or a combination of these two processes. This causes inflammation to spread to the bone-implant interface. This can impair the typical operations of orthopaedic implants by causing aseptic osteolysis and eventual implant loosening [10].

Therefore, excellent corrosion and wear resistance in the physiological environment and the chosen alloying components in the implant materials are essential for good biocompatibility [11].

Permanent and temporary orthopaedic implants are the two categories of orthopaedic implants.

13.3 Orthopaedic Implants That Are Permanent

The hip, knee, ankle, shoulder, elbow, wrist, and finger joints are only a few of the total joints that can be therapeutically replaced [12]. Metals, ceramics, and polymers are frequently utilised in these permanent orthopaedic implants, which are meant to last the patients' whole lives. In particular, the development and acceptance of joint prosthesis for the hip and knee have accelerated recently [13].

13.4 Orthopaedic Implants Used for Temperary Period

Temporary orthopaedic implants are another type that are required to mend damaged or fractured bones while they heal. Plates, screws, pins, wires, and intramedullary nails are examples of temporary orthopaedic implants that are meant to last only long enough for bones to heal [5].

13.5 Surgical Screws Used in Orthopaedics

One of the current tenets of orthopaedic fixing is that pressing the fracture fragments tightly together promotes bone healing. Along with their primary purpose of keeping the fracture in anatomic alignment, several orthopaedic devices are made to do just that. One of the most common hardware components is the screw [14].

Without orthopaedic screws, orthopaedic procedures are not possible. One of the principles of the orthopaedic fixing technique is the use of orthopaedic screws. If the fractures are pushed together tightly by orthopaedic screws, the bone heals better and more quickly [15]. The benefit of these screws is that they narrow the space between the bones, which reduces pressure on the orthopaedic implant as a result. Orthopaedic screws are a common type of surgical hardware that can be used alone to provide fixation or in conjunction with other devices to complete the surgery [16].

13.6 Types of Screws Include

1. *Cortical screws*, which are intended to secure a cortical bone and typically have very thin threads running the length of their shaft. Compared to cancellous screws, these sorts of screws have a smaller pitch. Consequently, a cortical orthopaedic screw has many more threads than a cancellous screw [15, 16].
2. *Cancellous screws* have smoother, coarser threads with an unthreaded region that allows them to function similarly to lag screws. Cortical screws are shorter than cancellous screws [14, 15].
3. Another often used screw type is *cannulated screws*. Due to their hollow shaft, cannulated screws are what they are called. It should be emphasised that compared to other screws, these orthopaedic screws have many more benefits [14, 15].
4. *Herbert orthopaedic screws* can be compared with Acutrak screws. It is better to implant the headless screw directly beneath the bone's surface. An Acutrak screw is not entirely similar to Herbert orthopaedic screws, though. Because it is fully threaded, the fracture can be located anywhere along the length of the screw while still holding the internal power well [17].
5. *Acutrak screws* are an additional kind of screw that are utilised to treat the majority of scaphoid fractures. This screw is cannulated, just like the Herbert screw. Its lack of a head makes it possible to implant it underneath the bone's surface [17]. Although it utilises the same variable thread pitch theory as the Herbert screw, it is completely threaded. A fracture or osteotomy site may be located anywhere along the length of the screw thanks to this feature, which may also increase internal holding power [18].
6. In order to achieve interfragmental compression, *lag screws* are required. Other devices should be employed in place of such screws to provide bending, rotation, and axial loading force protection against fractures [14].

Orthopaedists mostly employ all of these orthopaedic screws during surgical implantation.

An orthopaedic screw is made of three materials, specifically:

- *Stainless steel*: Stainless steel screws are widely used. These screws range in size from 1.5 to 4.5 mm for cortical bones and from 3.5 to 6.5 mm for cancellous bones [14].
- *Titanium*: Titanium is always the best material for implants. The mandibular fractures are best treated with titanium screws. The rate of infection in titanium screws is significantly lower than that of stainless screws [19].
- *Biodegradable*: Polyglycolic acid, poly-L-isocitic acid, and polylactic acids make up the majority of these screws. There is no need to eliminate them from the body because they eventually are absorbed by the body. They do, however, occasionally cause bodily reactions to be stimulated [20].

The following are the advantages of bioabsorbable screws [21]:

- Causes no MRI interference
- Does not obstruct potential future revision surgery
- A decline in the frequency of graft laceration
- No need to remove the implant

The following are the drawbacks of bioabsorbable screws [22]:

- A significant drawback is screw failure while being inserted.
- Some people may experience a foreign body reaction.
- Special screw drivers that span the entire length of the screw lessen incidence of screw breakage.

There are two more types of implants besides orthopaedic screws, which are primarily employed in the process of orthopaedic implants: plates and prostheses.

13.7 Plates

Since 1886, plates have been in use. The plates are required to have proper breadth and thickness because they were historically solely used to hold fractures together. All forms of forces and motions, such as bending, twisting, and compression, must be controlled and countered by these plates [23]. These plates may also be used in the tibia in addition to the femur. Orthopaedic plates are now frequently utilised to fixate fractures of the diaphyseal long bones. Plates come in a variety of flavours and are given names according to their uses.

Compression, neutralisation, buttress, reconstruction, and blade plates are the most common types [14, 15].

13.8 Washers

There are two common uses for washers. In order to prevent thin cortical bone from splitting, they are utilised to disperse stresses beneath screw heads [14].

Avulsed ligaments, minor avulsion fractures, and comminuted fractures are fixed to the remaining bone using serrated washers [14].

13.9 Wires

A tiny krischner wire, or K-wire, as it is more often called, is drilled across the targeted location in order to implant these orthopaedic screws in the body. The orthopaedic screw with a cannula is then positioned over the K-wire and lowered to the bone's surface. Orthopaedic surgeons employ a range of wires. One popular kind is cerclage wire, which is wrapped around the bone's perimeter to bring together separate fracture fragments [14, 24]. Tension band wire is yet another type of wiring used in orthopaedic surgery. As seen in the patellar fracture below, this form of wiring can be implanted either alone or in conjunction with a screw or Kirschner wire [25].

13.10 Nails and Rods

A wide range of devices, from Kirschner wires to substantial femoral nails, are inserted into the intramedullary canal of bones. These devices can generally be categorised according to whether intramedullary reaming is required before insertion. The medullary gap had to be reamed out before the initial nails were hammered down the femoral shaft to prevent the huge nail from shattering the bone [26].

13.11 Complications of Reaming

Reaming has drawbacks because it is an intrusive treatment that may endanger the medullary space's already precarious blood supply. Thermal osteonecrosis can also result from reaming, especially if the medullary canal is narrow, a tourniquet is applied while reaming, or there is obvious soft tissue damage. Fat emboli to the lungs are conceivable during reaming if intraosseous pressure rises. These factors have led to the development of numerous unreamed devices. The Rush rod, which frequently treats fibular shaft fractures and sporadically treats other tubular bones as well, has a chisel-like tip [14, 27].

The Ender nail is a different variety of unreamed nail. The end of these nails is chisel-like as well. Typically, three or four of these nails are pushed through at a time a cortical hole created by fluoroscopy that extends up or down the bone's shaft and across the fracture [28].

13.12 Devices for Spinal Fixation

The legendary Harrington rod is the standard spinal fixing tool. These typically come in two varieties: compression and diversion. The device is extended or compressed to the correct position using hooks that are intended to be positioned beneath the lamina or transverse processes. In the same spine, both kinds of rods may occasionally be used. Other more modern devices have largely replaced Harrington and Edwards rods [14].

13.13 External Fixators

Generally speaking, orthopaedists favour treating fractures in a closed manner. If that doesn't work, they would rather fixate on them inside. However, there are occasionally exceptional situations that make the use of internal fixation impossible. In certain situations, external fixators can be very beneficial [14, 29].

External fixation is appropriate in the following situations: The only option to treat fractures with insufficient bone stock or infection (external fixation provides simple access to wounds), open fracture with significant soft tissue damage, polytrauma patients, and to offer immediate fixing [14, 30].

13.14 Common Spinal Instrumentation

Spinal hardware is used to keep the spine stable as a patient recovers from surgery. It is made up of several plates, rods, cages, wires, spacers, hooks, and screws that are intended to be left in the body for an extended period of time. He clarifies that your surgeon will probably advise removal if the hardware becomes loose, contaminated, or feels noticeable under the skin [31].

"Most of the contemporary instrumentation is made of titanium," despite some gear being made of cobalt-chrome or stainless steel [32].

If a patient has a spinal deformity, degenerative condition, fracture or breaks, or any other issue causing back discomfort, spinal fusion is a popular type of surgery that incorporates this hardware. Hardware might be a complication during spinal fusion failures [32]. Excessive discomfort is frequently a sign of a loose screw following spinal fusion or other hardware issue. "If the hardware becomes loose or is irritating the nearby tissue and nerves, the patient may experience discomfort or may feel and hear crepitus—a crackling sound or popping sound—if the hardware is noticeable under the skin [32].

When the bones in the spine don't heal or fuse properly, there is a high likelihood that the instrumentation will become loose. Hardware may also shift or break as a result of improper healing [34]. This hardware failure may also be more common in patients who lift large objects, engage in high-impact activities, or sustain trauma soon after surgery. Several medical diseases, such as osteoporosis (weak bones) and osteopenia, can also enhance the possibility of shifting, breaking, or loosening [35].

Though it is very rare, some persons experience infections immediately following or in the weeks following device implantation, which could increase the risk of loosening, breaking, or shifting. The patient can experience pain as well as the development of a draining wound and fevers if the hardware gets infected. Chills and redness, swelling, or discomfort near your incision are additional crucial warning signs [32].

Patients may occasionally experience an allergic reaction to the hardware itself, but these occurrences have grown progressively less common in recent years. Before the usage of titanium, this was a regular occurrence with stainless steel hardware [32].

13.15 The Dangers and Difficulties of Removing Hardware

All surgical procedures carry a risk of potential complications, such as those related to anaesthesia, infection, harm to the nerves and blood vessels, bleeding, or blood clots. Infection, nerve damage, re-fracture (breaking the bone again), and anaesthesia hazards are the most frequent risks after hardware removal [36].

It may be difficult to identify nerves or other structures in the area during the surgical exposure if there is scar tissue left over from the initial surgery. This raises the possibility of these structures being damaged, even only slightly. When the hardware is covered in scar or bone, it can be challenging to recognise it. It could need more dissection in order to be discovered. It might need to be chiselled or drilled out because it is so firmly embedded in the bone. Removal might, at least briefly, impair the structure that the implant was stabilising. Hardware might occasionally malfunction while being removed. Deep within the bone, fractured hardware may be left in situ if the hazards of removing it outweigh the advantages [37].

13.16 Precautions Taken During Removal of Hardware

Small screws or wires could only require minimal or no activity restrictions. Exercises requiring jumping and other high-stress activities, such as athletic training, may be prohibited for a while. For up to 4 weeks after removal of the ankle fixation hardware, you may need to wear a walking cast boot or another type of brace [37].

13.17 Significance of Metals in Orthopaedics

13.17.1 Common Uses for Titanium and Titanium Alloys

Many orthopaedic implants are made of titanium as a common material. The majority of the titanium alloy is used to construct whole hip femoral stem parts. The majority of total shoulder arthroplasty stems can be stated to be similar. Additionally, titanium alloy makes up almost all intramedullary rods in use today [38]. For the stabilisation of fractures, conventional stainless steel plating techniques have been used for many years. Many surgeons have the option of using titanium or stainless steel plates and screws. Numerous studies have been conducted on reducing bacterial adhesion to titanium, which may be more prone to enabling adhesion. They are also used in various pedicle screws, rods, and interbody devices that are used in spine surgery [39].

Implanted materials, especially those meant to be kept in place for the duration of the host's life, must eventually become physiologically inactive or nearly so. For instance, the body might easily alter the material's mechanical characteristics if it were implanted with a substance that could be metabolised. This might have long-term effects associated with implantation that would be unforeseen and perhaps inevitable. Similar to this, it is undesirable to use materials that are prone to oxidation or release biologically active particles [38].

In contrast to these cases, titanium implants oxidise as expected. The implant is coated with a very thin coating of oxidised titanium that is produced by the oxidised titanium. This layer is not alive physiologically. This had several benefits, such as known and preserved mechanics of the material, no-host biologic reaction, durability, and material stability [40].

13.18 Stainless Steel

The first metals used in orthopaedic implants were stainless steels in 1926. Asthenic stainless steels, especially 316, and its variations are frequently used in orthopaedic implants due to their excellent corrosion resistance. Alloys based on iron (Fe) that typically include some Cr and Ni are known as stainless steels. Medical-grade stainless steels also contain smaller levels of silicon, carbon, nitrogen, phosphorus, molybdenum (Mo), and manganese as important alloying components. Stainless steels are generally biocompatible, but less so than other traditional metals because they corrode more quickly in physiological environments, releasing poisonous $Cr3+$ and $Ni2+$ [41].

Stainless steel has been and still is the material of choice for a variety of orthopaedic implants. Numerous stainless steel alloys have been created for both industrial and medicinal applications. These mixtures are used to change the biological reaction of alloy components and their structural qualities. The majority of medical grade stainless steel is an alloy known as 316L, while there are certain alloys that differ significantly from one another. Aside from iron, it also contains chromium, nickel, and molybdenum. Numerous historical and contemporary orthopaedic implants have been made of stainless steel. It is often used in the production of intramedullary early-generation rigid nails, some flexible nails, sliding hip screws, orthopaedic plates, and screws. Braided stainless steel is also commonly used to make collar cables wires. Under light loads, such as shoulders, permanent implants made of 316L stainless steel are possible [42, 43]. In the meantime, Orthinox stainless steel has been employed as stem materials in permanent hip replacements due to its increased resistance to pitting and crevice corrosion as well as fatigue strength that is greater than the maximal loading stress value of hip stems [23, 42].

Stainless steel allergy reactivity has been discussed in the literature. Nickel was a component of several stainless steel alloys, which occasionally but not always can result in an allergic reaction. Reactivity to implants is a fairly contentious subject that has generated discussion in the literature. Additionally, there have been examples of explantation (removing the nickel-containing components) that led to an improvement in symptoms [38].

13.19 Cobalt Chrome

13.19.1 Alloys Based on Cobalt

Orthopaedic surgery now uses cobalt chrome as a common bearing surface. This is why cobalt chrome is now most frequently utilised in arthroplasty. Orthopaedic surgery now frequently uses bearing surfaces made of cobalt chrome. In bearing applications involving metal on polyethylene (plastic), it is a widely utilised bearing surface. Cobalt chrome was a widely used bearing surface throughout the emergence and later decline of metal on metal total hip arthroplasty. A more ductile cobalt chrome alloy is also used to create some cerclage wires. Cobalt chrome in bulk appears to be physiologically inert. It is frequently placed inside of joints. It has nickel and could cause problems in people who are hypersensitive to metals. Despite being a contentious issue, many practitioners avoid cobalt, chromium in individuals undergoing arthroplasty who have a severe nickel allergy. Total knee and total hip literature both provide documentation for this differentiation [38].

CoCr alloys are the standard name for co-based alloys. Similar to stainless steels, when Cr is present in high concentrations, a passive Cr_2O_3 layer spontaneously forms in the fluid environment of the human body. There are numerous surgical complications linked with cobalt chrome particles from metal-on-metal wear, and they are well-documented to be immunogenic. These tiny particles cause a lymphocyte-mediated reaction when they come into contact with metal-on-metal wear. Pseudotumours can develop from unabsorbed metal particles. Widespread damage of bone tissue can result from the immune system's reaction to metal particles [11].

It's significant to note that metal-on-metal arthroplasty scenarios are not the only ones where pseudotumor and metal debris trigger osteolysis. A condition known as metal fretting (trunionosis) can develop where the total hip stems and total hip heads meet. Similar wear particles as well as cobalt chrome and titanium micro-particles are generated by relative motion at the head/stem junction of a complete hip morse taper [44].

13.20 Tantalum

Tantalum implants are less frequently utilised in orthopaedic surgery nowadays, although their applications may grow as a result of the high cost of the material and rising arthroplasty revision rates. Due to the inexpensive material cost and mechanical similarity to titanium, tantalum implants for fracture fixation are also taken into consideration. Tantalum is currently most frequently used as an augment. In situations involving arthroplasty, tumours, and various types of fractures, augmentations fill bone deficiencies. This substance may be manipulated to become extremely porous, and it appears to permit bone ingrowth and efficient biologic assimilation.

High levels of intraoperative workability go hand in hand with this biologic integration [45].

13.21 Polyethylene

Since the advent of contemporary arthroplasty, various types of polyethylene have been used. The first long-lasting total hip arthroplasty used a polyethylene-bearing surface and was developed by arthroplasty pioneer Sir John Charnley. The orthopaedic community's most popular usage of polyethylene has mostly been in arthroplasty. The most typical application is as a synthetic joint surface. The majority of knee and hip replacement systems contain this plastic-like substance. Joint replacements made of plastic (polyethylene) and metal (cobalt chromium) are frequently subject to heavy loads. From the perspective of host reaction, bulk polyethylene has proven to be largely inert. However, in vivo testing has shown that wear debris made of polyethylene is extremely reactive [46].

Early reports of host responses to wear debris were characterised as "cement disease" and incorrectly attributed to a cement-driven process of bone resorption. These studies were actually host responses to polymethyl methacrylate bone cement. According to current thinking, polyethylene wear particles (debris) can cause implant loosening. The clinical manifestation of this is bony resorption at and around implant interfaces, which is caused by polyethylene wear debris that can cause a macrophage-mediated response that can result in bone resorption.

Complications from this resorption include pelvic discontinuity, subsidence, periprosthetic fracture, and other issues connected to catastrophic bone loss [46].

13.22 Polyethylene Infused with Vitamin E

Numerous bearings have been altered to include vitamin E, which is still widely used today. Free radicals are a problem because of the harm they do to the bearing. In order to scavenge free radicals created during or after the cross-linking process, vitamin E was added. Although the impact on long-term results is up for debate, employing this bearing surface over normal polyethylene has been shown to increase the longevity of polyethylene bearings [47].

13.23 Ceramics

In arthroplasty, ceramics are playing an ever-expanding role. The most widespread use today is in hip replacement bearing surfaces. For total knee arthroplasty, zirconium-based implants have also found a market. When compared to other bearing surfaces now employed in orthopaedic surgery, ceramic bearing surfaces have some of the best wear qualities [48]. In particular, metal on metal bearing surfaces, which have a poor track record in the majority of hip arthroplasty applications, were replaced by ceramic as a viable option. The lowest wear particle volumes of any arthroplasty surface are produced by ceramic on ceramic. It has not yet been consistently demonstrated that the immune response to a properly positioned and intact ceramic on the ceramic-bearing surface is harmful. On top of ceramic given its better wear properties, ceramic on ceramic-bearing surfaces has been employed extensively in younger arthroplasty patients.

Ceramic head complications are well known. Some patients with total hip arthroplasties made of hard or hard ceramic had an audible squeak when walking [49]. In an otherwise symptom-free arthroplasty, this consequence results in extremely low patient satisfaction and is a known cause for revision. Additionally, porcelain is very fragile and can break in real time, especially in a hard-on-hard environment, leading to catastrophic joint failure and a joint filled with microscopic ceramic debris.

13.24 Metal Poisoning and Metallosis

Metal-on-metal hip replacements and other metal implants are examples of joint replacement devices having metal components that can result in metallosis, a type of metal poisoning. These gadgets are constructed from a mixture of metals, including titanium, molybdenum, nickel, cobalt, and chromium. Microscopic metal particles are released into the blood and surrounding tissues as the metal components brush against one another. As these metal ions accumulate in the surrounding bone, muscle, and other tissue, metallosis starts to occur. This may cause bone or other tissue to die.

The presence of metal in the blood and surrounding tissues can cause a number of further problems. The brain, heart, eyes, and other organs can be impacted by a buildup of cobalt or other metals. The metal implant must typically be removed and replaced during revision surgery to treat metallosis and related diseases. A metal ball and cup were utilised in some metal-on-metal hip implants while others employed a metal neck and stem. Hips built of metal-on-metal had cobalt and chromium components [50].

13.25 Symptoms of Metallosis After Hip Replacement

According to the US Food and Drug Administration, local symptoms of metallosis include hip or groyne discomfort, numbness, swelling, weakness, and a change in one's capacity to walk.

Before exhibiting local symptoms, a patient may have issues with the skin, heart, kidneys, neurological system, or thyroid. Early signs of metallosis include "a feeling of instability, an increase in the audible sounds from the hip, and pain that was not present immediately after surgery." Metallosis can progress to bone loss and tissue death around the implant if individuals don't treat these symptoms. Surgery is the only option for treating metallosis [51].

There are times when pain after the initial hip replacement surgery is not a sign of metallosis. For instance, a study of 116 patients having their hips resurfaced discovered that 18% of them experienced groyne pain after the procedure, but claimed that the "cause is most likely multifactorial, ranging from iliopsoas tendinitis to an adverse soft tissue reaction."

13.26 Cancer and Metal-on-Metal Hips

It has not been demonstrated that metal fragments from hip implants cause cancer. Total 10,728 patients who had gotten a metal-on-metal hip were compared to 18,235 patients who had received a more traditional hip replacement in a 2012 study published in the British Medical Journal. Researchers discovered that both groups' cancer risk was comparable. Cobalt and trivalent chromium, a subtype of chromium, are both listed by the World Health Organisation International Agency for Research on Cancer as having the potential to cause cancer.

The London Implant Retrieval Centre disclosed test data demonstrating that trivalent chromium ions were discharged from the implants after hundreds of metal-on-metal implants failed. Metal-on-metal hip implants may therefore produce particles that could potentially cause cancer, despite the fact that no higher cancer risk has been identified [50].

13.27 Chromium and Cobalt Hip Implants

Cobalt and chromium were frequently used in the construction of metal-on-metal hip implants. For healthy cellular activity, the body naturally retains a specific amount of cobalt and chromium. A component of vitamin B12, cobalt is required for the creation of red blood cells. But just trace amounts are required for the survival of plants, animals, and people. Cobalt overdose can result in metal toxicity. Too much cobalt can be swallowed, inhaled, or come into contact with the skin for an extended period of time to cause cobalt poisoning. Cobalt is a mineral that is used in the production of machine tools, batteries, alloys, and colours.

Metal implant components can release cobalt and chromium ions above those found naturally when they brush against one another. If you consume or breathe too much cobalt, you could develop thyroid issues, nerve issues, heart issues, and hearing loss. If there are too many metal ions in the blood, hip implant patients may experience similar issues. "Cobalt poisoning might start anywhere between 9 months and 4 years following surgery.

High cobalt levels in the blood can cause the following symptoms: depression, anxiety, and other mental health issues; cardiomyopathy (heart issues), including heart abnormalities and failure; thyroid issues; cognitive impairment; nerve issues, including peripheral neuropathy; visual impairment that could result in blindness; tinnitus, hearing loss, and deafness; eczema; headaches; groyne pain; irritability; thickening of the blood; implant loosening; skin rashes; vertigo and vertigo-like symptoms; and shortness of breath [52].

The body may swiftly eliminate some cobalt when it is released in excess. What is not eliminated is taken up by the blood and transported to various body parts, primarily the kidney, liver, and bones. In the case of metal implants, the tissue surrounding the device absorbs any extra cobalt that may otherwise accumulate.

13.28 Metallosis Diagnosis

According to the US Food and Drug Administration, doctors may use metal ion testing or soft tissue imaging to assess whether a patient has an excessive amount of metal ions in their system. Both can show that metal ions have damaged tissue, but neither should be the sole indicator of whether a hip is deteriorating from metallosis. The terms "soft tissue imaging" refer to MRIs, CT scans, and ultrasounds. Each approach has benefits and drawbacks. MRIs and CT images may exhibit distortions due to the metal in metal-on-metal hips. While ultrasounds give medical professionals access to tissue devoid of metal interference, they have a poorer resolution and can't image as deeply beneath the skin as MRIs or CT scans.

Additionally, cobalt, chromium, molybdenum, and titanium—all metals used in metal-on-metal hips—may be examined in a blood test by a physician. Any test should, according to the FDA, be able to quantify ions at concentrations as low as one part per billion, or ppb. Accordingly, there are one part metal ions for every billion parts blood. The criterion was seven times greater in the past. However, as of 2019, the FDA suggests using a significantly lower quantity as a criterion. The FDA discovered that while some patients below the 7.0 ppb criterion had already had metal-on-metal hip problems, some individuals over the threshold had no symptoms at all. The FDA additionally provides instructions for physicians on picking a lab to conduct the blood samples and how to interpret the results [53].

13.29 Subjects Who Will Undergo Metallosis Testing

There is a chance that metallosis will manifest in anyone who currently has a metal-on-metal hip implant. However, not everyone with a metal-on-metal hip will experience metallosis or other wear-related symptoms. If patients display signs of any of the following six illnesses, the US Food and Drug Administration advises that doctors test for high ion levels:

Skin rash or another hypersensitive reaction; cardiomyopathy, a chronic condition of the heart tissue; neurological changes, such as loss of hearing or vision; psychological changes, such as depression and memory loss; kidney function issues; thyroid dysfunction, such as neck discomfort, fatigue, weight gain, or a cold sensation [53].

For persons who have not yet displayed any symptoms of these disorders, the FDA does not advise testing as long as their orthopaedic surgeon determines that the hip is healthy.

13.30 Treatment for Metallosis

Metallosis can only be treated surgically by replacing the worn metal-on-metal. It prevents new metal ions from being released. Around the implant, the doctor will remove unhealthy bone and tissue. The degree of tissue and bone necrosis, or death, dictates the surgical result in cases of severe metallosis. Because of their weak bones and severe tissue damage, some patients experience fractures as a result of revision surgery. To reduce potential issues with metal ions in the future, doctors replace the metal-on-metal implant with a ceramic-on-metal or plastic-on-metal implant during revision surgery. According to case studies published in the State of Epidemiology Bulletin, after revision surgery, some patients may have a marked improvement in their symptoms within 3–6 months [52].

13.31 Orthopaedic Implant Materials

A thorough understanding of the fundamental requirements of orthopaedic materials and the ensuing biological response is essential for the design and optimisation of implants under physiological settings in the human body since orthopaedic implants must perform under various working situations in vivo. The optimal material choice for an orthopaedic implant depends on its intended use. Orthopaedic implants frequently use ceramics, polymers, and metallic alloys.

Different physical, chemical, and biological characteristics of these materials allow them to be used in various applications. Despite the success of conventional materials, new and improved biomaterials are constantly being created to meet the rising need.

In the majority of medical disciplines, metals and metal alloys have been used for several medical implant purposes. Orthopaedic surgery has evolved during the past 30 years. The quality of life, mobility, and pain relief offered by orthopaedic surgery have all improved. The most popular devices are those to replace arthritic joints with prostheses and those to treat fractures and stabilise the spine, hip, and knee joints The best clinical outcomes have been seen with knee and hip joint replacements. As a result, joint replacements have proven to be a very effective treatment, even over a lengthy period of time. Nevertheless, it is crucial to enhance and give a better quality of life despite the success of this kind of surgery. Given that problems and negative side effects are widespread in this type of orthopaedic surgery and prosthetic materials, which often result in the replacement of the prosthesis, longer-lasting implants are required. One of the risks of joint replacement is the degeneration of the prosthetic parts, which is followed by a biological response from the body to the implant's material. The issue of negative implant reactions.

13.32 Immunological Reaction to Metallic Implants

The most typical immunological reaction to implants is metal sensitivity [56]. The most popular metals for orthopaedic and dental implants are titanium, cobalt, chromium, and stainless steel (with nickel) [57, 58]. Typically, metal alloys such as titanium, zirconium, nickel, cobalt, chromium, and molybdenum are used to create medical implants [59]. It is not unexpected that the immunological response to medical implants, particularly hypersensitivity to orthopaedic hardware (such as joint replacement prostheses, fracture fixation devices, and pain-relief stimulators), is frequently reported in the literature.

Following the implantation of orthopaedic devices, metal sensitivity is a common side effect [59–63]. Osteolysis and aseptic loosening of implanted metal devices are associated with metal sensitivity [60, 61, 63–74]. It is unclear whether metal sensitivity or a failed implant—which can trigger a stronger immune response and possibly more clinical testing for metal reaction—is the initiating event. The immune system reacts to surface alterations and breakdown products as implants corrode or degrade [75]. According to Huber et al., there may be a connection between corrosion and implant-related hypersensitivity if patients have both corrosion products and a hypersensitivity reaction.

While some studies have found that patients with metal hypersensitivity prior to implantation can actually become desensitised and anergic after implantation, other studies have taken an opposing stance, concluding that hypersensitivity fails to develop [76]. While some writers have hypothesised that metal hypersensitivity may be linked to bone loss and aseptic loosening of implanted devices, others have countered that even in the case of a metal allergy, no harmful effects take place. Demehri et al. described a rare case of squamous cell skin cancer (more particularly, Marjolin ulcer) linked to a contact allergy to superficial metal implants, most likely owing to chronic inflammation [77, 78]. Additionally, metal sensitivity may be linked to fibromyalgia, autoimmune disorders, and chronic fatigue syndrome. When faced with symptomatic or failing devices, applying diagnostic criteria may be helpful in directing the decision-making process [79]. Although it may be expensive and of questionable clinical value to prescreen all patients for metal hypersensitivity, a number of specialised laboratory assays, such as the lymphokine migration inhibition factor (MIF) test, seem to be confirming. The differential diagnosis of metal hypersensitivity should be taken into consideration when a patient exhibits symptoms like recurrent pain and aseptic loosening connected to implanted hardware, which is extremely uncommon. In spite of the drawbacks and the worry that the immune environment and reactive immune cells are different in the musculoskeletal tissues than the skin. Before the cause of the issue is presumed to be an allergic reaction, other possible causes such as infection, nonunion, aseptic loosening, other inflammatory disorders, mechanical failure of the implant, and alignment problems must be ruled out first. The likelihood of an allergic reaction to the metal must be taken into consideration, assessed, and addressed once the more frequent causes of implant failure have been ruled out [80, 81].

Patients with metal-on-metal bearing surfaces are considered to be special cases because tribocorrosion, or the release of metal ions or particles into the joint, causes the immune system to react negatively, leading to pseudotumours, adverse local tissue reactions, and possibly prosthetic failure. Serum cobalt or chromium levels that are extremely high can cause chromosomal abnormalities, cardiotoxicity, and neurotoxicity. Serum tests are utilised for screening, and metal artefact reduction sequence (MARS) MRI can be used to see local soft-tissue reactions. Ion concentrations that are very high (>7 ppb) are a sign that improved imaging is needed. High cobalt ion concentrations have the potential to be both cardiotoxic and neurotoxic, whereas high chromium ion concentrations may be carcinogenic [82]. Systemic symptoms often require concentrations much greater than 7 ppb. Patients who have metal-on-metal prostheses may also experience metal ion leak as a result of surgical procedures. Patients who experienced adverse responses to metal debris (ARMD) and the position of the prosthetic cup were linked by Koutalos et al. but not by metal ion levels [83]. Modern implant designs have been created precisely to reduce metal ion release. The use of nickel-free bearing materials (such as hardened titanium, ceramic, or ceramicised metal) has been minimised, modular necks in femoral prostheses have been eliminated, and ceramic heads rather than metal heads have been used to lessen reactions at the trunnion of a stem. According to Markel et al., a dual-mobility cobalt-chromium hip replacement prosthesis was used, and after 1–2 years, metal ion levels were either undetectable or very low. Additionally, percentages of B cells and T cells were normal, and there was no rise in CD16 inflammatory monocytes, indicating no immune reaction to the implant [84].

13.33 Immune Reaction to Non-metallic Implants

Implant non-metallic components may also trigger an immunological reaction. Organic or inorganic materials can be non-metallic. Natural and artificial polymers, polysaccharides, and proteins, such as chitosan, glycosaminoglycans, hyaluronic acid, collagen, and silk, are examples of organic biomaterials. Bioactive glass and calcium phosphate are examples of inorganics.

Metal ions are technically metals, although the immune response to nonmetal implants frequently includes them as well. Total knee arthroplasty patients who have acrylic bone cement or its polymerisation additives (benzoyl peroxide and N,N-dimethyl-p-toluidine) may experience significant hypersensitivity responses [85]. Acrylates, benzoyl peroxide, toluidine, and antibiotics are examples of bone cement ingredients that may have immunogenic effects. Multiple types of coatings for implants have been created in an effort to reduce bacterial adhesion, inhibit the formation of biofilms, or directly kill bacteria in response to the recognition of the potentially disastrous effects of implant-associated infection [86, 87]. These coating materials, however, ironically, can also trigger an immune response to the coating materials themselves. These coating materials can promote host cell attachment and a local immunological response against the infectious organism. The following are examples of coating materials in use [86, 88]: anti-adhesive materials Silver ions in hydrogels Tungsten Dioxide Ions of selenium, copper, or zinc Antibiotics derivatives of chitosan Cytokines peptide antimicrobials Neutrophils and bacteria are both cytotoxic to silicon polyurethane silver ions, which reduces the immunological response to both the implant and the bacterium [87].

13.34 Aetiology of Immune Response to Metal Implants

A real systemic hypersensitivity reaction or local implant damage may be the source of the immune response to an implant. There is a type IV delayed cell-mediated response during allergic responses. Sensitised T cells react to an implanted device with a foreign body reaction after being exposed to metals or implanted devices repeatedly or for an extended period of time. Inflammation caused by the production of cytokines by activated lymphocytes, such as interferon [IFN] gamma [89], stimulates macrophages as a step in the inflammatory cascade. However, some metals, like cobalt, can directly irritate soft tissues without inducing an allergic reaction through toxicity of the metal ions. A medical device that has been surgically inserted triggers an immunological reaction right away. When immune cells, platelets, and coagulation mechanisms are activated by blood contact, a thrombus forms at the interface, which serves as the temporary provisional matrix. Growth factors, cytokines, and matrix metalloproteinases are abundant in this matrix, which aids in the immune response and draws neutrophils. Monocytes are propelled to migrate by activated platelets as they develop into macrophages. M1 proinflammatory macrophages arrive as part of the early inflammatory response [90, 91]; these macrophages discharge injured tissues and produce proinflammatory mediators such as interleukin (IL)-1, IL-6, IL-16, and tumour necrosis factor (TNF) [92]. Excessive inflammation results from these cytokines remaining in the body. The M1 macrophages often differentiate into M2 macrophages, which aid in controlling tissue remodelling. Osteoclast activation is critical in the initial stages of bone remodelling around the implant as well as for the clearance of necrotic tissue. Through the action of three major cytokines, the immune system regulates the osteoclastogenic process as follows: M-CSF, or macrophage colony-stimulating factor RANKL, or receptor activator of NF-B ligand OPG, or osteoprotegerin Increased osteoclast activity, rapid bone resorption, and excessive bone loss are caused by an elevated RANKL-to-OPG ratio. Chronic inflammation, sluggish tissue repair, and poor biomaterial integration are all consequences of a persistently increased M1 response with a diminished or non-existent M2 response. When an implant is present, a condition known as "frustrated phagocytosis" may occur, characterised by a pro- and anti-inflammatory state that is imbalanced, causing persistent inflammation [91]. Immune cells, including T cells of different subsets, cytokines, and microRNAs (miRNAs), control the transition from M1 to M2. The anti-inflammatory cytokine IL-10 is essential for preserving immunological homeostasis and controlling inflammation [93]. Dendritic cells use pattern recognition receptors to sense their local surroundings and communicate the characteristics of antigens to T cells in lymph nodes, which controls the immune response to the foreign body [93]. The severity of this foreign body reaction varies, and study is ongoing to determine why some patients respond excessively compared to other patients. However, it is obvious that particular osteomodulatory features of the implant contribute. Metal ions have cytotoxic haptenic potential and are thought to cause a delayed-type hypersensitivity response as well as a negative reaction to metallic debris [94]. Cobalt ions emitted by failed hip implants have been shown to trigger mitochondrial stress and cytokine production by synovial fibroblasts [92].

13.35 Patient Reactions to Metal Implants

Patients with metal implant responses may appear with a non-specific clinical picture. Patients may appear with systemic eczematous dermatitis as well as localised dermatitis or rashes.

Infection-like symptoms include swelling, discomfort, draining sinuses, and inflammation at the implant site. Dermatitis and other skin responses, joint discomfort, joint effusions, and sluggish wound healing are possible presentations. Although the indications and symptoms of a nickel or other metal hypersensitivity to an implanted orthopaedic device can vary, they typically include the typical complaints of a patient experiencing hardware failure. Loss of motion may be experienced by patients with joint replacements. A nickel allergy is more common in dermatitis patients than a metal allergy to another substance [59]. It is common for metal hypersensitivity to present as a skin rash at the site of the implant, especially with superficial implants such as plates at the ankle. Patients with joint replacements typically have symptoms of loosening, including pain and instability. (For example, with a total hip replacement, the patient often has groin pain radiating to the medial thigh.) Patients with hardware for fractures have symptoms of nonunion, including pain and motion at the fracture site. Local inflammatory symptoms similar to the symptoms of infection are also possible, including warmth, erythema, and swelling over the implant, though systemic complaints (e.g. fever) are unlikely. A skin rash may develop over the metal device but is not always present [95–97] Osteolysis and aseptic loosening should always be included in the differential diagnosis. Despite the introduction of highly crosslinked polyethylene in the mid-2000s, decreasing the incidence, these are historically common causes of local reaction, bone resorption, pain, and implant loosening. Fujishiro et al. identified an association between the extent of inflammation and the amount of visible metal particles and concluded that this relation implied the occurrence of an immune response to the metal [98]. However, the typical morphologic features of an immune inflammatory reaction, including loss of the surface synovial lining, fibrin deposits, and lymphocytes in diffuse and perivascular distributions, were not consistently present. More likely, the mechanism of osteolysis is primarily a local reaction to particulate debris [99, 100], which leads to a cascade of cellular reactions (including activation of monocytes/macrophages, phagocytosis, and release of cytokines) that eventually lead to increased osteoclastic activity around the prosthesis. Whereas the radiographic and clinical symptoms overlap with those of metal immune reaction, osteolysis is a reaction to local irritation from wear debris, not an immune hypersensitivity response [101]. The causes of these different patterns of inflammation are unknown, but the association between the extent of inflammation and visible metal particles (but not zirconium particles) supports the concept of an immune reaction to metal, and it illustrates that the process is not specific to metal-on-metal constructs.

13.36 Metal Wear Debris Produced from Prostheses

Wear debris can take both particulate and soluble forms and is produced by mechanical wear, surface corrosion, or a combination of the two. All metal surfaces are susceptible to corrosion, which can either cause the disintegration of the bulk metal alloy or the creation of a passive protective coating. During the corrosion of metal alloys, solutions including cobalt (Co(II)), titanium (Ti(V)), aluminium (Al(III)), iron (FE(III)), nickel (Ni(II)), and chromium (Cr(III)) have all been found. The release of Cr(VI) from the CoCrMo (molybdenum) alloy is supported by evidence, however this is still debatable. Within the synovial environment, metal oxides (Cr_2O_3, CoO, TiO_2, Al_2O_3, etc.) and hydroxides ($Cr(OH)_3$, $Co(OH)_2$, etc.) make up the majority of corrosion products. In non-synovial settings, calcium phosphate deposits and the subsequent production of metal phosphates ($CrPO_4$, $Co_3(PO_4)_2$, etc.) take place. The biological and chemical characteristics of free particle metals outside of the effective joint space may be considerably altered as a result [102]. Patients who have had arthroplasty have substantial metal wear products in their synovial fluid and peri-prosthetic tissues. Additional buildup was found postmortem in the local lymph nodes, liver, and spleen. The full degree of spread is yet unknown since metal particles are so tiny (nano size). Wear particles are moved through the lymphatic system either freely or after being phagocytosed. 2As ions or particles, metallic debris may also travel through the circulatory system. Blood and urine metal concentrations are utilised in occupational biomonitoring as biomarkers to determine exposure. The average metal levels found in exposed workers and patients who have joint replacements are frequently comparable. Health and safety groups like the Health and Safety Executive and the Deutsche Forschungsgemeinschaft have given advisory values for Cr and Co based on biological and atmospheric factors. In the United Kingdom, the exposure equivalents of carcinogenic substances (EKA values) for Co are 5.0 and 60 gL^{-1} in whole blood and urine, respectively, and for Cr are 17 and 20 gL^{-1} in erythrocytes and urine, respectively. Patients with biological metal levels higher than one or more of these values have been identified in several investigations in the field of orthopaedics. However, it is obvious that joint replacement patients, particularly those with metal-on-metal articulations, are likely to experience increased metal levels over the course of the prosthesis' lifespan. Biological reactions to metal wear detritus in cells and cellular uptake [102].

13.37 Biological Reactions to Metal Wear Detritus in Cells and Cellular Uptake

Endocytotic mechanisms, particularly non-specific receptor-mediated endocytosis and pinocytosis, are responsible for the uptake of metal nanoparticles (150 nm) by cells. In specialised cells like macrophages, larger particles (>150 nm) can trigger phagocytosis. Metal particles that have been ingested can cause cytotoxicity, chromosomal damage, and oxidative stress [103]. Passivation and particle size affect a particle's toxicity. Both of these variables affect how easily metal dissolves from surfaces, which could explain biological activity. The physical characteristics of the particles themselves may cause signs of cell injury, such as crooked cell membranes and swollen mitochondria. Because of the chromate anion's structure, Cr(VI) is easily absorbed by anionic channels, whereas Cr(III) builds up at the plasma membrane. With the brief synthesis of Cr(V) and Cr(IV), Cr(VI) is quickly reduced to Cr(III) and disseminated throughout the cell coupled to peptide and/or protein ligands. Co(II) and Ni(II) uptake may be facilitated by 3Divalent metal transporter ((DMT)-1), which is expressed in a variety of organs, and natural resistance-associated macrophage protein (NRAMP)1, which is situated on the phagosomal membrane. Cell-surface transferrin receptors have the ability to internalise transferrin-bound Fe(III), A1(III), Cr(III), or vanadium (V). A variety of cells are affected by metal ions released from orthopaedic implants, with Co(II) and V(III) being among the most cytotoxic. CoO, Cr_2O_3, and $CrPO_4$ are examples of corrosion products that exhibit some cytotoxicity [102, 104].

Redox metals Cr, Ni, Co, and Ti can produce reactive oxygen species such the superoxide radical (O_2) and the hydroxyl radical (OH) by reacting with hydrogen peroxide (H_2O_2) in a Fenton-driven manner. DNA, proteins, and lipids may become oxidatively damaged as a result of reactive oxygen species. A variety of orthopaedic metal ions, including Ni(II), Cr(VI), and Co(II), have been found to inhibit DNA repair, affect signal transduction, and change gene expression. regional tissue reactivity. Despite the reintroduction of metal-on-metal bearings as an alternative to metal-on-polyethylene articulations, aseptic loosening and osteolysis continue to be the main causes of implant failure. Aseptic loosening in patients with metal-on-polyethylene bearings is hypothesised to be caused by macrophages' reaction to particle wear debris. In contrast, the ability of particles from metal-on-metal bearings to activate macrophages is restricted, and they have the potential to cause osteolysis through an immune reaction including hypersensitivity. The perivascular infiltration of lymphocytes and the buildup of plasma cells in the peri-prosthetic tissue of loose metal-on-metal articulations distinguish the pattern of inflammation from that of metal-on-polyethylene articulations. Orthopaedic metals may have immunological effects that support a cell-mediated hypersensitive response, according to experimental evidence. A few number of epidemiological and experimental investigations using in vitro and in vivo models have been conducted to learn more about the toxicity caused by metals. On the systemic consequences of metal in patients who have had arthroplasty, there are sadly few data available. Currently, the following harmful reactions have been identified: Its blood. Both A1 and Cr(VI) have the ability to impair cellular iron consumption, which is associated with variations in haematocrit and haemoglobin levels. The development of microcytic anaemia in renal patients is a result of poor A1 clearance. Although in vitro oxidative effects, primarily lipid peroxidation, at high concentrations have been documented, no substantial Ni(II) effect has been found in vivo. the defence mechanism. Through a number of immunostimulatory or immunosuppressive mechanisms, metals control the actions of immunocompetent cells. The effects of orthopaedic metal ions typically include changes in cytokine release, altered T-cell, B-cell, and macrophage activity, the production of immunogenic chemicals, and direct immunotoxicity. Patients with metal-on-metal articulations had a considerable decrease in circulating lymphocytes, especially CD8+ T-cells; however, there was no direct relationship between this and serum metal concentrations. A threshold value of 5 ppb combined Co and Cr, though, was found below which no appreciable reduction was noticed. In patients with metal-on-polyethylene articulations, there has been evidence of an inverse relationship between the concentration of Cr and the numbers of circulating

CD4+ T-cells and CD20+ B-cells, although myeloid cells and CD8+ T-cells continuously reduced independent of metal levels. Despite the fact that lymphoid populations in experimental animals exposed to metal alloy solutions underwent a considerable change, these effects were not replicated [102, 105, 106].

13.38 Systemic Toxicity of Metals

The liver: Hepatocellular necrosis frequently happens in reaction to extremely high metal levels in the body, as shown in humans after acute Cr(VI) consumption. After exposure to A1, portal inflammation and oxidative stress have been noted, while pathogenic alterations in experimental animals were not noticeable [107, 109, 110].

The kidney: In experimental animals and humans, Cr can impair renal function, promote tubular necrosis, and create noticeable interstitial alterations because it is concentrated in the epithelial cells of the proximal renal tubules. Human participants who had occupational exposure to Cr(VI) showed signs of tubular dysfunction. Since Al, Ni, and Co are all quickly eliminated by the kidneys, renal toxicity typically calls for much higher doses [108, 111, 112].

The respiratory system: Due to the frequent occupational exposure to Co, Ni, and Cr, the effects on the respiratory system have been extensively studied. These effects include an increased risk of developing asthma and inflammatory diseases [107]. Welders of stainless steel who are frequently exposed to metal vapours including Cr and Ni experience these effects frequently. It is challenging to generalise toxic responses of the respiratory system to a circulatory pathway since they are mostly connected to inhalation exposure [108, 113, 114].

The nervous system: Al poisoning in humans has been linked to a number of neurological symptoms, including memory loss, jerking, ataxia, and neurofibrillary degeneration. A1 buildup in the brain may be associated with the onset of several neuropathological disorders, such as amyotrophic lateral sclerosis, Parkinsonian dementia, dialysis encephalopathy, and senile plaques of Alzheimer's disease. Al is typically linked to modifications that could lower nerve conductivity, hasten neuronal degeneration, and heighten Fe-induced oxidative damage. A1 significantly affects the synthesis and accumulation of proteins linked to Alzheimer's disease, including -amyloid, whose release is boosted in vitro by Co(II). In particular, in reaction to Fe, oxidative stress may have a key role in the onset and/or progression of neurodegenerative diseases [108, 115–118].

The brains of laboratory animals exposed to Cr(VI) and V(V) have exhibited signs of oxidative damage. With a mean serum level of 14.4 ppb of V from occupational exposure, male workers have shown significant changes in their visuospatial ability and attention span [119].

The heart and vascular system: The accumulation of Co in the myocardium can cause cardiomyopathy, which became especially clear following the 1966 occurrence of "beer-drinkers' cardiomyopathy," in which Co was utilised as a foam-stabilising ingredient in beer. Despite the lack of clinically relevant cardiac impairment, a small series of cobalt industrial employees exposed to an average of 0.40 mg Co year-1 showed altered left ventricular function relaxation. After it was established that breathing in fine ambient particulate matter greatly raised the mortality rate from cardiovascular disease, Ni and V were assumed to have contributed to alterations in heart function in experimental animals [120–122].

The musculoskeletal system: Chronic exposure results in the deposition of A1 in the bone, which has been related to osteomalacia, bone pain, pathological fractures, proximal myopathy, and a failure to respond to vitamin D3 therapy. Osteoblast function is negatively impacted by orthopaedic metal particles and soluble metal compounds, which may have an impact on bone remodelling [123, 124].

Endocrine system: Cellular oestrogen receptors can bind to A1, Cr(II), Co, Ni, and V, which may help explain abnormal oestrogen signalling. In experimental models, Ni(II), Cr(VI), A1, and Co(II) have the ability to change the synthesis or circulation of sex hormones. This is typically because these substances have a direct impact on reproductive cells, as is the case with Cr(VI). By inhibiting the enzyme tyrosine iodinase, which can cause hypothyroidism, Co(II) prevents the uptake of iodine into the hormone thyroxine. Despite signs of a changed thyroid metabolism, occupational exposure in a limited sample of Danish ceramic painters had no impact on normal thyroid function. A1 is known to interfere with parathyroid hormone levels, which may explain why some dialysis patients get bone abnormalities [125–128].

Visual and auditory system: A1, Co, and Ni can severely degenerate the retina in test animals when they are present in high quantities in the visual and auditory systems. A case of a man with significant wear of a CoCrMo femoral head and elevated levels of Co in the serum (398 μg L^{-1}) and cerebrospinal fluid (3.2 μg L^{-1}) was recently reported. He experienced dermatitis, hearing loss, vision loss, and foot numbness [129–131].

The skin: Urticaria, vasculitis, and/or contact dermatitis are a few examples of cutaneous reactions to metals. In patients with complete joint replacement, with stable and loose prostheses, the incidence of cutaneous responses and positive skin-patch tests to Co, Ni, and Cr rises by 15% and 50%, respectively, over that of the general population [132, 133].

The reproductive system: In experimental animal models, chronic exposure to Cr(VI) causes a number of consequences that are harmful to fertility [134, 135]. These include a lower sperm count, epithelial deterioration, sperm abnormalities, fewer follicles and ova, and a higher proportion of atretic follicles. In contrast, workers in the production of chromium sulphate had a significant positive correlation between the incidence of morphologically abnormal sperm and blood Cr levels. This correlation was found in a large epidemiological study of stainless-steel workers, which found no significant causal link between exposure to Cr and reduced sperm quality [136]. It has been demonstrated that exposure to Ni(II), V, A1, and Co(II) causes aberrant spermatogenesis and some limited reproductive toxic effects in male experimental animals.

However, there appears to be a notable paucity of information about how these metals affect female animals [137–141].

Developmental toxicology: In a study of ten women who had metal-on-metal resurfacing and afterwards became pregnant, Co and Cr levels increased. This finding raises the possibility that orthopaedic metals may translocate from the maternal to the foetal circulation [142]. Numerous metals, including Cr, Co, Ni, V, and Al, may cause developmental toxicity, according to experimental animal research [143]. For instance, male and/or female mice exposed to Cr(VI) either before or during gestation may experience fewer implantations and live births. Numerous metals, including Cr, Ni, and V, can also cause teratogenic deformities.

Some metals, such as Cr(III), have been linked to transgenerational carcinogenesis, which is the transmission of the risk of cancer to untreated offspring of parents exposed to carcinogens before mating [144]. A study using V suggested that in addition to the transplacental pathway, metals may also be transferred from the mother to the growing child during nursing [145]. An extensive study found no appreciable increase in the likelihood of congenital defects or cancer in the offspring of male stainless-steel workers, but a considerable rise in the probability of spontaneous abortion in the partners of these men [146]. Additionally, epidemiological research has linked parental work exposure to a higher incidence of childhood cancer, while the precise aetiological agent is yet unknown [147]. The incidence of pregnancy-related problems did not differ from that in the general population in a relatively small study of 13 female arthroplasty patients [148].

Carcinogenesis: Both welders and patients who had arthroplasty had an elevated incidence of chromosomal abnormalities in their peripheral lymphocytes [149, 150]. There is a growing consensus that metal-induced DNA damage may contribute to carcinogenesis, but the importance of this discovery and its connection to an elevated risk of cancer are yet uncertain. The risk of developing cancer has been linked to occupational metal exposure, particularly to Cr [151].

Studies on THR patients in Norway have found a slight but significant increase in the incidence of malignant melanoma, endometrial, and prostate cancer [152, 153]. Cr(VI) and Ni(II) have been categorised as carcinogenic by the International Agency for Research on Cancer, which also lists metallic Ni and soluble Co as potentially carcinogenic. Metallic Cr has been classified as Cr(III) [154].

13.39 Clinical Evaluations of the Body's Reactions to Implanted Medical Devices

Table 1 shows the toxic effects of metals and their associated implants [54]. One of the most crucial endpoints to be addressed in a biological evaluation of an implanted metallic device or biomaterial is the toxic systemic effects of chemicals released from the device materials. The discharge of metal ions, specific wear debris, or both from implanted metallic devices is usually what causes the harmful effects. Therefore, determining the likelihood that metal ions or wear debris released from alloys will cause negative toxicological reactions systemically after being transported to target tissues far from the implant is a crucial step in determining the biocompatibility of a device or biomaterial.

The intensity of the toxic response to the implanted device is often determined by the amount of metal ions or worn debris discharged from the device. Metallic implants corrode, which can cause metal ions to be released or wear debris to be produced. When a device corrodes, metal ions or wear debris may be released, which could have harmful toxicological effects or cause the gadget to malfunction. The cellular and systemic levels of toxicity are frequently considered.

Cytotoxicity, which is the term used to describe harmful effects on cells as a result of inflammatory and immunological reactions.

However, systemic toxicity may be brought on by overactive or severe immune and inflammatory responses, as well as by the direct chemical toxicity of wear, corrosion, and degradation products. However, this is not always the case. Systemic toxicity may be easily diagnosed because the damage to the target organs presents visible signs and symptoms. One of the usual causes that produce both no immune and immunological systemic toxicity is the degradation products of implant materials. Only a few epidemiological and experimental studies using in vitro and in vivo models have been done to study the toxicity caused by metals [155].

13.40 Fatigue Failure

For people aged 45–54, the rate of hip replacement more than doubled in just 10 years, 13 accounting for about 20% of all hip replacements, according to the Centres for Disease Control and Prevention (CDC). According to the World Health Organization's (WHO) most recent report on average life expectancy, 72 years was the average life expectancy in 2016. This means that the average lifetime requirement for implants is higher than 22 years. The most frequent type of mechanical failure that implants face is fatigue failure. The number of stress cycles a material can tolerate under fatigue circumstances is inversely correlated to the magnitude of the applied stress, meaning that the number of stress cycles a material can withstand rises as the stress intensity decreases. The endurance limit is the tension at which a material may survive ten million stress cycles without failing. The phases of fatigue-induced failure are catastrophic failure, crack propagation, and crack initiation [156, 157].

13.41 Metals That Undergo Biodegradation

The aforementioned nondegradable metallic biomaterials, which act as bone fixation devices like plates, screws, and pins, must be removed by a second surgery following adequate tissue recovery. An alternative to conventional fixation implants has arisen in the form of a novel class of biodegradable metals. According to Zheng et al. [158, 159], biodegradable metals are anticipated to corrode gradually in vivo while eliciting the proper host response in reaction to the released corrosion products and to dissolve completely once their goal of promoting tissue repair is achieved. Mg-based alloys, Fe-based alloys, and zinc (Zn)-based alloys are the three primary categories of biodegradable metals [160]. Among them, Mg-based alloys have undergone the most in-depth in vivo, in vitro, and clinical research.

Although pure magnesium does not contain or release alloying elements, the development of magnesium alloys is driven by the high corrosion rate brought on by the unavoidable presence of impurities and the relatively low strength of pure magnesium. Al is a typical alloying element in Mg alloys to improve both the mechanical qualities and corrosion resistance. Al has a high maximum solubility of 12.7 wt% in Mg. A lot of research has been done on typical Mg-Al-based alloys like AZ31, AZ61, AZ91, and AM60 as biodegradable metals. However, due to the probable neurotoxicity caused by Al, Mg-Al-based alloys are not advised for orthopaedic implants [161].

In order to create Mg-Zn, Mg-Zr, Mg-Ca, Mg-Sr, and Mg-Sn alloys for orthopaedic applications, various nontoxic or low toxic components, such as Zn, Zr, Ca, strontium (Sr), and tin (Sn), were chosen [158, 159]. Mg-based alloys hold more promise for temporary orthopaedic implants than conventional alloys such as stainless steels, Co-based alloys, and Ti-based alloys. Mg naturally breaks down in the human body, eliminating the need for a second procedure to remove the implants once the tissue has healed. In the salty environment of human tissues, magnesium tends to deteriorate. Due to their organic breakdown behaviour, magnesium and related alloys are useful as absorbable implant materials.

The density of magnesium, a light metal, is 1.74 g cm^{-3}, which is also quite comparable to the density of bone (1.8–2.1 g cm^{-3}). Mg has strong biocompatibility because it is a necessary component of the human body [162]. Due to the low standard electrode potential of Mg2(aq) 2e/Mg(s) (−2.37 V at 25 °C), the primary problem with Mg alloys in biomedical implant applications is the quick corrosion in the physiological environment during healing. Because of mechanical integrity loss before appropriate healing, released alloying elements, excessive hydrogen evolution, and a local alkaline environment, the significant degradation rate of magnesium alloys has so far prevented a widespread clinical application. Therefore, it is desired for Mg alloys to degrade at a rate that permits complete healing, and surface treatment is a useful and affordable tactic to improve the corrosion resistance of Mg-based alloys [163, 164].

13.42 Scaffolds for Tissue Engineering

Even though fracture fixation technology has evolved over the past century, it wasn't until the 1970s that cells and scaffolds were mixed in an effort to "engineer" tissue. Since then, the science of tissue engineering—or, as it is now more frequently known, regenerative medicine—has grown at an astounding rate. Similar to bone, 5–13% of fractures take longer than expected to heal or don't heal at all, necessitating a secondary surgery like bone grafting. Sadly, current reconstructive techniques still don't have a high enough level of clinical predictability, and they also have other problems like immunological rejection and a significant lack of tissue donors.

In order to solve this problem, research attention has mostly switched to degradable and nondegradable synthetic and natural matrices. The materials used in scaffolds for bone tissue engineering can be roughly categorised as temporary (degradable) or permanent (non-biodegradable). Materials can be ceramics like hydroxyapatite and/or tricalcium phosphate (in a variety of shapes and chemical proportions), natural or artificial polymers like polylactic acid and polyglycolic acid, or more recently composite or hybrid materials. There are numerous research that have produced bone by combining the methods mentioned above [165]. The use of osteoconductive materials, which facilitate cell adhesion but do not promote differentiation, is a drawback of this method.

The addition of "osteoinductive" growth factors, such as bone morphogenetic protein (BMP)-2 and BMP-7, has given tissue engineering the capacity to induce osteoblast development and function.

Preclinical data show promise for cell-free [166] and cell-seeded scaffolds for de novo bone formation, the major clinical limitation of current bone tissue constructs. Despite limited success, the new generation of bone tissue constructs tends to favour multiphase materials that improve overall mechanical strength and allow different rates of degradation and growth factor release to improve tissue integration. Without the presence of the local vascular system, bone cannot live. The need for angiogenesis, the creation of blood vessels, is thus incorporated into the recently revised approach for bone tissue engineering. The majority of the research in this field is exploratory, although some interesting findings are starting to emerge [165].

13.43 Suture Anchors in Orthopaedic Surgery Made of Biomaterials

Suture anchors have transformed orthopaedic surgery because they make it easy and effective to fixate soft tissue (such as tendons and ligaments) to the bone during open and arthroscopic procedures at the shoulder, elbow, wrist, and lower limb joints. The types of procedures utilised in shoulder surgery, for example, have changed significantly from open repair of the rotator cuff and labrum utilising screws, washers, transosseous sutures, and staples to arthroscopic repair employing suture anchors.

The fundamental job of the suture anchor is to secure tissue at the appropriate location and hold it there without letting go or applying too much strain until physiologic healing is complete. Suture anchors should be simple to handle, maintain adequate pull-out strength, guard against suture abrasion, and be absorbable without causing any responses as they break down. Different types of anchors have been created, and during the past 10 years, anchor designs have changed to maximise their ability to produce a solid tendon-to-bone repair [167].

13.44 Suture Anchor Made of Metal

The original suture anchor designs were metallic and non-absorbable. A particular metal can be used alone or as an alloy with other metals. Titanium and stainless steel are the two metal anchors that are most frequently utilised. Titanium is a strong, lightweight material that is frequently used in orthopaedic applications. It can be mixed with iron or aluminium to increase strength and lightness. An alloy of iron, chromium, and carbon makes up stainless steel. It is stronger than pure iron and more corrosion-resistant than standard steel.

While titanium creates a surface layer of calcium and phosphate that connects directly to the bone without leaving any signs of this fibrous layer and with little indications of an inflammatory response, stainless steel anchors get enclosed by a fibrous membrane rich in inflammatory cells. Calcium and phosphate spontaneously precipitate on an oxide layer that has formed. Once attached, osteoblasts actively secrete osteoid matrix [168, 169].

Metallic suture anchors offer firm fixation and have been used successfully for a very long period, although they have a number of drawbacks, particularly when used near the shoulder. The risks include interference with diagnostic imaging, such as computed tomography scans and magnetic resonance imaging (MRI), migration, loosening, imprisonment of the metal anchors within the joint cavity, cartilage injury, and more. Additionally, if metal anchors are discovered, revision surgery is more challenging since the anchors must be avoided or removed and they discharge metal ions into the neighbouring tissues [170].

13.45 Suture Anchors That Are Biodegradable in Nature

Biodegradable suture anchors, created to help avoid the documented issues with metallic anchors, have recently become much more routinely utilised. When compared to metallic anchors, biodegradable anchors have the following advantages: (1) less difficult revision surgery; (2) improved postoperative imaging; (3) improved biocompatibility; and (4) absence of removal operation. Natural, synthetic, or biosynthetic polymers are biodegradable materials used in orthopaedic applications. These are created to be immune-suppressive and biocompatible. Over a hundred new polymers have recently been created for use in surgery. These polymers are long-chain macromolecules made up of a number of covalently bound monomers, which may be arranged in combinations or as a single repeating monomer [171, 172].

13.46 Polyglycolic Acid, Poly-Lactic Acid, Poly-L-Lactic Acid

One of the earliest biological sectors to conduct research on a degradable polymer was polyglycolic acid (PGA). PGA has been utilised by surgeons as a biodegradable suture since the 1970s. Although it was first employed as a biodegradable anchor as well, it was discontinued due to PGA's quick deterioration and subsequent lack of strength. Within the first week of anchor placement, PGA begins to degrade; as the glycolic acid products are released, they may result in synovitis, bursitis, or lytic bone abnormalities [3]. Later, poly-L-lactic acid (PLLA) was used to create anchors. It has been demonstrated that it dissolves very gradually and could last for up to 5 years. As a result of this characteristic, PLLA, in particular, is not as problematic as PGA; However, extremely slow rates of deterioration would prevent full bone replacement and would result in reactions to foreign bodies inside the osseous cavity. Bioabsorbable anchors' mechanical characteristics have been strengthened and the degradation time has been controlled with the help of copolymers like poly (D, L-lactide), which is made of L-lactide and D-lactide, and PLLA with PGA. Notably, there are drawbacks to using biodegradable suture anchors, such as issues during surgery or in the first few days afterward, such as (1) implant breakage during anchor insertion, (2) initial fixation loss, (3) incomplete anchor burial within a bone, which could harm articular cartilage, and (4) potential anchor migration [173]. Following surgery, there may be difficulties with (1) the onset of inflammatory responses leading to osteolysis, (2) cyst formation, and (3) the development of intra-articular granulomas associated with eodema. After rotator cuff repair, cyst development and osteolysis are two of these issues that need to be addressed. After the tendon that connects the bone to the bone heals, doctors anticipate the biodegradable anchors will be absorbed and replaced by bone. Grades for the fluid signal around the anchor on a T2-weighted MRI scan were given, with grade 0 meaning there was no fluid signal, grade 1 meaning there was little fluid collection, grade 2 meaning there was local fluid collection, grade 3 meaning there was fluid collection along the entire length of the anchor and grade 4 meaning the cyst diameter was larger than grade 3 [167, 174, 175].

13.47 Implants and Cancers in Humans and Animals

Despite the high number of implants used in clinical settings over a long length of time, neoplasms at the location of implanted medical devices are uncommon. There have been about 100 cases of tumours linked to orthopaedic and other surgical implants, and there have also been cases of tumours linked to other foreign substances, such as bullets, shrapnel, other metal fragments, sutures, bone wax, and surgical sponge. Sarcomas (i.e., mesenchymal tumours) make up the vast majority of malignant neoplasms connected to clinical fracture fixation devices, total joint replacements, vascular grafts, breast implants, and experimental foreign bodies in both people and animals. They include a number of histologic subtypes and are characterised by rapid and locally infiltrative growth, such as fibrosarcoma, osteosarcoma (osteogenic sarcoma), chondrosarcoma, malignant fibrous histiocytoma, and angiosarcoma. Both short- and long-term tumours associated with therapeutic implants have been documented after implantation. In a more recent case series and study that was limited to orthopaedic implant sarcomas [176], most tumours that were related to implants manifested as early as 0.5 years after surgery and as late as 30 years later. Three years after the installation of a clinical titanium total hip replacement for osteoarthritis in a 68-year-old man, an osteosarcoma was observed to grow nearby. Since the earliest signs of the tumour were mistaken for those of the underlying orthopaedic issue, the diagnosis was delayed. Carcinomas have been described in conjunction with foreign bodies far less frequently, and are typically only present when an implant has been inserted into the lumen of an organ that is lined with epithelium. According to several studies [176], lymphomas have also been linked to the capsules that enclose breast implants. It has also been documented that a primary tumour unrelated to implants—gastric cancer—metastasised to a total knee replacement.

It is still debatable whether implanted medical devices have an incidental influence in local or distant cancer in general or in certain circumstances. There is no evidence in humans for the tumorgenicity of non-metallic and metallic surgical implants, according to extensive epidemiological research and evaluations of the available data [176]. Some studies indicate that there may be a modest increase in the risk of haematological malignancies, such as leukaemias and lymphomas, despite the fact that cancers at the majority of other (systemic) sites do not appear to be affected. In this regard, one study suggests enhanced surveillance in total hip recipients with metal-on-metal articulations, where the number of metallic particles is estimated to be much higher than with metal-on-plastic bearings and the size of the particles is smaller, with a correspondingly high reactive surface area. Such cancers occur in organs where exogenous elements and particles may accumulate (e.g. lymph nodes) [176].

13.48 Conclusion

The toxicity of the substance can be significantly influenced by the physical and chemical form of the metal discharged from a metallic implant, whether it be as metal ions or as particulate metal. Therefore, while evaluating the risk caused by patient exposure to metals discharged from metallic implants, it is crucial to take this into consideration. When determining the risk posed by patient exposure to metals released from implanted metallic medical devices, specific considerations should be made, including the need to take into account the form (particles vs. ionic) and valence of the compound released from the device, the ability to estimate the dose of the compound released from the device using biomonitoring data, and the need to take into account systemic effects on target organs far from the implant. To determine the proper goals for the studies that manufacturers are now required to do as well as the proper clinical monitoring of patients, direct data is needed for the effects of chronic exposure to metal on systemic organ function in patients with implants that are functioning well.

References

1. https://pacificaorthopedics.org/orthopedics/importance-of-orthopedics-in-medicine/
2. Meakins JL. Site and side of surgery: getting it right. Can J Surg. 2003;46(2):85–9.
3. El-Tallawy SN, Nalamasu R, Salem GI, LeQuang JAK, Pergolizzi JV, Christo PJ. Management of musculoskeletal pain: an update with emphasis on chronic musculoskeletal pain. Pain Ther. 2021;10(1):181–209. https://doi.org/10.1007/s40122-021-00235-2.
4. Wood AM, Brock TM, Heil K, Holmes R, Weusten A. A review on the management of hip and knee osteoarthritis. Int J Chronic Dis. 2013;2013:845015. https://doi.org/10.1155/2013/845015.
5. Kim T, See CW, Li X, Zhu D. Orthopedic implants and devices for bone fractures and defects: past, present and perspective, engineered regeneration. Eng Regener. 2020;1:6–18. ISSN: 2666-1381
6. Moghadasi K, et al. A review on biomedical implant materials and the effect of friction stir based techniques on their mechanical and tribological properties. J Mater Res Technol. 2022;17:1054–121. ISSN: 2238-7854
7. Biały M, Hasiak M, Łaszcz A. Review on biocompatibility and prospect biomedical applications of novel functional metallic glasses. J Funct Biomater. 2022;13(4):245. https://doi.org/10.3390/jfb13040245.
8. Hallab NJ, Anderson S, Stafford T, Glant T, Jacobs JJ. Lymphocyte responses in patients with total hip arthroplasty. J Orthop Res. 2005;23:384–91.
9. Kheder W, Al Kawas S, Khalaf K, Samsudin AR. Impact of tribocorrosion and titanium particles release on dental implant complications – a narrative review. Jpn Dent Sci Rev. 2021;57:182–9. https://doi.org/10.1016/j.jdsr.2021.09.001.
10. Hallab NJ, Jacobs JJ. Biologic effects of implant debris. Bull NYU Hosp Jt Dis. 2009;67:182–8.
11. Eliaz N. Corrosion of metallic biomaterials: a review. Materials (Basel). 2019;12(3):407. https://doi.org/10.3390/ma12030407.
12. Park J, Lakes RS. Biomaterials: an introduction. Springer; 2007.
13. Navarro M, Michiardi A, Castaño O, Planell JA. Biomaterials in orthopaedics. J R Soc Interface. 2008;5(27):1137–58. https://doi.org/10.1098/rsif.2008.0151.
14. https://rad.washington.edu/about-us/academic-sections/musculoskeletal-radiology/teaching-materials/online-musculoskeletal-radiology-book/orthopedic-hardware/
15. https://monib-health.com/en/post/17-orthopedic-screws
16. https://orthopaedicprinciples.com/2013/06/bone-screws-in-orthopaedic-surgery/
17. Loving VA, Richardson ML. Scaphoid fracture fixation with an Acutrak(®) screw. Radiol Case Rep. 2015;1(2):58–60. https://doi.org/10.2484/rcr.v1i2.13.
18. Gereli A, Nalbantoglu U, Sener IU, Kocaoglu B, Turkmen M. Comparison of headless screws used in the treatment of proximal nonunion of scaphoid bone. Int Orthop. 2011;35(7):1031–5. https://doi.org/10.1007/s00264-010-1129-y.

19. Sahoo NK, Anand SC, Bhardwaj JR, Sachdeva VP, Sapru BL. Bone response to stainless steel and titanium bone plates: an experimental study on animals. Med J Armed Forces India. 1994;50(1):10–4. https://doi.org/10.1016/S0377-1237(17)31029-8.
20. Narayanan G, Vernekar VN, Kuyinu EL, Laurencin CT. Poly (lactic acid)-based biomaterials for orthopaedic regenerative engineering. Adv Drug Deliv Rev. 2016;107:247–76. https://doi.org/10.1016/j.addr.2016.04.015.
21. Ramos DM, Dhandapani R, Subramanian A, Sethuraman S, Kumbar SG. Clinical complications of biodegradable screws for ligament injuries. Mater Sci Eng C. 2020;109:110423. ISSN: 0928-4931
22. Kramer DE, Kalish LA, Kocher MS, Yen YM, Micheli LJ, Heyworth BE. Complications of bioabsorbable tibial interference screws after anterior cruciate ligament reconstruction in pediatric and adolescent athletes. Orthop J Sports Med. 2020;8(2):2325967120904010. https://doi.org/10.1177/2325967120904010.
23. Uhthoff HK, Poitras P, Backman DS. Internal plate fixation of fractures: short history and recent developments. J Orthop Sci. 2006;11(2):118–26. https://doi.org/10.1007/s00776-005-0984-7.
24. Harasen G. Orthopedic hardware and equipment for the beginner: Part 1: Pins and wires. Can Vet J. 2011;52(9):1025–6.
25. John J, Wagner WW, Kuiper JH. Tension-band wiring of transverse fractures of patella. The effect of site of wire twists and orientation of stainless steel wire loop: a biomechanical investigation. Int Orthop. 2007;31(5):703–7. https://doi.org/10.1007/s00264-006-0238-0.
26. Wood GW 2nd. Intramedullary nailing of femoral and tibial shaft fractures. J Orthop Sci. 2006;11(6):657–69. https://doi.org/10.1007/s00776-006-1061-6.
27. SPRINT Investigators, Bhandari M, Guyatt G, Tornetta P 3rd, Schemitsch E, Swiontkowski M, Sanders D, Walter SD. Study to prospectively evaluate reamed intramedually nails in patients with tibial fractures (S.P.R.I.N.T.): study rationale and design. BMC Musculoskelet Disord. 2008;9:91. https://doi.org/10.1186/1471-2474-9-91.
28. https://www.siiora.lk/blog/functioning-of-rods-and-nails/
29. Hadeed A, Werntz RL, Varacallo M. External fixation principles and overview. In: StatPearls [Internet]. Treasure Island: StatPearls Publishing; 2022.
30. Cross WW 3rd, Swiontkowski MF. Treatment principles in the management of open fractures. Indian J Orthop. 2008;42(4):377–86. https://doi.org/10.4103/0019-5413.43373.
31. https://www.healthcentral.com/condition/back-pain/spinal-fusion-instrumentation-removal-pros-cons
32. https://www.healthcentral.com/condition/back-pain/spinal-hardware-removal
33. Nouh MR. Spinal fusion-hardware construct: basic concepts and imaging review. World J Radiol. 2012;4(5):193–207. https://doi.org/10.4329/wjr.v4.i5.193.
34. Mjöberg B. Hip prosthetic loosening: a very personal review. World J Orthop. 2021;12(9):629–39. https://doi.org/10.5312/wjo.v12.i9.629.
35. Sözen T, Özışık L, Başaran NÇ. An overview and management of osteoporosis. Eur J Rheumatol. 2017;4(1):46–56. https://doi.org/10.5152/eurjrheum.2016.048.
36. Reith G, Schmitz-Greven V, Hensel KO, Schneider MM, Tinschmann T, Bouillon B, Probst C. Metal implant removal: benefits and drawbacks – a patient survey. BMC Surg. 2015;15:96. https://doi.org/10.1186/s12893-015-0081-6.
37. https://www.footcaremd.org/conditions-treatments/injections-and-other-treatments/hardware-removal
38. Tapscott DC, Wottowa C. Orthopedic implant materials. [Updated 2022 Jul 25]. In: StatPearls [Internet]. Treasure Island: StatPearls Publishing; 2023.
39. Chouirfa H, Bouloussa H, Migonney V, Falentin-Daudré C. Review of titanium surface modification techniques and coatings for antibacterial applications. Acta Biomater. 2019;83:37–54.
40. Yılmaz E, Gökçe A, Findik F, Gulsoy HO, İyibilgin O. Mechanical properties and electrochemical behavior of porous Ti-Nb biomaterials. J Mech Behav Biomed Mater. 2018;87:59–67.

41. Prasad K, Bazaka O, Chua M, Rochford M, Fedrick L, Spoor J, Symes R, Tieppo M, Collins C, Cao A, Markwell D, Ostrikov KK, Bazaka K. Metallic biomaterials: current challenges and opportunities. Materials (Basel). 2017;10(8):884. https://doi.org/10.3390/ma10080884.
42. Gyaneshwar T, Nitesh R, Sagar T, Pranav K, Rustagi N. Treatment of pediatric femoral shaft fractures by stainless steel and titanium elastic nail system: a randomized comparative trial. Chin J Traumatol. 2016;19(4):213–6.
43. Bostrom MP, Asnis SE, Ernberg JJ, Wright TM, Giddings VL, Berberian WS, Missri AA. Fatigue testing of cerclage stainless steel wire fixation. J Orthop Trauma. 1994;8(5):422–8.
44. Meekes C, Schouten BJM, Nix M, Ongkiehong BF, Wolterbeek R, Van der Wal BCH, Nelissen RGHH. Pseudotumor in metal-on-metal hip arthroplasty: a comparison study of three grading systems with MRI. Skeletal Radiol. 2018;47(8):1099–109. https://doi.org/10.1007/s00256-018-2873-0.
45. Ling TX, Li JL, Zhou K, Xiao Q, Pei FX, Zhou ZK. The use of porous tantalum augments for the reconstruction of acetabular defect in primary total hip arthroplasty. J Arthroplasty. 2018;33(2):453–9.
46. Devane PA, Horne JG, Ashmore A, Mutimer J, Kim W, Stanley J. Highly cross-linked polyethylene reduces wear and revision rates in total hip arthroplasty: a 10-year double-blinded randomized controlled trial. J Bone Joint Surg Am. 2017;99(20):1703–14.
47. Lambert B, Neut D, van der Veen HC, Bulstra SK. Effects of vitamin E incorporation in polyethylene on oxidative degradation, wear rates, immune response, and infections in total joint arthroplasty: a review of the current literature. Int Orthop. 2019;43(7):1549–57.
48. Fisher J, Jin Z, Tipper J, Stone M, Ingham E. Tribology of alternative bearings. Clin Orthop Relat Res. 2006;453:25–34.
49. Gillespie JA, Kennedy JW, Patil SR, Meek DR. Noise production in ceramic-on-ceramic total hip arthroplasty is associated with lower patient satisfaction and hip scores. J Orthop. 2016;13(4):282–4.
50. https://www.drugwatch.com/hip-replacement/metallosis/
51. Pritchett J. Adverse reaction to metal debris: metallosis of the resurfaced hip. Curr Orthop Pract. 2012;23(1):50–8. https://doi.org/10.1097/bco.0B013E3182356075.
52. Alaska Epidemiology Bulletin, May 28, 2010.
53. https://www.fda.gov/medical-devices/metal-metal-hip-implants/information-all-health-care-professionals-who-provide-treatment-patients-metal-metal-hip-implant
54. Villero Suárez J, Farak Gómez J, Pérez García M, et al. Poisoning by metals used in prothetic materials in orthopedics and its current management. Rev Cuba J Ortop Traumatol. 2021;35(2):e420.
55. Drummond J, Tran P, Fary C. Metal-on-metal hip arthroplasty: a review of adverse reactions and patient management. J Funct Biomater. 2015;6(3):486–99. https://doi.org/10.3390/jfb6030486.
56. Wawrzynski J, Gil JA, Goodman AD, Waryasz GR. Hypersensitivity to orthopedic implants: a review of the literature. Rheumatol Ther. 2017;4(1):45–56.
57. Pacheco KA. Allergy to surgical implants. Clin Rev Allergy Immunol. 2019;56(1):72–85.
58. Xie Y, Hu C, Feng Y, Li D, Ai T, Huang Y, et al. Osteoimmunomodulatory effects of biomaterial modification strategies on macrophage polarization and bone regeneration. Regen Biomater. 2020;7(3):233–45.
59. Haddad SF, Helm MM, Meath B, Adams C, Packianathan N, Uhl R. Exploring the incidence, implications, and relevance of metal allergy to orthopaedic surgeons. J Am Acad Orthop Surg Glob Res Rev. 2019;3(4):e023.
60. Merritt K, Rodrigo JJ. Immune response to synthetic materials. Sensitization of patients receiving orthopaedic implants. Clin Orthop Relat Res. 1996;326:71–9.
61. Deutman R, Mulder TJ, Brian R, Nater JP. Metal sensitivity before and after total hip arthroplasty. J Bone Joint Surg Am. 1977;59(7):862–5.
62. Elves MW, Wilson JN, Scales JT, Kemp HB. Incidence of metal sensitivity in patients with total joint replacements. Br Med J. 1975;4(5993):376–8.

63. Waterman AH, Schrik JJ. Allergy in hip arthroplasty. Contact Derm. 1985;13(5):294–301.
64. Merritt K. Role of medical materials, both in implant and surface applications, in immune response and in resistance to infection. Biomaterials. 1984;5(1):47–53.
65. Brown GC, Lockshin MD, Salvati EA, Bullough PG. Sensitivity to metal as a possible cause of sterile loosening after cobalt-chromium total hip-replacement arthroplasty. J Bone Joint Surg Am. 1977;59(2):164–8.
66. Evans EM, Freeman MA, Miller AJ, Vernon-Roberts B. Metal sensitivity as a cause of bone necrosis and loosening of the prosthesis in total joint replacement. J Bone Joint Surg Br. 1974;56-B(4):626–42.
67. Goldring SR, Clark CR, Wright TM. The problem in total joint arthroplasty: aseptic loosening. J Bone Joint Surg Am. 1993;75(6):799–801.
68. Merritt K, Brown SA. Metal sensitivity reactions to orthopedic implants. Int J Dermatol. 1981;20(2):89–94.
69. Huber M, Reinisch G, Trettenhahn G, Zweymüller K, Lintner F. Presence of corrosion products and hypersensitivity-associated reactions in periprosthetic tissue after aseptic loosening of total hip replacements with metal bearing surfaces. Acta Biomater. 2009;5(1):172–80.
70. Merritt K, Brown SA. Biological effects of corrosion products from metals. In: Fraker A, Griffin C, editors. Corrosion and degradation of implanted materials: second symposium. ASTM STP 859. Philadelphia: American Society for Testing and Materials; 1985. p. 105–16.
71. Panigutti MA, Merritt K, Bruner RJ, et al. Correlation of allergy, metal levels, implant alloy, and implant damage in patients undergoing revision joint arthroplasties. Trans Soc Biomater. 1992;15:7.
72. Rostoker G, Robin J, Binet O, Blamoutier J, Paupe J, Lessana-Leibowitch M, et al. Dermatitis due to orthopaedic implants. A review of the literature and report of three cases. J Bone Joint Surg Am. 1987;69(9):1408–12.
73. Szliska C, Raskoski J. Sensitization to nickel, cobalt and chromium in surgical patients. Contact Derm. 1990;23(5):378–9.
74. Akil S, Newman JM, Shah NV, Ahmed N, Deshmukh AJ, Maheshwari AV. Metal hypersensitivity in total hip and knee arthroplasty: current concepts. J Clin Orthop Trauma. 2018;9(1):3–6.
75. Negrescu AM, Cimpean A. The State of the Art and prospects for osteoimmunomodulatory biomaterials. Materials (Basel). 2021;11:14.
76. https://emedicine.medscape.com/article/1230696-overview
77. Washington University in St Louis. Unusual skin cancer linked to chronic allergy from metal orthopedic implant. ScienceDaily. 9 Oct 2014. Available at https://www.sciencedaily.com/releases/2014/10/141009153817.htm. Accessed 5 Apr 2022.
78. Demehri S, Cunningham TJ, Hurst EA, Schaffer A, Sheinbein DM, Yokoyama WM. Chronic allergic contact dermatitis promotes skin cancer. J Clin Invest. 2014;124(11):5037–41.
79. Schalock PC, Thyssen JP. Patch testers' opinions regarding diagnostic criteria for metal hypersensitivity reactions to metallic implants. Dermatitis. 2013;24(4):183–5.
80. Amini M, Mayes WH, Tzeng A, Tzeng TH, Saleh KJ, Mihalko WM. Evaluation and management of metal hypersensitivity in total joint arthroplasty: a systematic review. J Long Term Eff Med Implants. 2014;24(1):25–36.
81. Lachiewicz PF, Watters TS, Jacobs JJ. Metal hypersensitivity and total knee arthroplasty. J Am Acad Orthop Surg. 2016;24(2):106–12.
82. Sampson B, Hart A. Clinical usefulness of blood metal measurements to assess the failure of metal-on-metal hip implants. Ann Clin Biochem. 2012;49(Pt 2):118–31.
83. Koutalos AA, Toms AP, Cahir JG, Smith EJ. Correlation of MARS MRI findings with cup position, metal ion levels and function in metal-on-metal total hip arthroplasty. Hip Int. 2020;30(1):64–70.
84. Markel D, Bou-Akl T, Rossi M, Pizzimenti NM, Wu B, Ren WP. Response profiles of circulating leukocytes and metal ions in patients with a modular dual-mobility hip implant. Hip Int. 2019;21:1120700019865530.

85. Pahlavan S, Hegde V, Bracey DN, Jennings JM, Dennis DA. Bone cement hypersensitivity in patients with a painful total knee arthroplasty: a case series of revision using custom cementless implants. Arthroplast Today. 2021;11:20–4.
86. Gallo J, Holinka M, Moucha CS. Antibacterial surface treatment for orthopaedic implants. Int J Mol Sci. 2014;15(8):13849–80.
87. Li T, Wang N, Chen S, Lu R, Li H, Zhang Z. Antibacterial activity and cytocompatibility of an implant coating consisting of TiO2 nanotubes combined with a GL13K antimicrobial peptide. Int J Nanomed. 2017;12:2995–3007.
88. Brown A, Mandelberg NJ, Munoz-Mendoza D, Palys V, Schalock PC, Mogilner A, et al. Allergy considerations in implanted neuromodulation devices. Neuromodulation. 2021;24(8):1307–16.
89. Thomas P, von der Helm C, Schopf C, Mazoochian F, Frommelt L, Gollwitzer H, et al. Patients with intolerance reactions to total knee replacement: combined assessment of allergy diagnostics, periprosthetic histology, and peri-implant cytokine expression pattern. Biomed Res Int. 2015;2015:910156.
90. Özçelik H, Vrana NE, Gudima A, Riabov V, Gratchev A, Haikel Y, et al. Harnessing the multifunctionality in nature: a bioactive agent release system with self-antimicrobial and immunomodulatory properties. Adv Healthc Mater. 2015;4(13):2026–36.
91. Kzhyshkowska J, Gudima A, Riabov V, Dollinger C, Lavalle P, Vrana NE. Macrophage responses to implants: prospects for personalized medicine. J Leukoc Biol. 2015;98(6):953–62.
92. Eltit F, Noble J, Sharma M, Benam N, Haegert A, Bell RH, et al. Cobalt ions induce metabolic stress in synovial fibroblasts and secretion of cytokines/chemokines that may be diagnostic markers for adverse local tissue reactions to hip implants. Acta Biomater. 2021;131:581–94.
93. Alobaid MA, Richards SJ, Alexander MR, Gibson MI, Ghaemmaghami AM. Developing immune-regulatory materials using immobilized monosaccharides with immune-instructive properties. Mater Today Bio. 2020;8:100080.
94. Goodman SB, Gallo J, Gibon E, Takagi M. Diagnosis and management of implant debris-associated inflammation. Expert Rev Med Devices. 2020;17(1):41–56.
95. Whittingham-Jones PM, Dunstan E, Altaf H, Cannon SR, Revell PA, Briggs TW. Immune responses in patients with metal-on-metal hip articulations: a long-term follow-up. J Arthroplasty. 2008;23(8):1212–8.
96. Granchi D, Cenni E, Tigani D, Trisolino G, Baldini N, Giunti A. Sensitivity to implant materials in patients with total knee arthroplasties. Biomaterials. 2008;29(10):1494–500.
97. Hallab NJ, Caicedo M, Finnegan A, Jacobs JJ. Th1 type lymphocyte reactivity to metals in patients with total hip arthroplasty. J Orthop Surg Res. 2008;3:6.
98. Fujishiro T, Moojen DJ, Kobayashi N, Dhert WJ, Bauer TW. Perivascular and diffuse lymphocytic inflammation are not specific for failed metal-on-metal hip implants. Clin Orthop Relat Res. 2011;469(4):1127–33.
99. St John K, editor. Particulate debris from medical implants. ASTM STP 1144. Philadelphia: American Society for Testing and Materials; 1992.
100. Goodman SB. Wear particles, periprosthetic osteolysis and the immune system. Biomaterials. 2007;28(34):5044–8.
101. Wright TM, Goodman SB, editors. Implant wear in total joint replacement: clinical and biologic issues, material and design considerations. Rosemont: American Academy of Orthopaedic Surgeons; 2001. p. 61–70.
102. Keegan GM, Learmonth ID, Case CP. Orthopaedic metals and their potential toxicity in the arthroplasty patient. J Bone Joint Surg Br. 2007;89-B(5):567–73. https://doi.org/10.1302/0301-620X.89B5.18903.
103. Bommala VK, Krishna MG. 14 Magnesio metal. Compuestos en aplicaciones biomédicas. 2020;273
104. Bitar D, Parvizi J. Biological response to prosthetic debris. World J Orthop. 2015;6(2):172–89. https://doi.org/10.5312/wjo.v6.i2.172.

105. Xiong P, Huang X, Ye N, Lu Q, Zhang G, Peng S, Wang H, Liu Y. Cytotoxicity of metal-based nanoparticles: from mechanisms and methods of evaluation to pathological manifestations. Adv Sci (Weinh). 2022;9(16):e2106049. https://doi.org/10.1002/advs.202106049.
106. Manke A, Wang L, Rojanasakul Y. Mechanisms of nanoparticle-induced oxidative stress and toxicity. Biomed Res Int. 2013;2013:942916. https://doi.org/10.1155/2013/942916.
107. Theodoros B, Thomas G, Peter G. Granulomatous lung disease: a novel complication following metallosis from hip arthroplasty. Hip Pelvis. 2016;28(4):249–53.
108. Powell SK, Cruz RL, Ross MT, Woodruff MA. Past, present, and future of soft-tissue prosthetics: advanced polymers and advanced manufacturing. Adv Mater. 2020;32(42):e2001122.
109. Kurosaki K, Nakamura T, Mukai T, Endo T. Unusual findings in a fatal case of poisoning with chromate compounds. Forensic Sci Int. 1995;75:57–65.
110. Kametani K, Nagata T. Quantitative elemental analysis on aluminium accumulation by HVTEM-EDX in liver tissues of mice orally administered with aluminium chloride. Med Mol Morphol. 2006;39:97–105.
111. Oliveira H, Santos TM, Ramalho-Santos J, de Lourdes Pereira M. Histopathological effects of hexavalent chromium in mouse kidney. Bull Environ Contam Toxicol. 2006;76:977–83.
112. Barceloux DG. Chromium. Clin Toxicol. 1999;37:173–94. Bonde J, Vittinghus E. Urinary excretion of proteins among metal welders. Human Exp Toxicol. 1996;15:1–4.
113. Nemery B. Metal toxicity and the respiratory tract. Eur Respir J. 1990;3(202–19):72.
114. Antonini J, Lewis AB, Roberts JR, Whaley DA. Pulmonary effects of welding fumes: review of worker and experimental animal studies. Am J Indust Med. 2003;43:350–60.
115. Olivieri G, Novakovic M, Savaskan E, et al. The effects of beta-estradiol on SHSY5Y neuroblastoma cells during heavy metal induced oxidative stress, neurotoxicity and betaamyloid secretion. Neuroscience. 2002;113:849–55.
116. Youdim MB, Ben-Shachar D, Riederer P. Iron in brain functions and dysfunction with emphasis on Parkinson's disease. Eur Neurol. 1991;31(Suppl 1):34–40.
117. Travacio M, Polo JM, Llesuy S. Chromium (VI) induces oxidative stress in the mouse brain. Toxicology. 2001;162:139–48.
118. Garcia GB, Biancardi M, Quiroga A. Vanadium (V)-induced neurotoxicity in the rat central nervous system: a histo-immunohistochemical study. Drug Chem Toxicol. 2005;28:329–44.
119. Barth A, Schaffer AW, Komaris C, et al. Neurobehavioural effects of vanadium. J Toxicol Environ Health [Am]. 2002;65:677–83.
120. Barceloux D. Cobalt. J Toxicol Clin Toxicol. 1999;37:201–16.
121. Linna A, Oksa P, Groundstroem K, et al. Exposure to cobalt in the production of cobalt and cobalt compounds and its effects on the heart. Occup Environ Med. 2004;61:877–85.
122. Lippmann M, Ito K, Hwang JS, Maciejczyk P, Chen LC. Cardiovascular effects of nickel in ambient air. Environ Health Perspect. 2006;114:1662–9.
123. Jeffery E, Ebero K, Burgess E, Cannata J, Greger JL. Systemic aluminium toxicity: effects on bone, hematopoietic tissue, and kidney. J Toxicol Environ Health. 1996;48:649–65.
124. Vermes C, Glant TT, Hallab NJ, et al. The potential role of the osteoblast in the development of periprosthetic osteolysis: review of in vitro osteoblast responses to wear debris, corrosion products, and cytokines and growth factors. J Arthroplasty. 2001;16(Suppl 1):95–100.
125. Darbre PD. Metalloestrogens: an emerging class of inorganic xenoestrogens with potential to add to the oestrogenic burden of the human breast. J Appl Toxicol. 2006;26:191–7.
126. Murthy RC, Junaid M, Saxena D. Ovarian dysfunction in mice following chromium (VI) exposure. Toxicol Lett. 1996;89:147–54.
127. Brock T, Stopford W. Bioaccessibility of metals in human health risk assessment: evaluating risk from exposure to cobalt compounds. J Environ Monit. 2003;5:71–7.
128. Prescott E, Netterstrom B, Faber J, et al. Effect of occupational exposure to cobalt blue dyes on the thyroid volume and function of female plate painters. Scand J Work Environ Health. 1992;18(101–4):8.
129. Lu Z-Y, Gong H, Ameniya J. Aluminum chloride induces retinal changes in the rat. Toxicol Sci. 2002;66:253–60.

130. Khosla PK, Murthy KS, Tewari H. Retinal toxicity of trace elements. Indian J Ophthalmol. 1987;35:311–4.
131. Steens W, von Foerster G, Katzer A. Severe cobtalt poisoning with loss of sight after ceramic-metal pairing in a hip: a case report. Acta Orthop. 2006;77:830–2.
132. Hallab N, Jacobs J, Black J. Hypersensitivity to metallic biomaterials: a review of leukocyte migration inhibition assays. Biomaterials. 2000;21:1301–14.
133. Hallab N, Mikecz K, Jacobs J. Metal sensitivity in patients with orthopaedic implants. J Bone Joint Surg [Am]. 2001;83-A:428–36.
134. Aruldhas M, Subramaniam S, Sekar P, et al. Chronic chromium exposure-induced changes in testicular histoarchitecture are associated with oxidative stress: study in a non-human primate (Macaca radiata Geoffroy). Human Reprod. 2005;20:2801–13.
135. Elbetieha A, Al-Hamood MH. Long-term exposure of male and female mice to trivalent and hexavalent chromium compounds: effect on fertility. Toxicology. 1997;116:39–47.
136. Bonde J. The risk of male subfecundity attributable to weldng of metals: studies of semen quality, infertility, adverse pregnancy outcome, and childhood malignancy. Int J Androl. 1993;16(Suppl 1):1–29.
137. Kumar S, Sathwara NG, Gautam AK, et al. Semen quality of industrial workers occupationally exposed to chromium. J Occup Health. 2005;47:424–30.
138. Pandey R, Kumar R, Singh SP, Saxena DK, Srivastava SP. Male reproductive effect of nickel sulphate in mice. Biometals. 1999;12:339–46.
139. Domingo JL. Vanadium: a review of the reproductive and developmental toxicity. Reprod Toxicol. 1996;10:175–82.
140. Llobet JM, Colomina MT, Sirvent JJ, Domingo JL, Corbella J. Reproductive toxicology of aluminium in male mice. Fundam Appl Toxicol. 1995;25:45–51.
141. Anderson MB, Pedigo NG, Katz RP, George WJ. Histopathology of testes from mice chronically treated with cobalt. Reprod Toxicol. 1992;6:41–50.
142. Kanojia R, Junaid M, Murthy P. Embryo and fetotoxicity of hexavalent chromium: a long-term study. Toxicol Lett. 1998;95(165–72):1.
143. Domingo J. Metal-induced developmental toxicity in mammals: a review. J Toxicol Environ Health. 1994;42:123–41.
144. Yu W, Sipowicz MA, Haines DG, et al. Preconception urethane or chromium (III) treatment of male mice: multiple neoplastic and non-neoplastic changes in offspring. Toxicol Appl Pharmacol. 1999;158:161–76.
145. Morgan A, El-Tawil O. Effects of ammonium metavandate on fertility and reproductive performance of adult male and female rats. Pharmacol Res. 2003;47:75–85.
146. Hjollund N, Bonde JP, Jensen JK, et al. Male-mediated spontaneous abortion among spouses of stainless steel welders. Scand J Work Environ Health. 2000;26:187–92.
147. O'Leary LM, Hicks AM, Peters JM, London S. Parental occupational exposures and risk of childhood cancer: a review. Am J Int Med. 1991;20:17–35.
148. Meldrum R, Feinberg JR, Capello WN, Detterline AJ. Clinical outcome and incidence of pregnancy after bipolar and total hip arthroplasty in young women. J Arthroplasty. 2003;18:879–85.
149. Ladon D, Doherty A, Newson R, et al. Changes in metal levels and chromosome aberrations in the peripheral blood of patients after metal-on-metal hip arthroplasty. J Arthroplasty. 2004;19(Suppl 3):78–83.
150. Iarmarcovai G, Sari-Minodier I, Chagpoul F, et al. Risk assessment of welders using analysis of eight metals by ICP-MS in blood and urine and DNA damage evaluation by the comet and micronucleus assays: influence of XRCC1 and XRCC2 polymorphisms. Mutagenesis. 2005;20:425–32.
151. Cole P, Rodu B. Epidemiologic studies of chrome and cancer mortality: a series of meta-analyses. Reg Toxicol Pharmacol. 2005;43:225–31.

152. Visuri T, Pukkala E, Pulkinen P, Paavolainen P. Decreased cancer risk in patients who have been operated on with total hip and knee arthroplasty for primary osteoarthritis: a meta-analysis of 6 Nordic cohorts with 73,000 patients. Acta Orthop Scand. 2003;74:351–60.
153. Visuri T, Pukkala E, Polkkinen P, Paavolainen P. Cancer incidence and causes of death among total hip replacement patients: a review based on Nordic cohorts with a special emphasis on metal-on-metal bearings. Proc Inst Mech Eng H. 2006;220:399–407.
154. No Authors Listed. IARC monographs on the evaluation of carcinogenic risk to humans. 2004. http://monographs.iarc.fr/. Date last accessed 5 Oct 2006.
155. https://www.fda.gov/media/131150/download
156. Teoh SH. Fatigue of biomaterials: a review. In: Nanobioceramic coatings for biomedical applications introduction: metallic implants. Santosh Kumar; 2014.
157. Friis EA, DeCoster TA, Thomas JC. Chapter 7 – Mechanical testing of fracture fixation devices. In: Friis E, editor. Mechanical testing of orthopaedic implants. Woodhead Publishing; 2017. p. 131–41. ISBN: 9780081002865.
158. Jin W, Chu PK. Orthopedic implants. In: Narayan R, editor. Encyclopedia of biomedical engineering. Elsevier; 2019. p. 425–39. ISBN: 9780128051443. https://doi.org/10.1016/B978-0-12-801238-3.10999-7.
159. Zheng YF, Gu XN, Witte F. Biodegradable metals. Mat Sci Eng R Rep. 2014;77:1–34.
160. Li HF, Zheng YF, Qin L. Progress of biodegradable metals. Prog Nat Sci Mater Int. 2014;24:414–22.
161. Yokel RA. The toxicology of aluminum in the brain: a review. Neurotoxicology. 2000;21:813–28.
162. Staiger MP, Pietak AM, Huadmai J, Dias G. Magnesium and its alloys as orthopedic biomaterials: a review. Biomaterials. 2006;27:1728–34.
163. Zhao Y, Yeung KWK, Chu PK. Functionalization of biomedical materials using plasma and related technologies. Appl Surf Sci. 2014;310:11–8.
164. Wu G, Ibrahim JM, Chu PK. Surface design of biodegradable magnesium alloys – a review. Surf Coat Technol. 2013;233:2–12.
165. Jones JR. Chapter 19 – Scaffolds for tissue engineering. In: Hench LL, Jones JR, editors. Woodhead Publishing series in biomaterials, biomaterials, artificial organs and tissue engineering. Woodhead Publishing; 2005. p. 201–14. ISBN: 9781855737372. https://doi.org/10.1533/9781845690861.4.201.
166. Sheikh Z, Najeeb S, Khurshid Z, Verma V, Rashid H, Glogauer M. Biodegradable materials for bone repair and tissue engineering applications. Materials (Basel). 2015;8(9):5744–94. https://doi.org/10.3390/ma8095273.
167. Chen Y, Gan C, Zhang T, Yu G, Bai P, Kaplan A. Laser-surface-alloyed carbon nanotubes reinforced hydroxyapatite composite coatings. Appl Phys Lett. 2005;86:251905. https://doi.org/10.1063/1.1951054.
168. Lei T, Wang L, Ouyang C, Li N-F, Zhou L-S. In situ preparation and enhanced mechanical properties of carbon nanotube/hydroxyapatite composites. Int J Appl Ceram Technol. 2011;8:532–9. https://doi.org/10.1111/j.1744-7402.2010.02602.x.
169. Cho CH, Bae KC, Kim DH. Biomaterials used for suture anchors in orthopedic surgery. Clin Orthop Surg. 2021;13(3):287–92. https://doi.org/10.4055/cios20317.
170. de Medeiros WS, de Oliveira MV, Pereira LC, de Andrade MC. Bioactive porous titanium: an alternative to surgical implants. Artif Organs. 2008;32(4):277–82.
171. Schlegel P, Hayes JS, Frauchiger VM, et al. An in vivo evaluation of the biocompatibility of anodic plasma chemical (APC) treatment of titanium with calcium phosphate. J Biomed Mater Res B Appl Biomater. 2009;90(1):26–34.
172. Weiler A, Hoffmann RF, Stahelin AC, Helling HJ, Sudkamp NP. Biodegradable implants in sports medicine: the biological base. Arthroscopy. 2000;16(3):305–21.
173. Nho SJ, Provencher MT, Seroyer ST, Romeo AA. Bioabsorbable anchors in glenohumeral shoulder surgery. Arthroscopy. 2009;25(7):788–93.

174. Goradia VK, Mullen DJ, Boucher HR, Parks BG, O'Donnell JB. Cyclic loading of rotator cuff repairs: a comparison of bioabsorbable tacks with metal suture anchors and transosseous sutures. Arthroscopy. 2001;17(4):360–4.
175. Barber FA. Complications of biodegradable materials: anchors and interference screws. Sports Med Arthrosc Rev. 2015;23(3):149–55.
176. Barber FA. Biodegradable materials: anchors and interference screws. Sports Med Arthrosc Rev. 2015;23(3):112–7.
177. Kim SH, Oh JH, Lee OS, Lee HR, Hargens AR. Postoperative imaging of bioabsorbable anchors in rotator cuff repair. Am J Sports Med. 2014;42(3):552–7.
178. Schoen FJ. Chapter II.2.7 – Tumors associated with biomaterials and implants. In: Ratner BD, Hoffman AS, Schoen FJ, Lemons JE, editors. Biomaterials science. 3rd ed. Academic; 2013. p. 558–65. ISBN: 9780123746269. https://doi.org/10.1016/B978-0-08-087780-8.00049-8.

Chapter 14
Significant Risks Medical Devices – Radiology

Pugazhenthan Thangaraju, B. Aravind Kumar, Hemasri Velmurugan, Sajitha Venkatesan, Ripudaman Arora, and Soumitra Trivedi

14.1 Introduction

Medical devices in radiology play a vital role in diagnosing and treating medical conditions through the use of various imaging technologies. These devices include X-ray machines, CT scanners, MRI machines, ultrasound machines, and others. Radiological medical devices have significantly improved patient outcomes by providing non-invasive and accurate diagnostic information, enabling early detection and treatment of various diseases. However, these devices also pose risks to patients, such as radiation exposure, mechanical failure, software malfunctions, and other safety hazards. Therefore, it is essential to carefully assess and manage the risks associated with radiological medical devices throughout their life cycle, from design and development to post-market surveillance. Effective risk management can help ensure that radiological medical devices are safe and effective for patients and healthcare providers. The US Food and Drug Administration (FDA) has developed guidelines for managing the significant risks associated with medical devices used

P. Thangaraju (✉)
Department of Pharmacology, All India Institute of Medical Sciences, Raipur, Chhattisgarh, India

B. Aravind Kumar · H. Velmurugan
Department of Pharmacology, Pondicherry Institute of Medical Sciences, Pondicherry, India

S. Venkatesan
Department of Microbiology, All India Institute of Medical Sciences, Raipur, Chhattisgarh, India

R. Arora
Department of ENT, All India Institute of Medical Sciences, Raipur, Chhattisgarh, India

S. Trivedi
Department of Anatomy, All India Institute of Medical Sciences, Raipur, Chhattisgarh, India

P. S. Timiri Shanmugam et al. (eds.), *Significant and Nonsignificant Risk Medical Devices*, https://doi.org/10.1007/978-3-031-52838-5_14

in radiology. These guidelines provide a framework for identifying, evaluating, and mitigating the risks associated with imaging device. The guideline has listed Boron neutron capture therapy and hyperthermia systems and applicators as examples of significant risk devices in radiology.

14.2 Boron Neutron Capture Therapy

Boron Neutron Capture Therapy (BNCT) is a type of radiation therapy that uses boron-10, an isotope that readily absorbs low-energy thermal neutrons to deliver a high dose of radiation to cancer cells. Boron is delivered to cancer cells via a compound that selectively targets cells, such as boronated drugs. When the compound reaches the cancer cells, the cells are exposed to thermal neutrons, which causes the boron-10 to capture the neutron and release high-energy alpha particles, leading to cell death [1].

Boron Neutron Capture Therapy (BNCT) is primarily used for the treatment of:

- Brain tumours
- Skin cancers
- Head and neck tumours
- Other types of cancers (under investigation) [2].

14.2.1 Risks

- Normal tissue damage
- Development of secondary cancers
- Side effects of the boron delivery method
- Long-term effects on brain function
- Limited availability of treatment centrer and specialized expertise
- High cost of therapy and limited insurance coverage [2, 3].

14.2.2 Prior Investigations

Boron Neutron Capture Therapy (BNCT) is a type of radiation therapy that uses a combination of boron-containing compounds and neutron radiation to treat certain types of cancer. The USFDA has not yet approved BNCT for clinical use in the United States, but research is ongoing, and there have been investigations into the safety and effectiveness of BNCT for radiology use. Here are some examples of the investigations that have been conducted on BNCT for radiology use:

1. *Preclinical Studies*

Preclinical studies were conducted in animals to evaluate the safety and effectiveness of BNCT. These studies typically involve testing boron-containing compounds in animal models of cancer and evaluating the effectiveness of the therapy in reducing tumor growth.

2. *Clinical Trials*

Clinical trials were conducted to evaluate the safety and effectiveness of BNCT in humans. These trials typically involve testing BNCT in a group of patients with the condition for which the therapy is intended and comparing the results to those of a control group.

3. *Dosimetry Studies*

Dosimetry studies were conducted to determine the optimal doses of neutron radiation and boron-containing compounds that should be used in BNCT. These studies typically involve measuring the dose of radiation delivered to the tumor and surrounding tissue and adjusting the therapy accordingly.

4. *Toxicity Studies*

Toxicity studies were conducted to evaluate the potential side effects of BNCT. These studies typically involve monitoring patients for side effects during and after therapy and evaluating the severity and duration of adverse events.

5. *Quality Control and Assurance Studies*

Quality control and assurance studies were conducted to ensure that the equipment and procedures used in BNCT were consistent and produced reliable results. These studies typically involved monitoring the equipment and procedures used in BNCT and evaluating their performance over time.

14.3 Hyperthermia Systems and Applications

Hyperthermia is a clinically proven sensitizer to enhance the effectiveness of radiotherapy and chemotherapy in cancer patients. The aim of hyperthermia treatment is to raise the temperature of the tumour to 40–43 °C for one hour. Adding hyperthermia increases the tumour response rate typically by about 20%. The effect of hyperthermia is selective to tumours and radiation or chemotherapy-related toxicity is not increased [4].

Hyperthermia systems are used in healthcare for the following purposes:

1. Cancer Treatment

Hyperthermia is used in combination with radiation therapy and chemotherapy to enhance the efficacy of cancer treatment by increasing the temperature of tumor tissue, making it more vulnerable to the effects of radiation and chemotherapy.

2. Pain Management

Hyperthermia can be used to relieve pain in conditions such as arthritis, muscle strains, and sprains, by increasing blood flow to the affected area and reducing inflammation [6].

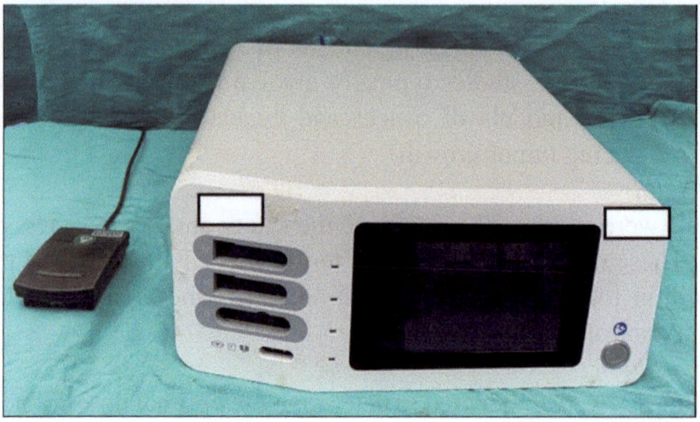

Fig. 14.1 Bipolar cautery machine

3. Wound Healing

Hyperthermia has been shown to promote wound healing by increasing blood flow and oxygenation in the affected area.

4. Physical Therapy

Hyperthermia can be used to enhance physical therapy by reducing pain and inflammation, improving flexibility, and promoting healing.

5. Cardiac Rehabilitation

Hyperthermia has been used to improve cardiovascular function and increase exercise tolerance in patients undergoing cardiac rehabilitation [5]. Figure 14.1 shows bipolar cautery machine.

In particular, hyperthermia systems are used in radiology for the following purposes:

1. Cancer treatment
2. Tumour ablation
3. Radiation sensitizer
4. Improved drug delivery
5. Non-invasive heating of tissue
6. Adjuvant to radiation therapy
7. Enhancement of immune response
8. Combination with chemotherapy
9. Improved blood flow to targeted area
10. Pain management.

The uses of radiology applicators include the following.

1. X-ray film holders
2. Intravenous (IV) contrast administration
3. Barium enema administration
4. CT and MRI contrast administration
5. Mammogram compression

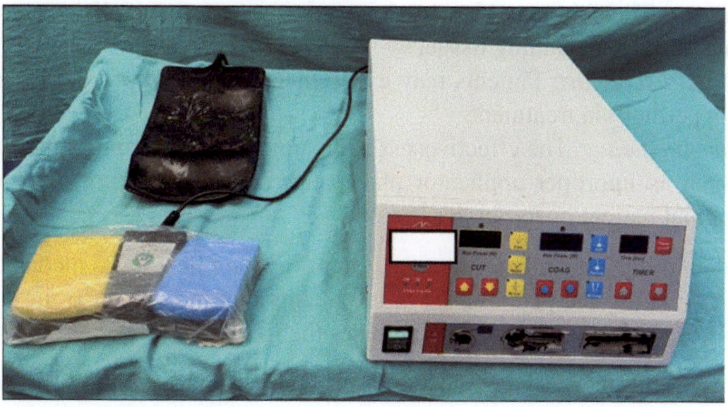

Fig. 14.2 Radiofrequency ablator

6. Radiation therapy applicators
7. Nuclear medicine administration
8. Interventional radiology procedures
9. Fluoroscopic guidance during procedures
10. Patient positioning for examination

These are some of the most common uses of hyperthermia systems in healthcare, but the technology is constantly evolving and new applications may emerge in the future. The toxicity of the heat generated during hyperthermia is generally low. Pain and burns represent typical hyperthermia-associated risks that may be minimized or avoided via correct heating techniques. When body temperature rises, the heat balance of the body is restored by increased blood flow to the skin and sweating. These responses increase the work of the heart and cause loss of salt and water from the body. These events impair the normal function of the heart and cause heart overload and hemoconcentration, which may further lead to coronary and cerebral thrombosis, particularly in elderly individuals with atheromatous arteries. Healthy people can tolerate body core temperature excursions of up to 40 °C when adequately hydrated and higher temperatures may lead to cell death. Figure 14.2 shows the radiofrequency ablator.

14.3.1 Risks

1. *Thermal injury*: Overheating of treated tissue can lead to pain, tissue damage, and scarring.
2. *Equipment malfunction*: Technical malfunctions of the hyperthermia system or applicators can result in incorrect or uneven tissue heating.
3. *Infection*: The use of hyperthermia systems and applicators may increase the risk of infection, particularly if proper infection control procedures are not followed.

4. *Skin burns*: The skin may burn by direct contact with the applicator or by overheating of the surrounding tissue.
5. *Pain and discomfort*: Patients may experience pain and discomfort during and after hyperthermia treatment.
6. *Treatment efficacy*: The effectiveness of treatment may be compromised by factors such as improper applicator placement, incorrect heating parameters, or suboptimal patient positioning [7].

Risk determination of hyperthermia systems and applicators as medical devices involves evaluating the potential harm to patients that may result from the use of these devices. The process typically includes the following steps:

1. Identifying the hazards associated with hyperthermia, such as tissue damage, thermal injury, and skin burns.
2. Analyzing the likelihood and severity of harm, taking into account factors such as the type of hyperthermia system, the patient's health status, and the intended use of the system.
3. Evaluating existing controls and their effectiveness, including the design of the hyperthermia system and applicators, user training and qualifications, and proper use of protective equipment.
4. Developing and implementing risk management plans, including proper maintenance and calibration of the system, regular monitoring of patients for adverse events, and establishing protocols for the emergency management of hyperthermia-related complications.
5. Monitoring and updating risk management plans as needed, taking into account new information and changes in medical practices or patient populations.

It is important to minimize the risk of harm associated with hyperthermia systems and applicators by following established guidelines and protocols, ensuring the proper training and qualifications of users, and maintaining and properly using protective equipment. Figure 14.3 shows the debrider machine in cancer therapy.

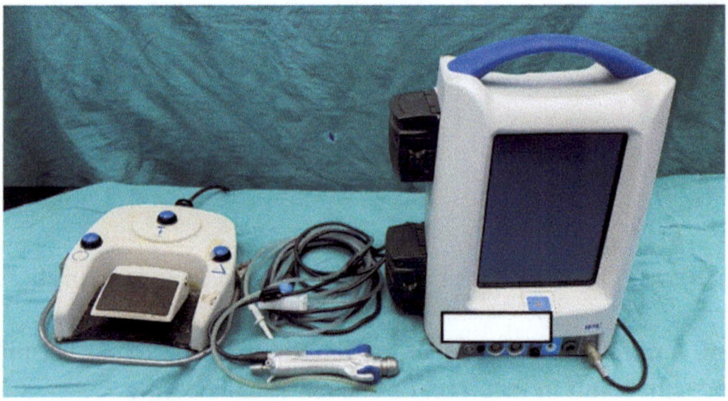

Fig. 14.3 Debrider machine in cancer therapy

14.3.2 Prior Investigations

Hyperthermia systems and applicators are medical devices that use heat to treat certain types of cancers. These devices are regulated by the USFDA and manufacturers are required to conduct extensive testing and clinical trials to ensure their safety and effectiveness. Here are some examples of the investigations that have been conducted on hyperthermia systems and applicators for radiology use:

14.3.2.1 Preclinical Studies

Preclinical studies are conducted in animals to evaluate the safety and effectiveness of hyperthermia systems and applicators. These studies typically involve testing the devices in animal models of cancer and evaluating the effectiveness of the therapy in reducing tumor growth.

1. *Clinical Trials*

Clinical trials are conducted to evaluate the safety and effectiveness of hyperthermia systems and applicators in humans. These trials typically involve testing the devices in a group of patients with the condition for which the therapy is intended and comparing the results to those of a control group.

2. *Temperature Monitoring Studies*

Temperature monitoring studies were conducted to ensure that the hyperthermia system delivered the correct temperature to the target tissue. These studies typically involve monitoring the temperature of the target tissue during treatment and adjusting the hyperthermia system accordingly.

3. *Applicator Studies*

Applicator studies were conducted to evaluate the design and performance of the hyperthermia applicator. These studies typically involve testing the applicator in various settings and evaluating its ability to deliver heat to the target tissue.

4. *Safety Studies*

Safety studies were conducted to evaluate the potential side effects of hyperthermia treatment. These studies typically involve monitoring patients for side effects during and after treatment and evaluating the severity and duration of adverse events.

5. *Quality Control and Assurance Studies*

Quality control and assurance studies are conducted to ensure that the equipment and procedures used in hyperthermia treatment are consistent and produce reliable results. These studies typically involve monitoring the equipment and procedures used in hyperthermia treatment and evaluating their performance over time.

References

1. Keshavarz S, Emamzadeh E, Sardari D, Darki SY, Kabirian M. Boron neutron capture therapy for the treatment of lung cancer and assessment of dose received by organs at risk. Arch Pathol Clin Res. 2022;6:027–31.
2. Malouff TD, Seneviratne DS, Ebner DK, Stross WC, Waddle MR, Trifiletti DM, et al. Boron neutron capture therapy: a review of clinical applications. Front Oncol [Internet]. 2021:11. Available from: https://www.frontiersin.org/articles/10.3389/fonc.2021.601820
3. Zhang X, Geng C, Tang X, Bortolussi S, Shu D, Gong C, Han Y, Wu S. Assessment of long-term risks of secondary cancer in paediatric patients with brain tumours after boron neutron capture therapy. J Radiol Prot. 2019;39(3):838–53.
4. Kok HP, Crezee J. Hyperthermia treatment planning: clinical application and ongoing developments. IEEE J Electromagn RF Microw Med Biol. 2021;5(3):214–22.
5. Jha S, Sharma PK, Malviya R. Hyperthermia: role and risk factor for cancer treatment. Achiev Life Sci. 2016;10(2):161–7.
6. Jeziorski K. Hyperthermia in rheumatic diseases. A promising approach? Reumatologia. 2018;56(5):316–20.
7. Habash RW, Krewski D, Bansal R, Alhafid HT. Principles, applications, risks and benefits of therapeutic hyperthermia. Front Biosci (Elite Ed). 2011;3(3):1169–81.

Chapter 15
Non-significant Risk Medical Devices

Thamizharasan Sampath, Sandhiya Thamizharasan, Vinod Patiram Bhalerao, Karnika Singh, and Prakash Srinivasan Timiri Shanmugam

Abbreviations

CDRH	Center for Devices and Radiological Health
CFR	Code of Federal Regulations
CGM	Continuous Glucose Monitoring Sensors
CTSI	Clinical and Translational Sciences Institute
DMIST	Digital Mammographic Imaging Screening Trial
ECT	Electroconvulsive Therapy
ECG	Electrocardiogram
EMA	European Medicines Agency
EU	European Union
FDA	Food and Drug Administration
FDC	Federal Food, Drug, and Cosmetic Act
GMP	Good Manufacturing Practices
HPPP	Human Research Protection Program
IDE	Investigational device exemption
IEC	International Electrotechnical Commission
IRB	Institutional Review Board
ISO	International Organization for Standardization
IV	Intravenous

T. Sampath (✉)
Department of Pharmacology & Toxicology, VMCGH, Dr. YSR University of Health Sciences, Kurnool, AP, India

S. Thamizharasan
The Tooth Doctor, Advanced Implant Centre, Vellapanchavadi, Chennai, India

V. P. Bhalerao
Department of Forensic Medicine, GMC, SEONI, MP, India

K. Singh
The Ohio State University, Comprehensive Cancer Center, Columbus, OH, USA

P. S. Timiri Shanmugam
Global Product Safety & Toxicology, Avanos Medical Inc., Alpharetta, GA, USA

P. S. Timiri Shanmugam et al. (eds.), *Significant and Nonsignificant Risk Medical Devices*, https://doi.org/10.1007/978-3-031-52838-5_15

IVDR	In Vitro Diagnostic Medical Device Regulation
IVD	In Vitro Diagnostics
IUD	Intrauterine Devices
IOL	Intraocular Lenses
LED	Lightemitting diodes
LLLT	Low-Level Laser Therapy
MDR	Medical Device Regulation
MRI	Magnetic Resonance Imaging
NEMA	The National Electrical Manufacturers Association
NSR	Non-significant Risk
PORP	Partial Ossicular Replacement Prosthesis
RF	Radiofrequency
SAE	Serious Adverse Events
SAR	Specific Absorption Rate
SOP	Standard Operating Procedures
SR	Significant Risk
TORP	Total Ossicular Replacement Prosthesis
TENS	Transcutaneous Electric Nerve Stimulation
TMJ	Temporomandibular Joint
TSS	Toxic Shock Syndrome
UTI	Urinary Tract Infections

15.1 Introduction

Non-significant risk (NSR) devices play an important role in medical practice, offering a range of benefits and applications. These devices are classified as low-risk devices and are intended for use in situations where the potential harm to the patient is minimal. One of the key advantages of NSR devices is their relative safety profile. Due to their low-risk nature, they typically undergo a simplified regulatory process for approval and can be brought to market more quickly compared to higher-risk devices. This expedited process allows for faster access to innovative technologies and medical advancements.

NSR devices are often used in diagnostic and monitoring procedures, providing valuable information to health-care professionals without posing significant risks to patients. These devices may include blood pressure monitors, certain types of imaging equipment, and various diagnostic tools. By facilitating accurate and timely assessments, NSR devices aid in early detection, diagnosis, and treatment planning.

There are three types of medical device studies described in 21 CFR Part 812: significant risk (SR) device studies, non-significant risk (NSR) device studies, and studies using devices that are exempt from the IDE requirements. The investigator and/or sponsor is responsible for initially determining the device risk. As part of its review, the IRB must agree with the assessment based on the proposed use of the device in the research and not on the device alone; however, the FDA has the ultimate decision in determining if a device is IDE exempt, NSR, or SR.

Federal regulations define a significant risk (SR) device investigation as a study of a device that presents a potential for serious risk to the health, safety, or welfare of a subject and:

- Is an implant
- Is used in supporting or sustaining human life
- Is of substantial importance in diagnosing, curing, mitigating or treating disease, or otherwise preventing impairment of human health
- Otherwise presents a potential for serious risk to the health, safety, or welfare of a subject

SR devices include such things as pacemakers, urological stints, electroconvulsive therapy devices, diaphragms, implantable prostheses, surgical lasers, and tracheal tubes.

A non-significant risk (NSR) device investigation is one that does not meet the above stated definition for a significant risk device investigation. NSR device investigations, however, should not be confused with the concept of "minimal risk," a term used in IRB regulations to identify certain studies that may be approved through an expedited review procedure.

NSR devices include such things as low power lasers for treatment of pain, daily wear contact lenses, jaundice monitors for infants, wound dressings, and Foley catheters. A more comprehensive lists of significant and non-significant devices has been compiled by the FDA's Center for Devices and Radiological Health.

To qualify as an NSR device, the device must not pose a serious risk to human subjects. Medical devices that do not meet IDE exempt criteria qualify as NSR devices provided they do not meet the SR device criteria. Medical devices are SR if they are:

- A banned device
- Intended as an implant and present a potential for serious risk to the health, safety, or welfare of a subject
- Purported or represented to be for use in supporting or sustaining human life
- Of substantial importance in diagnosing, curing, mitigating, or treating disease, or otherwise preventing impairment of human health

NSR devices must follow a portion of the regulations outlined in 21 CFR 812:

- Labeling of the device
- IRB approval and maintaining approval
- Informed consent from the subject (21 CFR 50)
- Monitoring of the trial
- Record keeping and reporting
- Prohibition from promotion, commercialization, and misrepresentation of the device

15.2 Monitoring of NSR Device Trials

The goal of trial monitoring is to ensure adequate protection of the rights, welfare, and safety of human subjects and the quality of data collected. With respect to monitoring an NSR medical device study's progress, the regulations focus on two concepts: identifying and investigating any unanticipated adverse device effects and ensuring compliance with the investigational plan, applicable regulations, and any conditions of IRB approval. The procedures taken to comply with NSR monitoring requirements, however, may vary from study to study.

Independent of the device-specific risks, the FDA encourages broader, risk-based monitoring of clinical trials based on other study characteristics. The FDA permits variation in the types and intensity of monitoring activities best suited to address a particular study's characteristics. While the regulations and guidance materials refer to monitoring as a "sponsor" responsibility, it is important to remember that in the absence of a formal external sponsor, the principal investigator (PI) assumes the role of the sponsor and must fulfill these requirements.

15.3 Developing a Monitoring Plan

The FDA provides helpful guidance in assisting sponsors and investigators in developing risk-based monitoring strategies and plans for investigational studies. When developing a monitoring plan, numerous considerations should be made related to timing, types, frequency, and extent of monitoring which should vary based on the complexity of the study design, inclusion of clinically complex or vulnerable study populations, relative safety of the investigational product, experience and qualifications of the PI and study team, and quantity of data and data collection methods.

15.4 Who Should Perform Monitoring?

Study monitoring does not necessarily require independent review by an individual external to the study team. Depending on the study characteristics, the required study monitoring might be accomplished through the routine and ongoing oversight by the PI and study team. Increased complexity of the trial or increased risk of participation, however, might justify a more formal approach, including selecting someone external to the study team to serve as the study monitor. In this situation, the Indiana CTSI may be able to provide monitoring services for more complex NSR device studies or those with more relative risk [1].

15.5 What Should Be Monitored?

The FDA guidance provides examples of data and processes that should ordinarily be identified as critical for monitoring including but not limited to:

Verification that informed consent was obtained appropriately
Adherence to protocol eligibility
Appropriate accountability and administration of the investigational device
Conduct and documentation of procedures and assessments related to study endpoints, safety assessments, and evaluating, documenting, and reporting SAEs and unanticipated adverse device effects

The extent of data review will vary from study to study. While the FDA's guidance highlights the importance of monitoring critical data and processes, it does not require that they be monitored across all subjects. While data could be randomly selected for review, you may instead wish to consider the pace of enrollment, the amount of data to be collected, and the number of personnel involved in study conduct when determining what data to review and how often. Monitoring critical data and processes soon after the first, or first few, subjects are enrolled may help to catch and fix potential problems early in the study's conduct. Additionally, monitoring may need to be increased—either in frequency or scope—in response to study events such as staff turnover, protocol amendments, and/or to satisfy corrective and preventive action plans.

Self-audit templates and tools are available on the HRPP Website. These templates may be helpful to assist study teams when conducting their own monitoring in identifying potential noncompliance, protocol deviations, and data management issues. While the monitoring process may be accomplished in a number of ways, regardless of the approach, all monitoring activities should be documented in the research record. Additionally, observations from monitoring activities should be discussed with the PI and study team to determine appropriate corrective and preventive actions and whether events must be reported to the FDA or IRB.

15.6 How Should Monitoring Be Documented?

While the FDA regulations do not require the monitoring plan to be a formal written document, in the event of an inspection, it will be critical for the PI and study team to be able to describe how they monitor the study's progress, how the PI is engaged in these activities, and to provide evidence that monitoring has occurred. For informal monitoring, this may include an explanation of team/department practices or SOPs and/or documentation related to study team meetings and study team correspondence. For more formal monitoring, this might include documentation of self-reviews and/or external monitor reviews, including dates of occurrence, the individual performing the review, the outcomes of the self-reviews, and communication of observations to the PI.

15.7 Assessing Study Characteristics

The plan for study monitoring should be based on the characteristics of the trial and the overall risks to subject health, safety, and welfare and to the quality and integrity of the study data [2].

Low Risk: This might involve open-label or single-arm trials, with minimal device-related safety risks anticipated and no intention to enroll individuals from vulnerable or medically complex study populations. The research will be overseen by a proficient principal investigator (PI) and a seasoned study team. Study monitoring could occur through informal means, ensuring adherence to routine researcher duties, responsibilities, and effective communication within the study team.

Moderate Risk: This might involve randomized, blinded, or cross-over designs, with moderately increased device safety risks. The study could include participants who are vulnerable or medically complex, and procedures may be conducted at multiple locations by a diverse team of staff with varying levels of experience. Study monitoring could be conducted in a more structured manner through regular self-review or self-auditing of specific data segments, both periodically and continuously.

High Risk: This might encompass intricate study designs, such as adaptive trial structures or those modifying device exposure based on subject response. There may be elevated safety risks, potentially necessitating safety-related stopping rules. The study could involve vulnerable populations or those with serious illnesses, requiring heightened safety monitoring throughout. Additionally, complex protocol-driven procedures may be employed, possibly overseen by a less experienced PI or study team. Extensive data transcription or conversions before analysis could also be part of the process. A structured study monitoring plan could be implemented, and the monitoring process may be carried out by an individual external to the study team.

15.8 Non-significant Risks of Medical Devices

When it comes to medical devices, it's essential to recognize that the vast majority of devices are designed, tested, and used safely without causing significant risks or adverse effects. However, like any medical intervention, there can be some non-significant risks associated with medical devices.

15.8.1 Discomfort or Pain

Some medical devices, such as catheters or braces, may cause temporary discomfort or mild pain during their use or application. This discomfort is generally manageable and subsides once the device is removed or adjusted.

Dental Braces: Dental braces are orthodontic devices used to correct misaligned teeth or jaw problems. During the initial stages of treatment and subsequent adjustments, patients may experience discomfort, soreness, or pain in their teeth and gums. This discomfort is often due to the pressure exerted by the braces on the teeth, and it typically subsides as the patient's mouth adjusts to the device.

Intravenous (IV) Catheters: IV catheters are used to administer fluids, medications, or draw blood from patients. The insertion of an IV catheter can cause a brief moment of pain or discomfort, especially when the needle punctures the skin. However, once the catheter is in place, the discomfort usually diminishes. Occasionally, patients may experience mild discomfort or tenderness around the insertion site during the catheter's use.

Urinary Catheters: Urinary catheters are thin, flexible tubes inserted into the bladder to drain urine. The insertion and presence of a urinary catheter can cause temporary discomfort or a sensation of pressure in the urethra or bladder. However, patients typically adjust to the presence of the catheter over time, and any initial discomfort usually resolves.

Compression Stockings: Compression stockings are elastic garments worn to improve blood circulation and prevent blood clots in the legs. While they provide therapeutic benefits, some individuals may find them slightly uncomfortable or tight, especially if they are not properly sized or worn for extended periods. Adjusting the fit or using stockings with lower compression levels can help alleviate any discomfort.

Prosthetic Limbs: Prosthetic limbs are artificial limbs designed to replace missing or amputated body parts. While modern prosthetic devices are carefully designed to maximize comfort and functionality, some individuals may experience initial discomfort or soreness as they adapt to wearing and using the prosthetic limb. This discomfort often lessens as they become accustomed to the device and receive proper adjustments or training.

15.8.2 Skin Irritation or Sensitivity

Certain medical devices, particularly those in direct contact with the skin, may occasionally cause skin irritation or sensitivity. This can manifest as redness, itching, or a rash in the area where the device is applied. In most cases, these reactions are temporary and resolve after discontinuing use or with appropriate skin care.

Adhesive Bandages: Adhesive bandages, commonly used for wound dressing, can sometimes cause skin irritation or sensitivity. Some individuals may develop redness, itching, or a rash in the area where the adhesive comes into contact with the skin. This can be due to an allergic reaction to the adhesive material or friction caused by repeated application and removal of the bandage.

ECG Electrodes: Electrocardiogram (ECG) electrodes are used to record the electrical activity of the heart. The adhesive used to secure the electrodes to the skin may cause skin irritation in some individuals. This can manifest as redness, itching, or a rash at the electrode sites. Hypersensitivity to the adhesive or prolonged contact with the skin can contribute to these reactions.

Transdermal Patches: Transdermal patches, such as nicotine patches or hormone patches, deliver medication through the skin. Occasionally, individuals may experience skin irritation or sensitivity at the application site. This can be due to the adhesive, the medication itself, or individual sensitivity to the patch components. Rotating the patch placement and using hypoallergenic adhesives can help minimize skin reactions.

Continuous Glucose Monitoring (CGM) Sensors: CGM sensors are used to monitor blood glucose levels continuously. The adhesive used to attach the sensor to the skin can sometimes cause skin irritation or sensitivity. This can result in redness, itching, or a rash around the sensor site. Proper skin preparation, using barrier creams or patches, and rotating sensor placement can help reduce skin reactions.

Ostomy Appliances: Ostomy appliances, such as ostomy bags or pouches, are used by individuals with a stoma to collect waste from the body. The adhesive used to secure the appliance to the skin can occasionally cause skin irritation or sensitivity. This can result in redness, itching, or a rash around the stoma or adhesive area. Using skin barrier products, ensuring proper fit, and addressing any allergies or sensitivities can help manage these reactions.

15.8.3 Device Malfunction

Although medical devices undergo rigorous testing and quality control, there is always a small risk of device malfunction or failure. This can occur due to manufacturing defects, improper use, or wear and tear over time. However, such malfunctions are relatively rare and can often be identified and addressed through regular maintenance and quality assurance protocols.

Device malfunctions can occur in medical devices, although they are relatively rare due to strict regulatory oversight and quality control measures. Here are a few examples of medical devices that can experience malfunctions:

Insulin Pumps: Insulin pumps are used by individuals with diabetes to deliver a continuous supply of insulin. Device malfunctions in insulin pumps can include issues with the pump's display, infusion sets, or insulin delivery mechanism. This can result in inaccurate insulin dosing, interruption of insulin delivery, or failure to detect high or low blood sugar levels.

Pacemakers: Pacemakers are implantable devices used to regulate and control abnormal heart rhythms. Malfunctions in pacemakers can involve battery depletion, electrode displacement, or programming errors. These malfunctions can lead to a disruption in the device's ability to deliver electrical impulses to regulate the heartbeat properly.

Implantable Defibrillators: Implantable cardioverter-defibrillators (ICDs) are devices implanted in individuals at risk of life-threatening heart rhythm disturbances. Malfunctions in ICDs can include battery failure, faulty leads, or inappropriate shocks. These malfunctions can result in either the device not delivering a necessary shock when needed or delivering inappropriate shocks when not required.

Surgical Tools: Surgical tools, such as electrocautery devices or robotic surgical systems, can experience malfunctions. This can include issues with the electrical or mechanical components of the devices, software glitches, or equipment failure during a surgical procedure. Malfunctions in surgical tools can impact the precision, control, or safety of the procedure.

Infusion Pumps: Infusion pumps are used to deliver medications, fluids, or nutrients intravenously. Malfunctions in infusion pumps can involve inaccuracies in medication dosing, occlusion alarms, or problems with the pump's user interface. These malfunctions can lead to incorrect medication delivery rates or interruptions in therapy.

15.8.4 Bruising or Minor Bleeding

Certain medical devices, such as needles used for injections or blood collection, may occasionally result in minor bruising or bleeding at the puncture site. These are typically self-limiting and resolve without intervention.

Certain medical devices can sometimes cause bruising or minor bleeding at the site of application or insertion. Here are a few examples:

Needles: Needles used for injections, blood draws, or intravenous (IV) catheter insertions can occasionally cause bruising or minor bleeding. This can happen due to factors such as needle size, technique, or fragile blood vessels. Applying pressure to the site after needle removal can help minimize bruising and bleeding.

Blood Pressure Cuffs: Blood pressure cuffs, commonly used to measure blood pressure, are inflated around the upper arm or wrist. In some cases, the cuff may be applied too tightly or removed abruptly, leading to minor bruising or small blood vessel rupture. Ensuring proper cuff placement and gradual release of pressure can help reduce the risk of bruising.

Compression Garments: Compression garments, such as stockings or sleeves, are used to apply pressure to the limbs to improve circulation or manage swelling. In some instances, these garments may be worn too tightly or for an extended duration, resulting in minor bruising or skin discoloration. Ensuring proper sizing and following recommended wearing guidelines can minimize this risk.

Tourniquets: Tourniquets are occasionally used to temporarily restrict blood flow during medical procedures, such as blood draws or intravenous (IV) line placements. If applied too tightly or for an extended period, tourniquets can cause bruising or minor bleeding at the application site. Proper application technique and timely release of the tourniquet can help prevent these complications.

Surgical Procedures: Some surgical procedures, particularly those involving invasive interventions, can result in minor bruising or localized bleeding. This can occur due to surgical instruments, manipulation of tissues, or blood vessel trauma during the procedure. Surgical teams take precautions to minimize these occurrences and control bleeding during and after the surgery.

While bruising or minor bleeding caused by medical devices is typically self-limiting and not a cause for significant concern, it's important to monitor the site for any signs of excessive bleeding, severe pain, or prolonged symptoms.

15.8.5 Psychological Impact

In some cases, the use of medical devices, especially those that are visible or require lifestyle adjustments, can have a psychological impact on individuals. This may include feelings of self-consciousness, anxiety, or emotional adjustment to living with a device. Patient education, counseling, and support can help individuals cope with these emotional challenges.

It's important to note that while these risks exist, they are generally considered non-significant in the broader context of medical device safety. Health-care professionals and regulatory bodies work diligently to evaluate, monitor, and address any potential risks associated with medical devices to ensure patient safety and well-being.

15.9 Non-significant Risk Devices

The following examples may help sponsors and IRBs in making SR and NSR determinations. The list includes many commonly studied medical devices. Inclusion of a device in the NSR list is not a final determination because the evaluation of risk must reflect the proposed use of a device in a study [3].

- Caries Removal Solution
- Contact Lens Solutions intended for use directly in the eye (e.g., lubricating/rewetting solutions) using active ingredients or preservation systems with a history of prior ophthalmic/contact lens use or generally recognized as safe for ophthalmic use
- Conventional Gastroenterology and Urology Endoscopes and/or Accessories

- Conventional General Hospital Catheters (long-term percutaneous, implanted, subcutaneous and intravascular)
- Conventional Implantable Vascular Access Devices (Ports)
- Conventional Laparoscopes, Culdoscopes, and Hysteroscopes
- Daily Wear Contact Lenses and Associated Lens Care Products not intended for use directly in the eye (e.g., cleaners; disinfecting, rinsing, and storage solutions)
- Dental Filling Materials, Cushions, or Pads made from traditional materials and designs
- Denture Repair Kits and Realigners
- Digital Mammography
- Electroencephalography (e.g., new recording and analysis methods, enhanced diagnostic capabilities, measuring depth of anesthesia if anesthetic administration is not based on device output)
- Externally Worn Monitors for Insulin Reactions
- Functional Noninvasive Electrical Neuromuscular Stimulators
- General Biliary Catheters
- General Urological Catheters (e.g., Foley and diagnostic catheters) for short-term use (< 28 days)
- Jaundice Monitors for Infants
- Low-Power Lasers for treatment of pain
- Magnetic Resonance Imaging (MRI) Devices within FDA-specified parameters
- Manual Image-Guided Surgery
- Menstrual Pads (Cotton or Rayon only)
- Menstrual Tampons (Cotton or Rayon only)
- Non-implantable Electrical Incontinence Devices
- Non-implantable Male Reproductive Aids with no components that enter the vagina
- Ob/Gyn Diagnostic Ultrasound within FDA-approved parameters
- Partial Ossicular Replacement Prosthesis (PORP)
- Total Ossicular Replacement Prosthesis (TORP)
- Transcutaneous Electric Nerve Stimulation (TENS) Devices for treatment of pain (except for chest pain/angina)
- Ureteral Stents
- Urethral Occlusion Device for less than 14 days
- Wound Dressings, excluding absorbable hemostatic devices and dressings (also excluding Interactive Wound and Burn Dressings that aid or are intended to aid in the healing process)

15.9.1 Daily Wear Contact Lens

Daily wear contact lenses are thin, soft lenses that are typically worn during the day and removed at night for cleaning and disinfection. They are designed to correct refractive errors, such as nearsightedness, farsightedness, and astigmatism, providing users with clear vision and improved quality of life.

In terms of regulatory oversight, daily wear contact lenses are subject to the guidelines outlined in 21 CFR Part 812 for NSR devices. These regulations ensure that the lenses are thoroughly evaluated for their safety, effectiveness, and overall performance before they are made available to the public.

Clinical investigations play a vital role in the evaluation of daily wear contact lenses as NSR devices. These investigations aim to assess factors such as lens fit, visual acuity, comfort, and the potential for adverse reactions. By conducting well-designed clinical studies, manufacturers and researchers can gather valuable data to support the safety and effectiveness of daily wear contact lenses.

Additionally, adherence to good manufacturing practices (GMP) is crucial in the production of daily wear contact lenses. GMP guidelines ensure that these devices are manufactured in a controlled environment with strict quality control measures to minimize the risk of contamination, maintain consistent lens quality, and ensure compliance with regulatory standards.

It is important to note that although daily wear contact lenses are classified as NSR devices, wearers must still follow proper hygiene practices and use them as directed by their eye care professionals. This includes regular cleaning, disinfection, and replacement of lenses to prevent potential eye infections and complications.

Overall, daily wear contact lenses such as NSR devices provide individuals with a convenient and comfortable vision correction option. By understanding the unique characteristics and regulatory landscape of daily wear contact lenses, eye care professionals, manufacturers, and wearers can make informed decisions that prioritize visual health and overall well-being.

15.9.2 Digital Mammography

Digital mammography is a critical tool in the early detection and diagnosis of breast cancer, providing high-resolution images for effective screening and assessment. As a non-significant risk (NSR) device, digital mammography systems are designed to offer improved diagnostic capabilities while posing minimal risk to patients.

Digital mammography is the technique by which the radiographic image is obtained with digital detectors and recorded electronically in a digital format. The image is further processed and displayed as a gray-scale image that can be displayed in multiple formats. Digital mammography has several advantages compared with conventional film screen mammography. Image acquisition, display, and storage are much faster. Image manipulation through adjustments in contrast, brightness, and magnification of selected regions enables radiologists to obtain superior views. This technology makes it possible to subtract various layers of computerized imagery to examine suspicious areas and improve the ability to detect and diagnose breast carcinoma. Greater contrast resolution allows better screening of women with dense breasts and breast implants. With the ability to manipulate and postprocess the images, subtle abnormalities are increasingly detected. Images can be stored easily for future reference and can be sent electronically to be read at multiple viewing stations, thereby allowing double reading when necessary. The main disadvantages of digital mammography include the cost of the equipment and the reduced spatial resolution compared with film.

Digital mammography has been compared with film screen mammography in various studies with little difference reported in cancer detection rates. In the Digital Mammographic Imaging Screening Trial (DMIST), 49,528 asymptomatic women underwent both film and digital mammography. Although there was no significant difference in overall diagnostic accuracy, digital mammography was more accurate for premenopausal and perimenopausal women. Furthermore, it was superior for women with dense breasts. Approximately 25,000 women aged 45–69 years were randomized to either digital or film screen mammography in the Oslo II Study. The breast cancer detection rate at 2 years was significantly higher in the full-field digital mammography group compared with film screen mammography (0.59% and 0.38%, respectively). In the United States, the majority of imaging centers use digital mammography. It may provide a small screening advantage in women younger than 50 years old. However, it must be noted that film mammography is an acceptable screening method for all women [4].

From a regulatory standpoint, digital mammography systems are subject to the guidelines outlined in 21 CFR Part 812 for NSR devices. These regulations ensure that these devices are evaluated for their safety, effectiveness, and performance characteristics before being used for clinical purposes. Digital mammography as an NSR device plays a significant role in the early detection and diagnosis of breast cancer. By adhering to regulatory guidelines, conducting thorough clinical investigations, and implementing quality control measures, digital mammography systems continue to evolve and improve, offering enhanced diagnostic capabilities for improved patient outcomes. By understanding the unique features and regulatory considerations associated with digital mammography, health-care professionals can effectively utilize this technology in the fight against breast cancer, ultimately saving lives and improving the quality of care for patients.

15.9.3 Menstrual Tampons (Cotton or Rayon Only)

Menstrual tampons have long been used as a convenient and discreet form of feminine hygiene product during menstruation. As a non-significant risk (NSR) device, menstrual tampons, specifically those made of cotton or rayon materials, are designed to offer effective absorption while posing minimal risk to the health and safety of users. Menstrual tampons are typically cylindrical-shaped absorbent materials inserted into the vaginal canal to collect menstrual flow. They provide an alternative to pads and other menstrual products, allowing women to engage in daily activities with ease and comfort.

Tampons are designed to be inserted into the vagina with or without an applicator. FDA-cleared tampons are made of cotton, rayon, or a blend of the two. The absorbent fibers used in FDA-cleared tampons sold today are made with a bleaching process that is free from elemental chlorine, which also prevents products from having dangerous levels of dioxin (a type of pollutant found in the environment).

Before any tampons can be legally sold in the USA, they must go through the FDA's review to determine whether they are as safe and effective as (substantially equivalent to) legally marketed tampons. As part of the FDA's review, manufacturers submit data including the results of testing to evaluate the safety of the materials used to make tampons and applicators (if present); tampon absorbency, strength, and integrity; and whether tampons enhance the growth of certain harmful bacteria or change normal bacteria levels in the vagina. Reusable tampons may carry additional risks of infections such as yeast, fungal, and bacterial infections. The only tampons cleared or approved by the FDA are designed for single use [5].

Toxic shock syndrome (TSS) is rare and is caused by a toxic substance that is produced by certain kinds of bacteria. The toxic substance produced by the bacteria can cause organ damage (including kidney, heart, and liver failure), shock, and even death. Rates of reported TSS cases associated with tampons have declined significantly over the years. One reason is that the FDA evaluates whether a tampon enhances the growth of the bacteria that causes TSS before the product can be legally marketed. Only tampons that have been cleared by the FDA can be legally marketed in the USA. In addition, more informative tampon labeling, as well as educational efforts by the FDA and manufacturers, may have contributed to the reduction in TSS cases.

Tampon Safety Advice

- Symptoms and signs of TSS may include a sudden fever (usually 102 °F or more), vomiting, diarrhea, fainting or feeling like you are going to faint when standing up, dizziness, or a rash that looks like a sunburn.
- Wash your hands before and after using a tampon. This will help reduce the spread of bacteria.
- Only use tampons when you have your period. Tampons are not intended to be used at any other time or for any other reason.
- Change each tampon every 4–8 hours. Never wear a single tampon for more than 8 hours at a time.
- Use the lowest absorbency tampon needed.

15.9.4 Jaundice Monitors for Infants

Jaundice monitors for infants are portable devices that use noninvasive methods, such as transcutaneous bilirubinometer, to measure the level of bilirubin in the baby's skin. These monitors provide health-care professionals with a quick and convenient way to assess jaundice severity and determine the need for further intervention or treatment. Jaundice can be treated easily by irradiating the infant with blue light that breaks bilirubin down to be excreted through urine. The treatment itself, however, can disrupt bonding time, cause dehydration, and increase the risks of allergic diseases. Neonatal jaundice is one of the leading causes of death and brain damage in infants in low- and middle-income countries.

To address the tricky balance of administering the precise amount of blue light needed to counteract the exact levels of bilirubin, researchers have developed the first wearable sensor for newborns that is capable of continuously measuring bilirubin. In addition to bilirubin detection, the device can simultaneously detect pulse rate and blood oxygen saturation in real time.

The real-time monitoring of jaundice is critical for neonatal care. Continuous measurements of bilirubin levels may contribute to the improvement of quality of phototherapy and patient outcome.

Currently, medical professionals use handheld bilirubinometer to measure bilirubin levels, but there is not a device that can simultaneously measure jaundice and vitals in real time.

Researchers succeeded in miniaturizing the device to a size that can be worn on the forehead of a newborn baby. By adding the function of a pulse oximeter to the device, multiple vitals can easily be detected. Held to the baby's forehead by a silicone interface, the device has a lens capable of efficiently transmitting lights to neonatal skin via battery-powered light-emitting diodes, commonly known as LEDs.

At the present stage, coin cell batteries are used, and the overall shape is very thick. In the future, it will be necessary to further reduce the thickness and weight by using thin-film batteries and organic materials. The researchers tested the device on 50 babies, and they found that the device is not currently accurate enough to suffice for clinical decision-making. According to them, they will reduce the thickness and increase the flexibility of the device, as well as improve the silicone interface to facilitate better skin contact. In the future, the researchers plan to develop a combined treatment approach that pairs a wearable bilirubinometer with a phototherapy device to optimize the amount and duration of light therapy based on continuous measurements of bilirubin levels.

15.9.5 General Urological Catheters

General urological catheters can potentially fall under the category of non-significant risk (NSR) devices. NSR devices are those that pose a low level of risk to the subjects involved in clinical investigations. In general, urological catheters are medical devices used for various purposes related to the urinary system, such as drainage, irrigation, or introducing substances. These catheters can include Foley catheters, intermittent catheters, suprapubic catheters, and others.

The design of the Foley catheter is simple. The catheter typically has two channels, the drainage channel for the passage of urine and the inflation channel, to allow the balloon at the end of the catheter to be inflated with sterile water from a syringe, to retain the catheter within the bladder. The smooth rounded tip of the catheter extends beyond the balloon and one or more eye-holes are cut in the tube adjacent to the tip to allow urine to drain. Were it not for its complications, it would often be desirable for a Foley catheter to be able to be in place for up to ~12 weeks before the possibility of mechanical failure of the balloon would dictate its replacement [6].

The principal reasons for indwelling catheterization are as follows:

- To permit urinary drainage in patients with neurological conditions that cause bladder dysfunction
- To manage urinary incontinence in patients lacking cognitive function
- To minimize skin breakdown and pressure ulcers in paralyzed, comatose, or terminally ill patients
- To irrigate the bladder
- To administer chemotherapy
- To aid in urological surgery or other surgery on contiguous structures
- To obtain accurate measurements of urinary output in critically ill or postoperative patients
- To empty the bladder during childbirth
- To undertake urodynamic studies (such as pressure measurements)

Bladder drainage may be performed by passing a Foley catheter through the natural urethral passage (termed transurethral catheterization) or by creating an artificial track between the lower abdominal wall and the bladder (suprapubic catheterization). Transurethral catheterization is the simpler and safer approach. The female urethra is short, being ~40 mm in length, muscular and straight. The male urethra is ~160 mm long, more sensitive and curved, which can give rise to complications. Passage of a catheter can be painful and, in the male, the curvature of the urethra introduces a risk that its tip may cause damage. Some designs of catheters are curved to minimize this risk. The main problem with suprapubic catheterization is the risk of perforating the bowel on insertion of the guidance cannula. Guidelines on minimizing morbidity associated with suprapubic catheter usage have been published by the British Association of Urological Surgeons.

Catheter size is usually expressed in French gauge (Fr or FG = circumference in mm). The normal practice is to use the smallest catheter compatible with good drainage: 12–16 Fr is usually adequate and only rarely is a catheter larger than 18 Fr necessary.

The Modern Foley Catheter

Foley's original catheter was made of latex, the mechanical properties of which are ideal for this purpose: it has a high stretch ratio, a high level of resilience and it is extremely waterproof. The main problem with latex is its cytotoxicity: for instance, in the 1980s, an epidemic of severe urethral strictures was recorded in patients as the result of using latex catheters. The cause was traced to cellular toxicity due to eluates from rubber. Latex catheters are now usually coated with silicone elastomer to reduce this risk. Many modern catheters are made entirely of silicone elastomer and hydrophilic coatings are used to provide a slippery surface to reduce friction. Silicone catheters are not only non-allergenic, but they also have superior resistance to kinking and better flow properties in comparison with latex catheters [7].

Emphasis has also been placed on the need for a smooth surface to the catheter and the drainage eyes. Rough surfaces encourage the deposition of bacterial biofilm, and sharp edges to the drainage eyes can cause bleeding from the urethral lining when the catheter is introduced or withdrawn.

Some catheter research over the last few years has focused on the development of antiseptic and antimicrobial coatings, with the aim of reducing the incidence of catheter-associated urinary tract infections, so far with negligible success. Thus, a randomized controlled trial performed to compare the ability and cost-effectiveness of an antiseptic- and antimicrobial-impregnated catheter versus a standard-coated catheter to minimize the risk of catheter-associated urinary tract infection revealed no evidence of benefit. Indeed, in an earlier randomized clinical trial, infection actually increased with a silver-impregnated catheter.

Some Foley catheters have a third channel, which can be used to infuse saline or other irrigating fluid into the bladder: this may be useful when there is a likelihood that blood clots may form in the bladder, perhaps as the result of postoperative bleeding. There is also a commercially available catheter that has two balloons at the end of the catheter. The balloon at the tip is intended to reduce the risk of trauma to the urothelium; the drainage eyes perforate a short section of catheter between the two balloons, the proximal of which serves as the retention device. A possible disadvantage of the dual-balloon catheter is that it may trap more urine in the bladder at the end of drainage, thus increasing the risk of bladder infection.

Adverse Events Caused by the Foley Catheter

Bacterial Colonization

The flow of urine through an indwelling catheter may be continuous or intermittent. The introduction of a Foley catheter without a valve results in continuous drainage and, thereby, suppresses the normal process by which the buildup of bacteria is inhibited by periodic flushing. Periodic flushing is usually facilitated by a manually operated pinch or rotary valve. It can also be provided by "tidal drainage," which allows the bladder to fill and empty automatically. By raising the height of the drainage tubing leading from the catheter to a few centimeters above the level of the bladder, the bladder fills to the corresponding hydrostatic pressure before a syphon is formed that empties the bladder, after which the cycle is repeated. In a series of 33 patients with neurologically damaged bladders following spinal cord injuries, tidal drainage reduced the rate of infection from 73% to 15%. Although this is a marked improvement and urethral damage can be avoided by suprapubic catheterization, tidal drainage is seldom used nowadays, possibly because of the level of nursing care required.

Bacteria can invade the bladder by migrating along the inside and the outside of the catheter. With short-term catheterization, the daily infection rate is ~5%, so that ~95% of catheterized patients suffer bacterial invasion after 1 month. Urinary tract infection necessitates the use of antibiotics, which are all too frequently untested against the specific bacteria and consequently often prove to be ineffective until the right one is found by a process of trial and error. This adds to the cost of clinical management, as well as being a burden for patients and carers [8].

Antibiotic Resistance

The use of antibiotics to control catheter-induced infections contributes significantly to the development of resistant strains, about which the World Health Organization (WHO) has expressed grave concern. The WHO referred in particular to seven bacteria: the first of these, *Escherichia coli*, is strongly associated with urinary tract infections. In five out of the six WHO regions, it was found that the antimicrobial drugs that were used failed in 50% or more of the cases investigated. The second bacterium in the list, *Klebsiella pneumonia* that is also found in urinary tract infections, was similarly resistant. In view of the increasingly serious threat to global public health identified by the WHO, the present pervasive lack of interest in research aimed at finding a better alternative to the Foley catheter is both disturbing and inexcusable.

Chronic Infection

The balloon of the Foley catheter occupies the base of the bladder, obstructing the internal urethral orifice, with the result that 10–100 ml of urine remains in the bladder when its flow has ceased. This sump of residual urine is likely to be infected, so that uninfected urine descending from the kidneys will also rapidly become infected, resulting in chronic infection of the bladder.

Kidney and Bladder Damage

Invasion of the bladder by urease-producing bacteria, particularly *Proteus mirabilis*, results in the conversion of urea in the urine into ammonia. The consequential increase in the alkalinity of the urine causes phosphates to nucleate out of solution, forming crystals of struvite (magnesium ammonium phosphate) and hydroxyapatite (an hydroxylated form of calcium phosphate in which some of the phosphate groups are replaced by carbonate). Increasing fluid intake with citrate-containing drinks increases the pH at which crystals form in the urine and there is evidence that this could be used to control the rate at which catheter encrustation occurs.

The nucleation of struvite and hydroxyapatite crystals on the biofilm on the catheter resulting from the activity of the bacterial urease causes encrustation to form around and within the catheter, blocking the drainage eyes and the lumen and preventing the flow of urine. This is a medical emergency that not only can be excruciatingly painful for the patient but that also and more importantly requires a rapid response (usually the replacement of the blocked catheter) if permanent damage to the bladder and the kidneys (due to ureteric reflux) caused by the high bladder pressure is to be avoided. These problems are exacerbated if associated with bladder spasm. Moreover, it is apparent that some patients are more likely to block their catheters than others (blockers tend to have urine that is more alkaline, which is consistent with other observations).

It has been reported that, in cases of *Proteus mirabilis* infection, the necessity for antibiotics might be avoided by adding the biocide triclosan to the sterile water used to inflate the balloon of the Foley catheter. This appears to prevent the rise in urinary pH that drives biofilm formation and catheter blockage, presumably by leaching into the urine. It is disappointing, however, that exposure to triclosan has also been shown to encourage the development of resistant strains of *Proteus mirabilis*, so this does not hold out much promise as a long-term solution.

Bladder Stones

The crystals of struvite resulting from *Proteus mirabilis* infection act as nuclei for stone formation within the bladder. Bladder stones entrap *Proteus mirabilis* bacteria and, thus, maintain the infection. Recurrent blockage of a catheter raises a high suspicion that bladder stones may be present. Endoscopic transurethral techniques are used to remove bladder stones: fragmentation by crushing (litholapaxy), shock-wave ultrasound, or laser probes may be required to break them into particles small enough to be washed out of the bladder through the urethra.

Pseudopolyps

Insertion into the urethra of a hard unyielding catheter, with its balloon and its protruding tip perforated by drainage eyes, transforms the natural process of intermittent drainage. When the drainage valve at the distal end of the catheter is opened, the low viscous drag of the catheter allows the urine to flow rapidly, driven by both the collapsing bladder and the negative pressure of the hydrostatic column to the open end of the catheter. Toward the end of the drainage process, when the bladder wall comes into what is frequently traumatic contact with the tip of the catheter, the mucosal lining can be sucked into the drainage eye. Patients may experience a sharp pain at this stage, sometimes accompanied by "stuttering" when the urine flow momentarily ceases as the result of the bladder wall being sucked into the drainage eye and then released. This suction can result in the formation of hemorrhagic pseudopolyps, with cumulative damage.

Septicemia

The physical trauma caused by the catheter tip and the suction at the drainage eyes can damage the normally impermeable bacterial barrier provided by the urothelial lining of the bladder. This provides direct access for bacteria into the bladder wall and the bloodstream (bacteremia), with a high risk of septicemia. Moreover, reflux of infected urine via the ureters can lead to renal infection (pyelonephritis) and septicemia. If inadequately treated, septicemia can prove to be fatal.

Urethral Trauma

The process of inserting the catheter requires skill and practice if urethral trauma is to be avoided. One of the problems with indwelling catheters with silicone balloons is that, when the water is removed from the balloon with a syringe prior to catheter withdrawal, a phenomenon known as creep may cause the balloon to fail to collapse completely. This may result in a small rim that can make it difficult or impossible to remove the catheter. (This presents a particular problem in the case of suprapubic catheters because these pass through a rigid fibrous track into the bladder, rather than through the urethra with its more elastic muscular walls.) Even more serious damage can occur if the catheter is deliberately pulled out when the balloon is still inflated, as can be done by disorientated or demented patients. In women with neurological conditions such as multiple sclerosis, the catheter can be expelled spontaneously by a sudden inappropriate contraction of the bladder. Under these circumstances, the urethra is dilated by the balloon and, if frequently repeated, the sphincter mechanism may become incompetent [9].

Balloon Fragments

There is the risk that the catheter balloon may burst, during either insertion or withdrawal (particularly by a demented or disorientated patient) or when it is indwelling. If this should happen, the fragments must be removed, usually with the aid of a cystoscope, as otherwise they can lead to stone formation or catheter blockage.

15.9.6 Transcutaneous Electric Nerve Stimulation (TENS)

Transcutaneous electrical nerve stimulation (TENS) is a therapy that uses low-voltage electrical current to provide pain relief. A TENS unit consists of a battery-powered device that delivers electrical impulses through electrodes placed on the surface of your skin. The electrodes are placed at or near nerves where the pain is located or at trigger points. The battery-powered TENS device is about the size of a small cell phone. The device comes with several sets of electrode wires and end pads. The electrodes connect to the device at one end and are attached to about 2 inch by 2 inch pads at the other end. Each pad has an adhesive backing and is positioned on the skin in specific areas along nerve pathways in the area to be treated. (Instead of direct contact with the skin, an acupuncturist may connect the TENS unit to acupuncture needles.)

The device delivers pulses of electrical energy. Pulses can be adjusted for intensity, frequency, duration, and type (burst or continuous). A doctor, physical therapist, or acupuncturist determines and adjusts the machine's settings. There are two theories about how transcutaneous electrical nerve stimulation (TENS) works. One theory is that the electric current stimulates nerve cells that block the transmission of pain signals, modifying the perception of pain. The other theory is that nerve stimulation raises the level of endorphins, which are the body's natural pain-killing chemical. The endorphins then block the perception of pain.

TENS therapy has been used or is being studied to relieve both chronic (long lasting) and acute (short-term) pain. Some of the most common conditions for which TENS has been used include:

- Osteoarthritis (disease of the joints)
- Fibromyalgia (aching and pain in muscles, tendons, and joints all over the body, especially along the spine)
- Tendinitis (an inflammation or irritation of a tendon)
- Bursitis (inflammation of the fluid-filled sacs that cushion joints)
- Labor pain
- Low back pain
- Chronic pelvic pain
- Diabetes-related neuropathy (damage to the nerves that connect the brain and spinal cord to the rest of the body)
- Peripheral artery disease ("hardening of the arteries" that circulate blood to the body)

TENS is a noninvasive method of pain relief. It can be used alone or in addition to prescriptions or over-the-counter pain-relieving medications. The amount of medication may be able to be reduced in some patients who use the therapy. Patients should not stop taking or make any adjustments in your dose of medications without discussing it with doctor. Another benefit of the TENS unit is that it is small and portable and therefore can be used at home or away, anytime pain relief is needed.

TENS therapy has few reported side effects. In rare cases, patients have reported burns at the sites where the electrodes are placed. Some patients may be allergic to the adhesive used to affix the pad to the skin or the materials in the pad itself (the skin may appear red, irritated, or a rash may break out). Some people may be sensitive to or feel uncomfortable with the prickling/tingling sensation generated by the TENS unit [10].

Do not use TENS therapy at these specific body locations if you have any of the following conditions:

- An implantable device: (cardioverter/defibrillator, neurostimulators, bone growth stimulator, indwelling blood pressure monitors). Do not use TENS therapy over or close to the areas where an electronic device is implanted. TENS could cause these devices to malfunction.
- Do not apply TENS therapy to the abdomen; pelvic area; lower back; or to acupuncture points at the knee, hand, or ankle. (However, it can be used for labor pain.)
- Do not apply electrodes to areas of the body where there is known or suspected cancer. Do not use TENS if you have undiagnosed pain and a history of cancer in the last 5 years.
- Do not apply electrodes to your head, neck, or shoulders. The impulses could cause seizures.
- Deep vein thrombosis or thrombophlebitis: It may increase blood circulation, which may increase the risk of dislodging a blood clot.
- Bleeding (hemorrhagic) disorder: This therapy could increase bleeding at the tissue site or increase the risk of bleeding in persons with bleeding disorders.
- Heart disease: Use TENS therapy to the chest if you have heart disease, heart failure, or arrhythmias.

In addition, TENS should not be applied

- To infected tissues, wounds due to osteomyelitis, and tuberculosis. TENS therapy may result in the spread of infections.
- To areas of tissue that have been recently treated with radiation.
- To damaged skin (except for open wounds where the intent is to use electrical stimulation to heal tissue, in which cases, therapy should be guided by a skilled therapist).
- Near or over eyes or mouth, front or side of neck, or on the head.
- Near reproductive organs or genitals.
- To areas of the body that lack or have reduced sensation.
- In persons who have trouble communicating or who have mental impairment and cannot provide feedback to ensure the safe use of TENS.

15.9.7 Caries Removal Solution

Chemo-mechanical caries removal involves the chemical softening of carious dentine followed by its removal by gentle excavation. The reagent involved is generated by mixing amino acids with sodium hypochlorite; N-monochloroamino acids are formed which selectively degrade demineralized collagen in carious dentine. The procedure requires 5–15 minutes but avoids the painful removal of sound dentine thereby reducing the need for local anesthesia. It is well suited to the treatment of deciduous teeth, dental phobics, and medically compromised patients. The dentine surface formed is highly irregular and well suited to bonding with composite resin or glass ionomer. When complete caries removal is achieved, the dentine remaining is sound and properly mineralized. The system was originally marketed in the USA in the 1980s as Caridex. Large volumes of solution and a special applicator system were required. A new system, Carisolv, has recently been launched on to the market. This comes as a gel, requires volumes of 0.2–1.0 ml, and is accompanied by specially designed instruments [11].

15.9.8 Dental Filling Materials

A filling is used to treat a small hole, or cavity, in a tooth. To repair a cavity, a dentist removes the decayed tooth tissue and then fills the space with a filling material. When decay-causing bacteria come into contact with sugars and starches from foods and drinks, they form an acid. This acid can attack the tooth's surface (enamel), causing it to lose minerals.

When a tooth is repeatedly exposed to acid, such as when you frequently consume food or drink high in sugar and starches, the enamel continues to lose minerals. A white spot may appear where minerals have been lost. This is a sign of early decay. Tooth decay can be stopped or reversed at this point. Enamel can repair itself by using minerals from saliva and fluoride from toothpaste or through the application of fluoride by a dentist or dental hygienist. If more minerals are lost than can be restored, the enamel weakens and eventually breaks down, forming a cavity.

More severe decay can cause a large hole or even destruction of the entire tooth. If tooth decay is not treated, it can cause pain, infection, and even tooth loss. There are several types of filling materials used to repair cavities, including tooth-colored (composite) fillings and silver-colored (amalgam) fillings.

Silver Amalgam Fillings

This is a popular filling. Silver amalgam filling contains more than silver. It is a combination of other minerals, including tin, zinc, copper, and mercury. It is a common choice because it is sturdy, long-lasting, and less pricey than other options. A normal silver amalgam filling can stand up to 12 years of use. Dentists find it easy to use because it is malleable.

The main drawback is that it is not aesthetically pleasing, so it is not the best choice for a visible tooth. The material can also respond to temperature changes by contracting and expanding, causing the tooth to crack. The fluctuations may create a gap between the filling and tooth, ushering in food and bacteria and the formation of new cavities. Despite the controversy surrounding mercury in silver amalgam, the filling material has been declared safe for use.

Composite Fillings
Composite fillings are created from plastic and resin material. It is placed inside the tooth while it is soft and hardened with a bright curing light. It is a common option because it can be customized to match the color of the patient's existing teeth, so it is not conspicuous as the silver amalgam filling. Also, composite fillings are not as long-lasting as some other types. They can survive from up to 5 to 10 years, after which a replacement might be required.

Ceramic Fillings
They are created using porcelain material, which makes it both durable and cosmetically appealing. Ceramic fillings cost more than other fillings, but they are tooth-colored and resist stains and abrasion better than composite resin.

The drawback of using ceramic filling over composite is that it is more brittle and, therefore, has to be used on large cavities to prevent breakage. The dentist can enlarge the area to make room for the extra bulk.

Glass Ionomer Fillings
This filling is made from a mix of glass and acrylic. They are usually used for children whose teeth are still forming. They release fluoride into the tooth to protect it from additional decay. However, they only survive a few years because they are weaker than composite resin and are bound to crack or wear out.

15.9.9 Magnetic Resonance Imaging (MRI) Scanners

Magnetic resonance imaging (MRI) scanners are both medical devices and radiation-emitting electronic products subject to the requirements of the Federal Food, Drug, and Cosmetic Act. As medical devices, MRI scanners are subject to the general controls of the Act, such as establishment registration and device listing, premarket notification, maintenance of records and reports, and quality system regulations including good manufacturing practices. The FDA takes a risk-based approach to medical device regulation, and MRI scanners are Class II (moderate risk) medical devices, meaning that an MRI manufacturer is required to submit a 510(k) notification prior to marketing their MRI system.

As radiation-emitting electronic products, MRI scanners are subject to the general requirements of the Electronic Product Radiation Control provisions of the Act, such as maintenance of records and reports, notification of defects, repurchase, repair or replacement, and importation.

The following are FDA-recognized voluntary consensus standards relevant to MRI scanners.

- IEC 60601-2-33—Medical Electrical equipment: Particular requirements for the basic safety and essential performance of magnetic resonance equipment for medical diagnosis
- EC 62464-1—Magnetic resonance equipment for medical imaging Part 1: Determination of essential image quality parameters.
- NEMA MS 1—Determination of Signal-to-Noise Ratio (SNR) in Diagnostic Magnetic Resonance Images
- NEMA MS 2—Determination of Two-Dimensional Geometric Distortion in Diagnostic Magnetic Resonance Images
- NEMA MS 3—Determination of Image Uniformity in Diagnostic Magnetic Resonance Images
- NEMA MS 4—Acoustic Noise Measurement Procedure for Diagnostic Magnetic Resonance Imaging Devices
- NEMA MS 5—Determination of Slice Thickness in Diagnostic Magnetic Resonance Imaging
- NEMA MS 6—Determination of Signal-to-Noise Ratio and Image Uniformity for Single-Channel, Non-volume Coils in Diagnostic Magnetic Resonance Imaging (MRI)
- NEMA MS 8—Characterization of the Specific Absorption Rate for Magnetic Resonance Imaging Systems
- NEMA MS 9—Characterization of Phased Array Coils for Diagnostic Magnetic Resonance Images
- NEMA MS 10—Determination of Local Specific Absorption Rate (SAR) in Diagnostic Magnetic Resonance Imaging Systems
- NEMA MS 12—Quantification and Mapping of Geometric Distortion for Special Applications
- NEMA MS 14—Characterization of Radiofrequency (RF) Coil Heating in Magnetic Resonance Imaging Systems

Manufacturers of Devices Intended to Enter the MR Environment

The MR environment presents unique safety hazards for patients with implanted and accessory medical devices. Investigator should consider all of the following hazards when evaluating the safety of a device that would reasonably be anticipated to enter the MR environment during clinical care, including the following:

- Strong, static magnetic field of the MRI scanner induces displacement forces and torques on magnetic materials that may cause unwanted movement of a device.
- Radiofrequency energy and magnetic fields that change with time may cause heating of the device and the surrounding tissue.
- Magnetic fields and radiofrequency energy produced by an MRI scanner may also cause electrically active medical devices to malfunction and may induce voltages in leads or other long conductive portions of the medical device, which can result in a failure of the device to deliver the intended therapy.

- Presence of the device may degrade the quality of the MR image and may make the MR scan uninformative or may lead to an inaccurate clinical diagnosis, potentially resulting in inappropriate medical treatment.

15.9.10 Continuous Glucose Monitor (CGM)

Externally worn monitors for insulin reactions, also known as continuous glucose monitors (CGMs), are medical devices used to continuously monitor blood glucose levels in individuals with diabetes. CGMs are generally considered non-significant risk (NSR) devices, but it's important to note that the classification may vary based on specific factors and regulatory requirements in different regions. They may require compliance with certain regulations, including the submission of a premarket notification (510(k)) or obtaining FDA clearance through other pathways [12].

A CGM works through a tiny sensor inserted under your skin, usually on your belly or arm. The sensor measures your interstitial glucose level, which is the glucose found in the fluid between the cells. The sensor tests glucose every few minutes. A transmitter wirelessly sends the information to a monitor. The monitor may be part of an insulin pump or a separate device, which you might carry in a pocket or purse. Some CGMs send information directly to a smartphone or tablet. Many CGMs have special features that work with information from your glucose readings:

- An alarm can sound when your glucose level goes too low or too high.
- Patient can note meals, physical activity, and medicines in a CGM device.
- We can download data to a computer or smart device to see your glucose trends more easily.
- Some models can send information right away to a second person's smartphone—perhaps a parent, partner, or caregiver. For example, if a child's glucose drops dangerously low overnight, the CGM could be set to wake a parent in the next room.

15.9.11 Low-Level Laser Therapy (LLLT)

Low-level laser therapy involves the use of relatively low light energy (less than 100–200 mW). The potential physiologic benefits of LLLT appear to be nonthermal; it can have a stimulating effect on target tissues. It is used to decrease pain and inflammation, stimulate collagen metabolism and wound healing, and promote fracture healing.

Indications: It may have a role in wound care, especially for diabetic foot ulcers. Following LLLT, pain is immediately reduced in patients with acute neck pain and over a period of 1–5 months in those with chronic neck pain. It also reduces pain in the wrists, fingers, knees, and temporomandibular joints, as well as the pain associated with lateral epicondylitis.

A single dose of IR irradiation after implant placement will reduce postoperative pain and edema. Repeated irradiation sessions will stimulate actual osseointegration. It is also a useful additional therapy in the control of periimplantitis.

LLLT can promote bone healing and bone mineralization and thus may be clinically beneficial in promoting bone formation in skeletal defects. It may be also used as additional treatment for accelerating implant healing in bone. It can modulate the primary steps in cellular attachment and growth on titanium surfaces. Multiple doses of LLLT can improve efficacy, accelerate the initial attachment, and alter the behavior of human gingival fibroblasts cultured on titanium surfaces [13].

Contraindications and Precautions: Precautions should be taken to ensure that the LLLT beam does not hit the eye directly or after reflection off a shiny surface. It should not be used in areas with cancerous tissue.

One of the key advantages of NSR devices is their relative safety profile. Due to their low-risk nature, they typically undergo a simplified regulatory process for approval and can be brought to market more quickly compared to higher-risk devices. This expedited process allows for faster access to innovative technologies and medical advancements.

However, it is essential to recognize that even though NSR devices are considered low risk, they are not entirely devoid of potential hazards. Adverse events can still occur, albeit at a lower frequency and severity compared to higher-risk devices. Therefore, appropriate training, adherence to guidelines, and close monitoring of patients are still necessary to ensure their safe and effective use.

In conclusion, non-significant risk devices are valuable tools in health care, offering a balance between safety, accessibility, and affordability. Their use in diagnostics, monitoring, research, and other medical applications contributes to improved patient care, early intervention, and advancements in medical knowledge.

References

1. https://research.iu.edu/compliance/human-subjects/guidance/non-significant-risk-medical-devices.html.
2. https://pubmed.ncbi.nlm.nih.gov/35048029/.
3. http://www.fda.gov/downloads/RegulatoryInformation/Guidances/UCM126418.pdf.
4. Smith JA, Jhingran A. Principles of radiation therapy and chemotherapy in gynecologic cancer: basic principles, uses, and complications, comprehensive gynecology. Elsevier; 2022. p. 618–36.
5. https://www.fda.gov/consumers/consumer-updates/facts-tampons-and-how-use-them-safely.
6. https://www.medicaldesignbriefs.com/component/content/article/mdb/pub/briefs/38937.
7. European Commission 2010Guidelines on Medical Devices: Serious Adverse Event Reporting. MEDDEV 2.7/3
8. Stickler DJ. Clinical complications of urinary catheters caused by crystalline biofilms: something needs to be done. J Intern Med. 2014;276:120–9.

9. Lawrence EL, Turner IG. Kink, flow and retention properties of urinary catheters part 1: conventional Foley catheters. J Mater Sci Mater Med. 2006;17:147–52.
10. Sampath T, SandhiyaThamizharasan MS, Srinivasan P, Shanmugam T. Neurology and psychiatry. In: Toxicological aspects of medical device implants; 2021. p. 261–79.
11. Beeley JA, Yip HK, Stevenson AG. Chemo-mechanical caries removal: a review of the techniques and latest developments. Ned Tijdschr Tandheelkd. 2001;108(7):277–81. Dutch.
12. https://www.fda.gov/radiation-emitting-products/mri-magnetic-resonance-imaging/mri-information-industry.
13. Wu C-H. Braddom's Rehabilitation Care: A clinical handbook. Philadelphia: Elsevier; 2018.

Index

P. S. Timiri Shanmugam et al. (eds.), *Significant and Nonsignificant Risk Medical Devices*, https://doi.org/10.1007/978-3-031-52838-5

If you have any concerns about our products,
you can contact us on
ProductSafety@springernature.com

In case Publisher is established outside the EU,
the EU authorized representative is:
Springer Nature Customer Service Center GmbH
Europaplatz 3, 69115 Heidelberg, Germany

Printed by Libri Plureos GmbH
in Hamburg, Germany

MIX
Papier aus verantwortungsvollen Quellen
Paper from responsible sources
FSC® C105338

If you have any concerns about our products,
you can contact us on
ProductSafety@springernature.com

In case Publisher is established outside the EU,
the EU authorized representative is:
Springer Nature Customer Service Center GmbH
Europaplatz 3, 69115 Heidelberg, Germany

Printed by Libri Plureos GmbH
in Hamburg, Germany